ELECTRON PROBE QUANTITATION

ELECTRON PROBE QUANTITATION

Edited by
K.F.J. Heinrich
and
Dale E. Newbury

Chemical Science and Technology Laboratory
National Institute of Standards and Technology
Gaithersburg, Maryland

PLENUM PRESS • NEW YORK AND LONDON

Library of Congress Cataloging-in-Publication Data

Electron probe quantitation / edited by K.F.J. Heinrich and Dale E.
Newbury.
        p.   cm.
    "Result of a gathering of international experts in 1988 at ...
National Institute of Standards and Technology"--Pref.
    Includes bibliographical references and index.
    ISBN 0-306-43824-0
    1. Electron probe microanalysis--Congresses. 2. Microchemistry-
-Congresses.   I. Heinrich, Kurt F. J.   II. Newbury, Dale E.
QD117.E42E44   1991
543'.08586--dc20   91-11807
                                                                        CIP

ISBN 0-306-43824-0

© 1991 Plenum Press, New York
A Division of Plenum Publishing Corporation
233 Spring Street, New York, N.Y. 10013

Printed in the United States of America

PREFACE

In 1968, the National Bureau of Standards (NBS) published Special Publication 298 "Quantitative Electron Probe Microanalysis," which contained proceedings of a seminar held on the subject at NBS in the summer of 1967. This publication received wide interest that continued through the years far beyond expectations. The present volume, also the result of a gathering of international experts, in 1988, at NBS (now the National Institute of Standards and Technology, NIST), is intended to fulfill the same purpose.

After years of substantial agreement on the procedures of analysis and data evaluation, several sharply differentiated approaches have developed. These are described in this publication with all the details required for practical application. Neither the editors nor NIST wish to endorse any single approach. Rather, we hope that their exposition will stimulate the dialogue which is a prerequisite for technical progress. Additionally, it is expected that those active in research in electron probe microanalysis will appreciate more clearly the areas in which further investigations are warranted.

Kurt F. J. Heinrich (ret.)
Dale E. Newbury

Center for Analytical Chemistry
National Institute of Standards and Technology

CONTENTS

EARLY TIMES OF ELECTRON MICROPROBE ANALYSIS

R. CASTAING

Université de Paris-Sud
Orsay, France

Introduction

In 1967 I gave in London an after-dinner talk about "the early vicissitudes of electron probe x-ray analysis." At this time, it was already an old story; now, it looks like prehistory. I hope the reader will pardon me for evoking here, as an introduction to a series of papers which will illustrate the present state of the art in that field, the reminiscences of an old-timer.

In January 1947, I joined the Materials Department of the Office National d'Etudes et de Recherches Aéronautiques (ONERA), in a little research center 30 miles away from Paris, near the village of Le Bouchet. My laboratory was equipped with an ill-assorted collection of scientific apparatus, most of which had been obtained from Germany at the end of the war; but in December I received the basic equipment I had been promised when engaging at ONERA: two electron microscopes, at that time a real luxury. One of them was a French electrostatic instrument manufactured by the C.S.F.; the other an R.C.A. 50 kV microscope built in 1945. I immediately turned my attention toward the R.C.A. instrument; it was very pretty indeed and I was fascinated—I am ashamed to say—by the American technology. I used that microscope, together with the oxide-film replica technique, for studying the wonderful arrangements of oriented precipitates—platelets or needles—which form when light alloys are annealed at moderate temperatures. That was the occasion for me to meet Professor Guinier who studied the same alloys with x rays in his laboratory at the Conservatoire des Arts et Métiers. I was astonished to learn from him, in the course of one of our conversations, that the composition of most of the precipitates or inclusions which appear on the light micrographs was in fact unknown, in spite of the art of the metallographers, if their number was too small for recognizing the phase by x-ray diffraction. On that occasion he asked me my opinion about identifying at least qualitatively the elements present in such minute individual precipitates by focusing an electron beam onto them and detecting the characteristic x rays so produced. I replied straightaway that to my mind it was very easy to do; I was surprised that no one had done it before. We agreed that I would try to do the experiment, even if it appeared a little too elementary a subject for a doctoral thesis. In fact, there were some slight difficulties as I was to find out during the next 2 or 3 years.

The plan I drew up for my work was simple. Electron probes less than 10 nm in diameter had been produced by Boersch just before the war; Hillier used 20-nm probes in his attempts to microanalyze thin samples by the characteristic energy losses of the electrons. With the self-assurance of youth, I planned using such probe diameters for exciting the x rays, so that scanning microscopy was necessary for locating the points to be analyzed. The only problem I was foreseeing was the intensity of the x rays; the expected electron beam current was much less than a thousandth of a microampere and, from measurements carried out with Guinier on his conventional x-ray tubes, we concluded that even with a Geiger Muller counter, the counting rates from a curved quartz spectrometer would be of the order of one pulse per minute. Surely we were too pessimistic, as we disregarded

Electron Probe Quantitation, Edited by K.F.J. Heinrich and
D. E. Newbury, Plenum Press, New York, 1991

the fact that a bent crystal does not reflect correctly the radiation from a broad source, but at that moment we concluded that we would have to fall back on nondispersive methods such as balanced filters. In short, the basic idea was to build an instrument provided with nondispersive spectrometry and with a scanning electron microscope for viewing the object. We had little hope, in view of the small number of counts obtainable in a reasonable time and the supposed complexity of the laws governing the emission of characteristic lines by compounds, to get any accuracy for quantitative analysis. By the way, the instrument planned was nothing more than a modern scanning microscope equipped with E.D.S. However, for the years to come, things were to be quite different.

First I realized that in massive samples which concerned the metallurgists I would have to give up the splendid spatial resolving power that I had cheerfully envisaged, as I became aware of the terrific path that my electrons would perform haphazardly in the sample before agreeing to stop. I had to limit my ambitions to analyzing volumes of a few cubic micrometers. That was a big disappointment despite the gain of several orders of magnitude in resolution over localized spark source light spectrometry, which at this time was the best technique available for microanalysis. That disappointment showed a little when I presented my first results at the first European Regional Conference on Electron Microscopy held in 1949 in the lovely city of Delft.

I have happy memories of that Delft Conference. It was the first time my wife and I left France. I had prepared my presentation in French and was astounded when the organizers told me, the day before my talk, that I had better give my paper in English. By chance, when arriving at the conference place, we had made friends with young British participants Alan Agar and his wife. Agar kindly proposed to help me by translating my paper, which took half of the night. He is perfectly right when he claims that the first paper on the microprobe, attributed to Professor Guinier and me, was in fact written by him.

The experimental arrangement that I presented in that paper was quite simple indeed. I had modified my C.S.F. microscope, replacing the projector lens with a probe forming lens; the objective operated as a condenser. The probe formed 6 mm below the lower electrode; the specimen could be moved by a crude mechanical stage, but in addition, an electrostatic deflector allowed controlled displacement of the probe across the sample surface. The experiment consisted of demonstrating the spatial resolving power by plotting the changes in the total x-ray emission, as registered by a Geiger Muller counter, when the probe passed over chemical discontinuities. The first sample I observed was a plating of copper on aluminum; from the steep change of the emission one could estimate the resolving power to 1 or 2 micrometers. Similar intensity drops were observed on a sample of cast iron when the probe passed over the graphite flakes.

At this time preliminary trials with a Johannson curved quartz crystal that Guinier had brought from his laboratory, where he used it as a monochromator, had shown that counting rates of several hundred per second could be obtained on the Cu $K\alpha$ line with that 1-μm, 30-kV probe, in spite of the weakness of the beam intensity, less than one hundredth of a microampere. That opened the way to quantitative measurements of the lines and I hastened to fit the instrument with a Johannson focusing spectrometer manufactured at the ONERA workshop. It was ready at the beginning of 1950 when the Materials Department moved to the present building of Châtillon-sous-Bagneux. This spectrometer was a nice piece of work; it could be adjusted to allow very good separation of the $K\alpha$ doublets. The wavelength range was limited; it gave access to the K radiation of the elements between titanium and zinc. The L radiation allowed detecting heavier elements between cesium and rhenium. I also equipped the probe forming lens with an electrostatic stigmator, whose operation could be controlled quite satisfactorily by looking at the symmetry of shadow images of extremely thin plexiglass filaments, coated by vacuum deposition with chromium to avoid their melting under the beam. All was ready for attacking the problem of quantitative analysis.

I had undertaken a theoretical calculation of the intensity of the x-ray lines; the principle generally accepted at this time for x-ray emission analysis consisted of comparing the concentrations of two component elements by measuring the intensities of the corresponding characteristic radiations. I was quite excited indeed when I realized, in the course of that calculation, that the number of parameters, instrumental and others, would be considerably reduced if I did away with the comparison of different lines and proceeded by comparing the same x-ray line when emitted by the sample and by a known standard such as the pure element. That amounted to comparing directly—apart from a correction for self-absorption in the targets—the total number of ionizations that an electron produces in the sample on a given atomic level to the corresponding number in the standard. It was clearly the key to absolute quantitative measurements.

The best way for checking that principle was to analyze diffusion couples where the pure elements and various intermetallic phases would be found together, and the first one I observed was a silver-zinc couple. In principle I had only to move the specimen under the beam and to plot the Zn $K\alpha$ emission while the probe crossed the diffusion zone. But the apparatus was not equipped at this time with a viewing device and my crude specimen stage could not ensure calibrated displacements. I was reduced to moving the sample more or less haphazardly, watching the abrupt changes of the zinc emission, then scanning the probe across the discontinuity for exploring locally the diffusion curve near the phase boundaries. I remember my delight when I noticed, by observing the diffusion couple under the microscope after hours of fight against the beam instabilities, that all the phase boundaries were covered with myriads of contamination spots. I realized that the discontinuities of emission I had observed really corresponded to changes of composition. This was a revelation and since then I really believed in electron probe microanalysis.

That exhausting experiment made clear to me that a viewing device was a pressing necessity. That was not a simple job; I had to insert a mirror in the 6-mm clearance between the probe forming lens and the sample where part of the space was occupied by the stigmator. I still remember the wonder I felt when I was able to see the specimen during the bombardment; the counter became crazy when I brought a precipitate under the cross wires of the eyepiece. Some of those precipitates were good enough to become fluorescent under the beam. That somewhat monstrous apparatus was the most splendid toy I ever had.

Then things became serious. Measurements on homogeneous samples showed that the simple ratio k of the two readings, on the sample and on a pure standard, was not far from giving directly the mass concentration, at least in the case of neighboring components; but estimations of the electron penetration from Lenard's law made clear that the self-absorption of the x-ray photons in the target was far from negligible. It was clear that if a simple relation was to be confirmed between the k-ratios and the concentrations, it would hold for the k-ratios of the generated intensities only, and not for those of the emerging ones which had to be corrected for their absorption in the target. Another correction had to be made for subtracting the contribution of the fluorescence, but it was generally low and calculating it was not too complicated. Estimating the self absorption required the knowledge of the depth distribution of the characteristic emission, which was at this time practically unknown, at least at the required level of accuracy. I realized that the correction curve could be obtained by plotting the emerging intensity as a function of the angle of emergence of the beam—an extension of a simple method originally proposed by Kulenkampff. Rotating the whole spectrometer around the bombarded point raised serious technical difficulties, so I simply used a polished cylindrical specimen whose lateral displacement permitted a continuous change of the emergence angle of the x rays, without departing too much from the normal incidence of the electron beam. The ratio of the emerging intensity to the generated one is plotted as a function of the χ parameter which controls the self absorption, normalized by a small extrapolation to $\chi = 0$.

It was time to move to the applications. The viewing device allowed accurate measurements of the probe displacements and drawing diffusion curves was quite easy. In the simple case of a copper-zinc diffusion couple the curve could be drawn by simply plotting the Zn $K\alpha$ or the Cu $K\alpha$ emission as a function of the abscissa. Identifying unknown phases was also quite exciting. The first alloy I examined in that respect was a lead-tin-antimony-copper alloy used for manufacturing metal type for printing work. I would not claim that such a sample was of paramount metallurgical interest, but I was very proud to find out that the central part of the needles was Cu_3Sn whereas the composition of edges was consistent with the formula Cu_5Sn_4.

As a general rule, those experiments confirmed the simple linear relations between the mass concentrations and the generated emissions obtained by correcting the emerging ones for absorption and fluorescence. Even in the case of aluminum and copper whose atomic numbers differed appreciably, the α coefficients that I had introduced as a second approximation for taking into account possible differences of behavior of the various elements turned out to be near unity. We were at the beginning of 1951 and we agreed with Guinier that I would defend my thesis before the summer vacations. On the other hand, I enjoyed very much at this time exploring the possibility of local crystallographic analyses by Kossel line patterns, so that I did not push too far some measurements I had done on iron oxides, which seemed to indicate a definite departure for the k-ratios of Fe $K\alpha$ from the simple proportionality law of the first approximation. I was pressed by time and I hated the idea of compromising my summer vacations, so I convinced myself that I was not sure at all of the stoichiometry of my oxides. Instead of undertaking the preparation of homogeneous samples of such oxides, large enough for chemical analysis, I noted only in my thesis that the first approximation was verified for neighboring elements, but that serious discrepancies could possibly be observed in some cases. So I lost the opportunity of demonstrating the atomic number effect, whose importance was established 10 years later by the theoretical calculations of Archard and Mulvey, the measurements of Scott and Ranzetta, of Kirianenko, and of Poole and Thomas, as well as by the Monte Carlo calculations of Green.

Another point was puzzling me: the $\phi(\rho z)$ distribution law of the characteristic emission. The calculation indicated that, due to the progressive scattering of the electron trajectories, the exponential decrease of the Lenard law should occur at large depths only. In the first surface layers the deviation of the trajectories could compensate that decrease, but it appeared from Bothe's formula that an initial increase of the ϕ function would occur for heavy elements only, such as gold. The absorption correction curve that I had obtained for iron was consistent with a ϕ curve nearly stationary at small depths. Now, an experiment I had done in the very last weeks before defending my thesis indicated a much stronger influence of the scattering than that which I had estimated from Bothe's formula. I had evaporated a thin layer of chromium onto aluminum and covered a part of it with a thin layer of aluminum. Despite the absorption, the emission of the copper layer was definitely higher on the part covered with aluminum, which indicated a steep initial increase of the ϕ distribution curve. I realized that I could get the whole of the distribution curve in such a way, by using aluminum coverages of increasing thicknesses. That would give nice material for another chapter of my thesis, but I estimated that 2 or 3 weeks would be necessary for completing the measurements. Time was too short, as I persisted in defending my thesis

before the summer vacations, and I decided to postpone the tracer experiment for postdoctoral work. It was a wise decision indeed, because we later spent a little more than 2 years
to do that job with Descamps.

After my thesis, I was appointed to a lectureship in Toulouse and I spent half of my
time there, the other half in my laboratory of ONERA. Preparing the sandwich samples for
the tracer technique was horribly tedious with the evaporating devices at our disposal and
it was not until the end of 1953 that we could publish in the Comptes Rendus, Acad. Sc.
Paris our first ϕ distribution curve, in copper. We published all our results in 1955 in the
Journal de Physique, together with measurements by a similar tracer technique of the
fluorescence induced by the continuous spectrum. We showed there rough ϕ curves plotted in such a way for aluminum (copper tracer), copper (zinc tracer) and gold (bismuth
tracer). A tail due to the fluorescence excited by the continuous spectrum was particularly
visible in the gold curve. The ϕ curves were obtained subsequently by subtracting the part
of the emission due to that fluorescence. In that paper I noted that the validity of the first
approximation (proportionality emission-concentration) was related to the equality of the
integrals of the ϕ functions for the various components, but I did not emphasize the possibility of large discrepancies for elements of widely different atomic numbers, and I disregarded the possibility of large atomic number effects. We had chosen for that work a
well-defined accelerating voltage, 30 kV, that we controlled accurately with an electrome-
ter; but when writing the paper I noticed that we had neglected to subtract the voltage
drop in the resistor which polarized the Wehnelt electrode, and the readers of the paper
were puzzled a bit by this curious choice we had done of a 29 kV accelerating voltage for
the electrons.

Between those attempts to clarify the physical bases of microprobe analysis I examined
various samples that metallurgists or other scientists brought to me for inspection. I had
given a short talk on my thesis at the American Physical Society's meeting in Chicago on
my first trip to the United States in 1951. The paper had been added to the program at the
last minute, and I was asked very few questions; the essential one concerned the commercial availability of the instrument and its price. Nevertheless, this possibility of point analysis raised interest and I received samples from various places, some of them quite
interesting. As far as I can remember, two different industrial laboratories had sent me
samples of an Al-Mn alloy, asking me to investigate confidentially the segregation at the
boundaries of a cellular structure; I suppose they were competing companies and their
samples turned to be identical. I had also the visit of L. S. Birks who brought with him a
small sample of what he said to be a tungsten alloy. I was quite confused when I did not find
any of the components he announced; in fact, it was a steel if my memory is right. He told
me he had brought the wrong sample with him, but I still suspect him, and other visitors,
of having misled me intentionally on the real composition of their alloys with the purpose
of setting a trap for me.

At this time we were building an improved model of the microprobe, including mag-
netic lenses, an axial mirror objective and vacuum spectrometers. Two copies were manufactured in the ONERA workshops. One of them was delivered to the IRSID Laboratory
in St.-Germain-en-Laye, where a brilliant young metallurgist, whose name was Philibert,
was to make splendid use of it.

Before delivering the instrument to IRSID I exhibited it in operation at the 1955
French Physical Society's exhibition, in the hall of the Sorbonne University. At that famous
place, essentially devoted to the humanities, we had not any water supply to our disposal;
we refrigerated the diffusion pump by a closed fluid loop including a cooling truck radiator.
The instrument operated surprisingly well; the visitors were quite interested. I remember

that after 10 minutes of explanations and demonstration on small precipitates, the scientific reporter of a newspaper asked me, as a concluding question, "Is there anything new in this instrument?"

The range of applications was extending rapidly. Insulating samples could be observed by coating them with a thin conducting film, a technique that I mastered with some difficulties. My cleaning of the sample before depositing an aluminum film was very poor indeed and the coating film produced splendid bubbles under the probe. But the technique was to be brought to a high degree of perfection by Capitant and Guillemin, of the Bureau de Recherches Géologiques et Minières, who extended enormously the field of the mineralogical applications which was to become the choice ground for microprobe analysis in the hands of pioneers such as Klaus Keil.

Curiously, the first application of the probe to minerals was on a nonterrestrial material, when Kurt Fredriksson brought a sample of cosmic spherules to my laboratory. Quite exotic, too, was the sample of deep sea sediments that Gustav Arrhenius brought with him for looking at barium. I could claim that the field of application of the microprobe extended from the bottom of the Pacific Ocean until outer space.

Other scientists in various countries became involved in probe analysis. A decisive step was accomplished by Peter Duncumb when he built with Professor Cosslett, between 1953 and 1956, the first scanning microprobe. I had considered the possibility of doing that and I had proposed it as a probation work to Pierre Lanusse, a young engineer of the Physics Department of ONERA; but the head of the department of this young engineer was horrified when he heard about it. He explained to me that it was a 5-year program, and not a job for a 1-year probation work. He was not far from being true, but the main reason why we abandoned the project was that the wonderful idea of Duncumb and Cosslett, which was the real key for success, did not occur to us: to control directly the beam of the oscilloscope by the pulses from the counter. That was the simplest and the safest way for ensuring a linear relation between the counting rate and the local brightness of the distribution image. Duncumb and Cosslett's work gave a strong impetus to probe analysis. In addition, it made feasible the transfer of electron probe analysis into the electron microscope, initiated by Duncumb and his group at the end of the sixties. They pushed the spatial resolving power on thin samples below a tenth of a micrometer, bringing to full realization our initial concept, inapplicable to massive specimens.

Let's return to the middle fifties. I heard at this time about the work in the U.S.S.R. of Igor Borovskii, who developed independently a microprobe he built in 1953; he was to give deep insights on some physical problems involved in the x-ray emission. Borovskii was a tremendous man, full of vitality. He was really a good friend, exactly 10 years older than I, with the same birthday. The last time I saw him was in Toulouse, in 1983. He departed from this life; we will no longer hear him at the IXCOM meetings. The same as Vic Macres, a very pleasant young man. The charming little microprobe he built during the sixties was a model of ingenuity.

I shall not attempt to quote all the veterans who took the crude technique I had initiated and opened the way to the modern ages. Dave Wittry examined in his thesis the problem of the spatial resolving power; his later work extended widely the modes of operation of the probe. Dolby introduced the matrix method for disentangling neighboring lines in energy dispersive analysis. Bob Ogilvie checked the validity of the second approximation and developed the use of the Kossel patterns for local crystallographic studies in the laboratory of Professor Norton. Green attacked the problem of the stopping power factor, followed by Duncumb and Reed. Philibert gave an analytical shape to the absorption correction, opening the way to the ZAF procedure and making quantitative analysis a routine job. He developed with Tixier the analysis of thin films. Le Poole introduced the mini lens. Scott and Ranzetta, and Ken Carroll pioneered with Philibert and Crussard the metallurgical applications. Tousimis opened the field of biological applications, followed by Pierre Galle and by Hall who pioneered the analysis of thin biological specimens. Shimizu and Shinoda founded the Japanese school; they have guided a new generation of young scientists whose contributions, from the sixties until now, were to bring the technique to its modern standard.

As for me, I was tempted again by my old flame, electron microscopy. I still launched students on some specific points: Henoc on the general problem of the fluorescence correction, Derian on the direct measurement of the "backscattering loss," later on Françoise Pichoir and Pouchou on the problem of in-depth analysis. But I was more and more fascinated by the possibility of getting distribution images directly, without any "scanning artifice," by secondary ions. Microprobe analysis was to fend for itself. It had entered the golden age; early times were over. So I think I'd better stop now evoking the good old days.

STRATEGIES OF ELECTRON PROBE DATA REDUCTION

KURT F. J. HEINRICH

National Institute of Standards and Technology
Gaithersburg, MD 20899

I. Introduction

The purpose of data reduction procedures in electron probe microanalysis (EPMA) is to transform the measured x-ray intensities into mass fraction estimates of the elements present. A variety of approaches are used at present, and in each approach, changes in constants, parameters and algorithms may affect the accuracy of the results. In what follows, I will first discuss the classical case of analysis of a flat, semi-infinite and homogeneous specimen. We will consider the goal of data reduction, the events in a target, and methods and procedures. I aim at giving a balanced overview of the present state-of-art, and I will try to rationalize my preferences.

II. Statement of the Problem

We are analyzing a specimen which is homogeneous within the precision of x-ray intensity measurement to a depth larger than the electron penetration. The intensities observed from all elements will be compared with those from elementary standards. Both specimen and standard have flat surfaces and conduct electricity to the degree that the target phenomena are not observably affected by static charges during analysis. The electron beam is normally incident upon the specimen. The electrons are accelerated to a potential typically varying between 5 and 30 kV.

Cases which do not comply with the above conditions will be discussed in other contributions.

III. Events Within the Target

The events within the target are in principle well known, although their quantitative description is affected by significant uncertainties. The events of significance for x-ray emission are:

- Slowing down of electrons due to inelastic collisions.

- Deflection of their path, mainly by elastic collisions. As a consequence, a significant fraction of electrons is backscattered from the target; hence the energy available for x-ray production is diminished.

- Ionization of the inner shells of atoms by the primary electrons, with possible subsequent x-ray emission.

- Indirect ionization of inner shells by characteristic and continuous x-ray emission and by fast secondary electrons, leading to fluorescent emission.

- Absorption of the x rays emitted towards the spectrometer, on their way to the specimen surface.

Any data reduction scheme must take into account, explicitly or implicitly, the above mentioned mechanisms.

Electron Probe Quantitation, Edited by K.F.J. Heinrich and
D. E. Newbury, Plenum Press, New York, 1991

9

IV. Definitions

Precision: Repeatability of a measurement under stable conditions.

Accuracy: The degree to which calculated results of the analysis agree with the true value of what is measured.

Both precision and accuracy can be defined in terms of estimates of the statistical parameters (e.g., estimate of standard deviation), or deduced from error distribution graphs.

Under the usual conditions x-ray intensities can be measured with an accuracy of 1 to 2 percent. The goal of quantitative analysis is to obtain an accuracy of the same order of magnitude for the concentration estimates.

Concentration: Unless otherwise defined, this term denotes mass fraction.

Relative intensity: The ratio of x-ray intensities of a line, corrected for background, dead-time, and line overlap where applicable, from specimen and standard under identical experimental conditions.

Constant: A number expressing an invariable magnitude, which can be mathematical, physical or instrumental (e.g., x-ray take-off angle).

Parameter: A variable expressing a magnitude which depends on one or more other variables (e.g., the x-ray mass absorption coefficient, for a given element and line energy).

Algorithm: An algebraic expression, or set of algebraic expressions, of a parameter.

Model: A generic approach to data reduction.

Procedure: A set of instructions for data reduction, based on a model and *including all relevant constants, parameters and algorithms*.

While the efficacy of a scheme can be discussed in general terms, a numerical expression of estimated accuracy can only be given for a complete procedure. Much confusion is generated when it is attempted to describe the accuracy of a scheme without presenting the complete list of relevant constants, parameters and algorithms, as well as characterizing the set of measurements used for the estimate.

V. Information Available for the Formulation of a Procedure

A particular model and procedure can use the following information:

(1) Proposed algorithms for the description of the individual events in the target.
(2) Constants and parameters to be used in the algorithms.
(3) Tracer experiments on depth distribution of x-ray generation

(4) Variable take-off-angle experiments to determine $f(\chi)$.

(5) X-ray intensity ratios obtained from measurements on specimens of known composition.

It would be ideal to use all information contained in the first four items to formulate and adjust a procedure, and to reserve the measurements on "known" specimens for an ultimate verification, but investigators will adjust their procedures so as to minimize the spread of error diagrams. To avoid circular arguments, it would be desirable that the error distribution tests be performed on a commonly agreed on set of data, and, if possible, as a cooperative action.

VI. Classification of Models

While a few years ago there were only two procedures used in routine analysis (empirical and ZAF), new models have been recently proposed, and the Monte Carlo model has become more powerful and more frequently used. In order to discuss the possible approaches, we must define the models, even at the risk of disagreement about the validity of the definitions. We will in this communication define models as following:

A. EMPIRICAL MODELS

These are based exclusively on relative intensities from "known" specimens, usually for a specific composition range. The hyperbolic approximation [1] has been widely and successfully in the analysis of silicate minerals. With the rapid development of powerful small on-line computers, this model is losing some of its appeal due to its simplicity and we will not discuss it further.

B. ZAF MODEL

The ZAF model [2] was the first generalized algebraic procedure for data reduction. It was influenced by the initial assumption of a linear relation between concentration and generated intensity (Castaing's first approximation), and it treats deviation from linearity by means of multiplicative factors for the effects of atomic number (Z), absorption (A) and fluorescence (F), which are computed separately.

C. $\phi(\rho z)$ MODEL

This model [3] is based exclusively on the information obtained from tracer experiments. In practice, some of the $\phi(\rho z)$ procedures are the result of accommodation to analysis of known specimens, or even influenced by algebraic models.

D. MONTE CARLO (MC) MODEL

The procedures in this category are based on the construction of individual trajectories in which the events at any point of the trajectory are determined by random numbers, weighed by the probability of these events. The MC model is particularly useful in the investigation of unusual specimen configurations (particles, layered specimens, etc.).

VII. Functions Describing Depth Distribution and Absorption

Castaing [4] investigated the distribution in depth of x-ray generation by means of experiments in which a tracer emitting the observed radiation was imbedded in a matrix at varying depth. He defined a function, called $\phi(\rho z)$, which describes the variation of gener-

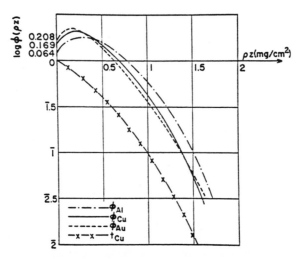

Figure 1. Distribution in depth of the primary characteristic emissions (ϕ curves) for samples of aluminum, copper, and gold, and transmission curve of electrons in copper (t_{Cu}). (Castaing).

ation of x rays in function of the depth, ρz; z being linear depth and ρ the density of the target material. To define the unit of the intensity scale, Castaing measured also the intensity from an unsupported layer of tracer material of the same thickness as that imbedded in the target. The value of $\phi(\rho z)$ at the depth z is thus the ratio between the x rays generated within a thin layer of tracer at this depth and that generated within a free-standing tracer layer of equal depth. The $\phi(\rho z)$ curve defines the x-ray emission from an element in the elementary target of another element, for a given electron beam energy (keV), choice of line, electron beam incidence angle and x-ray emergence angle (take-off angle). The experimental x-ray intensities observed in the tracer experiment, to represent generated primary emission, must be corrected for x-ray absorption (applying Beer's law), and for indirect x-ray emission. The area under the curve represents the total intensity of emission, and is thus related to the atomic-number correction, while the shape of the curve, indicating the depth distribution, relates to the absorption correction. Hence, all information needed to quantitatively describe the emission is contained in this curve (fig. 1), as is done explicitly in the $\phi(\rho z)$ model.

If one wishes to separate the effects of x-ray absorption from those of total generated intensity, it is preferable to use a normalized depth distribution, which we will call here $p(\rho z)$, in order to distinguish it clearly from $\phi(\rho z)$ as dimensioned by Castaing. The two intensities are simply related by

$$p(\rho z) = \phi(\rho z)/I_p \qquad (1)$$

where I_p represents the total intensity generated within a solid specimen; or

$$\int_0^\infty p(\rho z) = 1 . \qquad (2)$$

Let us now assume that the emitted x rays are observed by a detector or spectrometer covering the solid angle Ω. Since characteristic x rays are emitted isotropically, the intensity emitted towards the detector is equal to $I(\Omega/4\pi)$, where I is the total emitted intensity.

If the mean x-ray take-off angle is ψ, the ratio between generated and received intensities for this detector will be

$$f(\chi) = \int_0^\infty \phi(\rho z)dz \cdot e^{-\mu z \cdot \csc \psi} / \int_0^\infty \phi(\rho z) \cdot dz \equiv F(\chi)/F0 \ . \tag{3}$$

Curves of $f(\chi)$ as a function of $\chi \equiv \mu \cdot \csc \psi$ can therefore be obtained from the tracer experiment. The absorption function $f(\chi)$ is the basis of the absorption correction in the ZAF method; for moderately large values of χ, $1/f(\chi)$ is close to linear with respect to χ (fig. 2).

VIII. Models

A. THE ZAF MODEL

If the first approximation were to hold for the emerging intensities, we would obtain for the intensity ratio from specimen to standard:

$$k = I'^*/I'^s = c^* \tag{4}$$

where the star indicates parameters of the specimen, ^{s} those of the standard, c^* is the concentration of the emitting element in the specimen, and $I' = I \cdot f(\chi)$ is the emerging intensity. (We assume here that I is sampled over the same solid angle as I'.) However, the differences in absorption factors between specimen and standard are usually quite significant; therefore, Castaing applied the first approximation to the *generated* intensities:

$$k = c^*[f(\chi)^*/f(\chi)^s] = f_A \cdot c^* \ . \tag{5}$$

The factor f_A is called the absorption correction factor. With the recognition of the effects of specimen composition on the electron deceleration and backscatter, the atomic

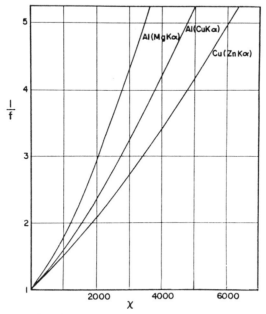

Figure 2. Inverse linear presentation of the absorption factors obtained from the tracer experiments of Castaing and Descamps [4].

number correction factor, f_Z, had to be added; finally, with the addition of a factor for fluorescence, f_F, the ZAF model was obtained:

$$k = c^* \cdot f_Z \cdot f_A \cdot f_F . \tag{6}$$

The use of multiplicative "correction" factors is due to the primitive concept of an ideally linear calibration curve; it is not justifiable on a physical basis; rather one should write:

$$k = [I_p^* \cdot f(\chi)_p^* + \Sigma I_f^* \cdot f(\chi)_f^*]/[I_p^s \cdot f(\chi)_p^s + \Sigma I_f^s \cdot f(\chi)_f^s] . \tag{7}$$

In this expression, the primary and fluorescent generated intensities of specimen and standard, each multiplied by its absorption factor, are ratioed. Multiplicative factors are, however, useful when the relative importances of the corrections are of interest. Obviously, a correction procedure expressed by means of multiplicative factors can be transformed into an algorithm following eq (7); hence; the essence of the ZAF procedure is not the use of such factors, but the separate estimation of the effects of primary generation of x rays, fluorescence and absorption. We therefore feel justified in including in this category all models and procedures in which these three effects (or more than three) are treated separately and explicitly.

The absorption correction was formulated in algebraic form by Philibert [2]; and a simplified version, with subsequent modifications, was used for many years. The shape of Philibert's simplified $\phi(\rho z)$ curve deviates significantly from the experimental curves: $\phi(0)$, i.e., the value of x-ray generation at zero depth, is zero (which is obviously inaccurate); to compensate, the maximum of the curve is displaced significantly towards the specimen surface. However, the shape of the absorption function $f(\chi)$ is insensitive to changes in the shape of $\phi(\rho z)$, so that the Philibert-type procedure proved useful until the determination of elements of very low atomic number was attempted.

In 1975, after a detailed study of the absorption correction [5] we concluded that experimental evidence then available did not warrant the use of a term for the atomic-number effects on absorption, and we proposed a very simple equation for $f(\chi)$, which we called the quadratic model:

$$f(\chi) = (1 + 1.2 \times 10^{-6} \gamma \chi)^{-2} ; \gamma = E_0^{1.65} - E_q^{1.65} . \tag{8}$$

The term γ expresses the dependence of $f(\chi)$ on the operating potential E_0 and the critical excitation potential for the line in question, E_q (both in kV). This algorithm was incorporated in the program FRAME [6] and proved to be as accurate as the more complicated Philibert-Duncumb-Heinrich formula. The corresponding expression for $\phi(\rho z)$ is

$$\phi(\rho z) = p(\rho z) \cdot I_p = \frac{z}{a^2 \gamma^2} \cdot e^{-z/a\gamma} \cdot I_p . \tag{9}$$

This expression again leads to a zero value for $\phi(0)$. Duncumb and Melford had previously proposed a "thin-film model" for very high absorption [7] based on the fact that in extreme cases the observed radiation comes from a very shallow layer on the specimen surface and that therefore be the shape of the function $\phi(\rho z)$ at greater depth becomes irrelevant. It must therefore be expected that both the simplified Philibert model and the quadratic model shown in eq (8) will become inaccurate for strong absorption such as occurs in the analysis for elements of low atomic number. The argument was brought

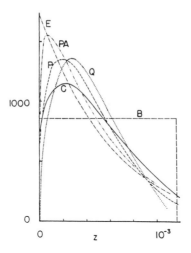

Figure 3. Models for $\phi(\rho z)$:

E	Exponential decay
C	Castaing and Descamps [4]
P	Philibert's complete model [2]
PA	Philibert's abbreviated model [2]
Q	Quadratic approximation [5]
B	Bishop's rectangular model [8]

The vertical scale is normalized so that areas under all curves are equal to one.

foreward strikingly with the proposed model of Bishop [8] in which a rectangular distribution is proposed for the function $\phi(\rho z)$. Although this model is grossly inaccurate, it provides reasonable values for strong absorption, when only the value of the function near the intercept [i.e., (0)] is important. Both the quadratic and the Bishop model have contributed to understand the limits of usefulness of the conventional Philibert model (fig. 3).

The rectangular model of Bishop was modified by Scott and Love [see 9]. In order to bring it more into line with the shape of the (ρz) curve, they replaced the rectangle by a quadrangle which had at its apices the point (0,0), the intensity at zero level, the point representing the maximum emission, and the range (fig. 1 of Scott and Love's paper in this publication).

Pouchou and Pichoir produced two models which aimed at presenting faithfully the experimental intensity distribution, so that the model would serve for the analysis of layered specimens as well as for the determination of low-atomic-number elements in homogeneous specimens (PAP models). The first model was quite rigorous; it uses a combination of two parabolas meeting at the point of inflexion. Although this model [10] gives very accurate results, its computation met with occasional failures in cases of very low absorption. For this reason, the authors produced the simplified version which is discussed in this publication [11].

Both the Scott-Love model and those of Pouchou and Pichoir include separate calculations of the effects of stopping power, backscattering, absorption and fluorescence. Although the presentation of the PAP models does not follow the formulation of the ZAF method, its transformation into a ZAF procedure is trivial; therefore we tend to include it among the ZAF-type approaches.

Brown and Packwood recognized the importance of accuracy in the $\phi(\rho z)$ model for cases of strong absorption and, following an early proposal by Wittry [12], they used a modified Gaussian curve to approximate the experiment [13-15]. They also performed a considerable number of tracer experiments which have enriched our experimental arsenal. Taking literally the observation that Castaing's formulation of the $\phi(\rho z)$ curve contains all the information needed for quantitation, they developed a model in which the effects of stopping power and backscattering are treated implicitly rather than separately. This method is used, with empirical adjustments which are in part related to the PAP model, by Bastin and Heijliger [16].

The use of a Gaussian distribution is not only a good empirical approach; it can be explained with the aid of a model first presented by Archard and Mulvey [17], in which it is assumed that the electrons penetrate to a point from which they diffuse in randomly fashion in all direction. This model would explain the fact that the volume of x-ray emission in the target is close to spherical. The point of contention with this method is its treatment of the atomic number correction. One may question if the accuracy of the intensity measurements from tracer layers and that of the estimates of layer thickness suffice to permit an adequate estimate of this correction. Certainly, the fact that the backscatter and stopping power effects are inseparable from those of absorption renders it more difficult to adjudicate residual errors to the correct mechanism.

The Monte Carlo method implies separate consideration of every single event of relevance in x-ray analysis. It can be modified to include phenomena not commonly considered such as excitation by high-energy secondary electrons; most importantly, it can be applied to all specimen configurations, including small particles and thin layers. The need for summation of a statistically meaningful number of trajectories is a disadvantage since it precludes its use on-line; in the past, the speed limitations of computers rendered necessary the use of approximation which introduced uncertainties and errors, and the accurate values of some of the parameters are still subject to controversy. It is not possible to rely on its result without experimental confirmation. The production of backscatter coefficients—which are accurately known [18,19]—is a necessary test of its fidelity. When data obtained by this method are reported, it is necessary to report all conditions and parameters that may have an effect on the result [20].

IX.   Miscellanea

Averaging of parameters for a non-elemental specimen: Most of the experimental evidence on which models are based was obtained on pure elements. It is not a trivial task to establish how the parameters involved in models should be averaged for such a case. For instance, there exist no experimental data on absorption in specimens containing more than one element, either by the tracer technique or by Green's variable take-off-angle technique. The choice of element for the tracer in the tracer experiment merely affects the minimum excitation potential provided that the tracer is very thin. An experiment with a tantalum tracer, observing the Ta *M* alpha radiation, for instance, should give exactly the same results as one using a silicon tracer, observing Si *K* alpha. Hence, the tracer experiment gives no information on the atomic number effect of elements not distributed in the matrix. Many averaging procedures in current and past models are based on unproven assumptions. In this area, measurements with the $f(\chi)$ instrument under construction at NIST [21] as well as Monte Carlo calculations may be useful.

The *analysis of specimens of known composition* should be the crucial test of performance of a procedure, and it is curious that this test is sometimes, and unjustly, belittled. The problem with the test arises when the same data obtained in an error distribution test have previously been used to adjust the procedure. It is well known that most such published tests show the author's procedure in a favorable light. If they did not, the author would modify the procedure to obtain the desired result. To really be useful and valid, such tests should be evaluated by a group, following previously established steps; it would also be necessary to carefully select the data to be used. On one hand, if a certain aspect, such as a backscatter correction, is to be tested, specimens in which such an effect is expected to be negligible or compensated should be excluded since they just contribute to the noise level of the test. On the other, data should only be used if they are part of a self-consistent set with systematic variation of either composition or operating voltage, so that inconsistencies of individual measurements can be spotted. It would also be wise to exclude all measurements made before modern equipment and deadtime correction procedures were available. All such exclusions must be done *prior to the evaluation*.

The role of computer development: At present we use on-line computers of ever increasing speed and power. Therefore, many of the simplifications used in the past are unnecessary and should be omitted. Even in the treatment of a large number of data, e.g., in the formation of quantitative composition maps, the time for data collection exceeds that required for data reduction. The criteria for what constitutes a good procedure have therefore changed. When numerical integration can be performed safely and rapidly, there is no need for formal integration which may have the elegance of a polished slide rule, but tends to obscure the nature of the procedure. In the same vein, it is no longer justifiable to use obsolete fluorescence procedures or to systematically ignore the fluorescence by the continuum.

The use of software provided by sources outside the laboratory (mainly the manufacturers of equipment) is a great advantage to the practical analyst. However, in conjunction with the equipment automation, it encourages the laboratory management to put into command operators who have no detailed knowledge of the analytical process. We have greatly benefitted from instrumental and computational innovations; yet, in microanalysis, as in so many other areas, there is no substitute for knowledge.

Expression of results: Many uncertainties concerning reported results are due to incomplete information on all pertinent conditions. These include as a minimum: Line and standards used, number of counts collected, description of background and deadtime correction, type of spectrometer system, treatment of line interferences where applicable, x-ray emergence angle, angle between electron beam and specimen; coating and other aspects of specimen preparation, and *a complete quantitative description of the data reduction procedure*. Data which do not provide such information should not be considered useful in the evaluation of procedures or in the characterization of the analyzed specimens.

X. References

[1] Ziebold, T. D. and Ogilvie, R. E. (1964), Anal. Chem. **46**, 322.
[2] Philibert, J. and Tixier, R. (1968), Quantitative Electron Probe Microanalysis, Heinrich, K. F. J., ed., NBS Spec. Publ. 298, p. 13.
[3] Brown, J. D. and Packwood, R. H. (1982), X-Ray Spectrometry **11**, 187.
[4] Castaing, R. and Descamps, J. (1955), J. Phys. Radium **16**, 304; Castaing, R. and Henoc, J. (1966), Proc. 4th Int. Congr. on X-Ray Optics and Microanalysis, Castaing, R., et al., eds., Hermann, Paris, p. 120.
[5] Heinrich, K. F. J. and Yakowitz, H. (1975), Anal. Chem. **47**, 2408.
[6] Yakowitz, H., et al. (1973), NBS Tech. Note 796.
[7] Duncumb, P. and Melford, D. A., Proc., Ref. 4, Proc. (1966), p. 240.
[8] Bishop, H. E. (1974), J. Phys. D: Appl. Phys. **7**, 2009.
[9] Sewell, D. A., et al. (1985), *ibid.*, **18**, 1245.

[10] Pouchou, J. L. and Pichoir, F. (1986), Proc. Int. Congress on X-Ray Optics and Microanalysis 11, London,
      Ontario, p. 249.
[11] Pouchou, J. L. and Pichoir, F., This publication, p. 31.
[12] Wittry, D. B. (1958), J. Appl. Phys. 29, 1543.
[13] Packwood, R. H. and Brown, J. D. (1981), X-Ray Spectrom. 10, 138; (1982), 11, 187.
[14] Brown, J. D., This publication, p. 77.
[15] Packwood, R., This publication, p. 83.
[16] Bastin, G. F. and Heijligers, H. J. M., This publication, p. 145.
[17] Aichard, G. D. and Mulvey, T. (1963), Proc. Int. Symp. on X-Ray Optics and Microanalysis 3, Stanford,
      CA, p. 393.
[18] Bishop, H. E. (1966), Proc. Int. Congr. on X-Ray Optics and Microanalysis 4, Paris, p. 153.
[19] Heinrich, K. F. J. ibid., p. 159.
[20] Henoc, J. and Maurice, F., This publication, p. 105.
[21] Small, J. A., et al., This publication, p. 317.

AN EPMA CORRECTION METHOD BASED UPON

A QUADRILATERAL $\phi(\rho z)$ PROFILE

V. D. SCOTT AND G. LOVE

University of Bath
Bath BA2 7AY
England

I. Introduction

Quantitative electron probe microanalysis (EPMA) results are obtained by comparing the intensity of a particular characteristic x-ray line from the specimen with that from a reference standard of known composition. The measured x-ray intensity ratio, k, is then related to the mass concentration, c, of the element provided that the instrumental settings are kept constant while the x-ray intensity readings are being taken. The ratio k is, however, not usually directly proportional to c and correction factors are required to take account of the different behavior of electrons and x rays in specimen and standard. We shall first identify these factors and, in doing so, provide the essential terminology.

II. The 'ZAF' or 'Matrix' Approach

The intensity of the primary x-ray emission may be expressed as

$$I = \phi(\Delta\rho z)\int_0^\infty \phi(\rho z)\exp(-\chi\rho z)\,d\rho z$$

(see, for example, [1]). The function $\phi(\rho z)$ describes the distribution of generated x rays with mass depth, ρz, in the target and typically has the form illustrated in figure 1; note that the actual shape of the $\phi(\rho z)$ curve depends upon the energy of the incident electrons and the atomic number of the target. The term $\exp(-\chi\rho z)$ takes account of x-ray absorption in the target, where $\chi = \mu/\rho\,\mathrm{cosec}\,\Psi$, μ/ρ being the relevant mass absorption coefficient, Ψ the take-off angle for the x-ray collection system, and ρ the target density. The term $\phi(\Delta\rho z)$ corresponds to the x-ray emission from an isolated thin film of the same material with mass thickness $\Delta\rho z$ and its significance will be shown below, but first we shall rewrite the emitted

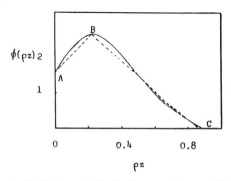

Figure 1. Typical curve showing distribution with depth of generated x-rays (solid line), upon which the quadrilateral profile ABC has been superimposed (broken line).

Electron Probe Quantitation, Edited by K.F.J. Heinrich and
D. E. Newbury, Plenum Press, New York, 1991

intensity as

$$I = \phi(\Delta\rho z) \int_0^\infty \phi(\rho z)\, d\rho z \frac{\int_0^\infty \phi(\rho z)\exp(-\chi\rho z)\, d\rho z}{\int_0^\infty \phi(\rho z)\, d\rho z} = \phi(\Delta\rho z) \int_0^\infty \phi(\rho z)\, d\rho z \cdot f(\chi)$$

(1)

where

$$\tilde{}(\chi) = \frac{\int_0^\infty \phi(\rho z)\exp(-\chi\rho z)\, d\rho z}{\int_0^\infty \phi(\rho z)\, d\rho z}$$

and represents the fraction of generated x rays that escapes from the specimen, i.e., the absorption factor.

Now consider a binary specimen containing elements A and B, where the mass concentration of A is to be measured by reference to a standard consisting of the pure element A. The ratio of emitted x-ray intensities, I_A^{AB}/I_A^A is given by

$$k_A = \frac{\phi(\Delta\rho z)_A^{AB}}{\phi(\Delta\rho z)_A^A} \frac{[\int_0^\infty \phi(\rho z)\, d\rho z]_A^{AB}}{[\int_0^\infty \phi(\rho z)\, d\rho z]_A^A} \frac{f(\chi)_A^{AB}}{f(\chi)_A^A}.$$

(2)

Next, it may be shown as follows that $\dfrac{\phi(\Delta\rho z)_A^{AB}}{\phi(\Delta\rho z)_A^A} = c_A$. The number of A atoms per unit area in the specimen AB is $\dfrac{Nc_A}{A} \cdot \Delta\rho z$ where N is Avogadro's number and A is atomic weight, and the number of ionizations produced by a given flux of electrons is, therefore, proportional to $\dfrac{QNc_A}{A} \Delta\rho z$, where Q is the ionization cross section. Thus the number of x rays generated will be proportional to $\omega\, QNc_A\, \Delta\rho z$, where ω is the fluorescence yield, and the intensity ratio of x rays from isolated films of AB and A of the same mass thickness will be

$$\frac{\omega\, QNc_A/A\ \Delta\rho z}{\omega\, QN/A\ \Delta\rho z} = c_A.$$

The integral terms in eq (2) correspond to the area under the respective $\phi(\rho z)$ curves and their ratio is referred to as the atomic number correction factor ('Z'). Further, by replacing the $f(\chi)$ terms by 'A', the absorption correction factor, eq (2) reduces to $k_A = c_A \cdot {}'ZA'$.

Thus far we have ignored the production of secondary x rays—fluorescence caused both by characteristic x rays and the continuum—but the factor may simply be incorporated to give $k_A = c_A \cdot {}'ZAF'$.

This particular notation was first introduced by Philibert and is mentioned in reference [2] in the proceedings of the first NBS workshop on quantitative EPMA. It may be seen that each correction factor can be treated separately, the approach adopted traditionally in quantitative EPMA and in the correction method to be described here.

After presenting our correction procedure, we shall examine in some detail its performance for a wide range of EPMA conditions likely to be met in practical situations.

III. The Atomic Number Correction

As mentioned above (see eq (2)), the atomic number correction is given by the ratio of the areas under the $\phi(\rho z)$ curves of specimen and standard, i.e., the intensity ratio of the primary generated x rays. However, the correction is not conventionally determined in this way and the more traditional approach as used in our model, is described below.

Considering an individual electron travelling distance s in the target, the generated x-ray intensity, I_g, is proportional to

$$\frac{N\rho}{A} \int_0^s Q \, ds \ .$$

Assuming that the electron remains in the target, the integration limits may be changed

$$I_g \propto \frac{N}{A} \int_{E_0}^{E_c} Q \, dE / (dE/d\rho s)$$

where E_0 is the incident electron energy, E_c is the critical excitation energy and $dE/d\rho s$ is the rate of electron energy loss. In practice, some of the incident electrons are backscattered from the target, resulting in a loss in x-ray intensity. This is accounted for by what is termed a backscatter factor, R, so that we have

$$I_g \propto \frac{RN}{A} \int_{E_0}^{E_c} Q \, dE / (dE/d\rho s) \ .$$

Hence for the binary alloy AB, the atomic number correction 'Z' is

$$\frac{R_A^{AB} \ [\int_{E_0}^{E_c} Q \cdot dE/(dE/d\rho s)]_A^{AB}}{R_A^A \ [\int_{E_0}^{E_c} Q \cdot dE/(dE/d\rho s)]_A^A} = \frac{R_A^{AB}}{R_A^A} \cdot \frac{S_A^A}{S_A^{AB}} \ .$$

We follow convention in treating the backscatter factor (R) and the stopping power factor (S) separately.

IV. The Stopping Power Factor

The stopping power factor is calculated using an empirical formula to describe the rate of electron energy loss. The formula was constructed from a study of the Bethe and Ashkin [3] energy loss law which is widely used in electron probe microanalysis. Earlier we showed [4] that at low electron energies the Bethe equation is unrealistic if the mean ionization energy J is taken to be independent of electron energy (as is generally assumed in microanalysis) because the electron never comes to rest in the target. It was further shown that, at higher electron energies, $d\rho s/dE$ as given by the Bethe expression varied in a parabolic fashion with electron energy. This fact has been utilized in the construction of the new energy loss formula which is in accord with the Bethe equation at high electron energies and is more realistic at lower energies. Rather than quote the equation for a binary system (AB), it is perhaps more useful to give the general form as applied to a specimen containing n elements.

$$\frac{dE}{d\rho s} = -\frac{1}{J} \left[\sum_{i=1}^{i=n} \frac{c_i Z_i}{A_i} \right] \bigg/ \left[1.18 \times 10^{-5} (E/J)^{0.5} + 1.47 \times 10^{-6} E/J \right] \ ,$$

where

$$J = \exp \left[\sum_{i=1}^{i=n} \frac{c_i Z_i}{A_i} \cdot \ln J_i \bigg/ \sum_{i=1}^{i=n} \frac{c_i Z_i}{A_i} \right] \text{ and } J_i = 0.0135 Z_i \ .$$

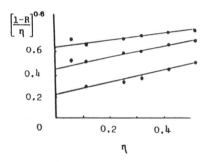

Figure 2. Plots of $(\frac{1-R}{\eta})^{0.6}$ versus η for different U_0 values.

If the form of the ionization cross section given by Bethe [5] is adopted, then

$$\int_{E_0}^{E_c} Q \, dE / (dE/d\rho s)$$

may be solved analytically. After several simplifications have been introduced (justified in Love et al. [4]), the stopping power factor reduces to

$$1/S = \left[1 + 16 \cdot 05(J/E_c)^{0.5} \left(\frac{U_0^{0.5} - 1}{U_0 - 1} \right)^{1.07} \right] / \sum_{i=1}^{i=n} \frac{c_i Z_i}{A_i} ,$$

where U_0, the overvoltage ratio, equals E_0/E_c.

Unlike the Duncumb and Reed [6] method, our formula is suitable for all overvoltage ratios and, furthermore, is not disadvantaged by utilizing the Bethe energy loss law (as, for example, in Philibert and Tixier [2]).

V.   The Backscatter Factor [1]

The backscatter factor has been established using data produced by Monte Carlo simulations of electron trajectories in targets [7]. R values have been obtained directly by summing the number of ionizations produced by each electron and then ratioing the value to the total number of ionizations produced had no electron backscattering taken place. Not surprisingly, the backscatter factor was found to be related to the backscatter coefficient η (the fraction of electrons incident upon the specimen that are backscattered) and also the overvoltage ratio. Based upon the physics of electron scattering and some experimental data given by Fitting [8], the following relationship was derived (Love et al. [4])

$$\left(\frac{1-R}{\eta} \right)^{0.6} = I(U_0) + \eta \, G(U_0) .$$

Plots of $\left(\frac{1-R}{\eta} \right)^{0.6}$ versus η, figure 2, showed that $I(U_0)$ and $G(U_0)$ were constants for a particular overvoltage (see table 1).

Dealing first with the intercept values, $I(U_0)$, these were plotted against $1/U_0$ (fig. 3), and the points shown to lie on a straight line at low values of U_0 but to diverge increasingly

[1] The authors draw attention to the fact that the expression for R differs from that given previously. Dr. K. F. J. Heinrich kindly pointed out that the original version, whilst performing satisfactorily for most systems, misbehaved when used at very high overvoltage ratios, e.g., when measuring carbon in tungsten carbide at 30 kV.

Table 1. Graphical data from plots of $\left(\dfrac{1-R}{\eta}\right)^{0.6}$ versus η

U_0	$I(U_0)$	$G(U_0)$
1.1	0.05	0.22
1.3	0.11	0.50
1.5	0.16	0.67
2.0	0.25	0.77
3.0	0.37	0.71
6.0	0.49	0.63
10.0	0.55	0.56
20.0	0.59	0.53
100.00	0.81	0.15
200.00	1.0	−0.08

at higher U_0 values. An equation of the form $a+b/U_0+c \exp(-d/U_0)$ might, therefore, seem appropriate, where a,b,c and d are constants; the exponential term is significant only at high U_0 values whilst the full equation is linear with $1/U_0$ at low overvoltages. The final equation was $I(U_0)=0.3\,[-1/U_0+\exp(1.5-1.5\ U_0^{-0.25})]$. It is shown as the solid line in figure 3 and is in good agreement with the experimental points.

After studying the $G(U_0)$ values, it was decided to plot them against $\ln U_0$ (fig. 4). The points are seen to lie close to a straight line at high overvoltages, but to fall away at low overvoltages according to an equation of the form $(a-b\ln U_0)\exp[f(U_0)]$, where a and b are constants. The final equation was

$$G(U_0)=(0.368-0.075\ln U_0)\exp(1-2.3U_0^{-4}),$$

and it is shown as a solid line in figure 4. Averaging R for a multicomponent specimen is carried out by weight averaging η according to $\displaystyle\sum_{i=1}^{i=n} c_i\eta_i$.

VI. The Absorption Correction

It is evident that a prerequisite for a good absorption correction is an accurate representation of the $\phi(\rho z)$ curve (fig. 1). There is now a large amount of experimental data available which illustrate how the $\phi(\rho z)$ curve varies as a function of incident electron energy, atomic number of the sample and x-ray line energy. A study of such data reveals that although the height of the $\phi(\rho z)$ curve and its extent along the mass depth axis varies

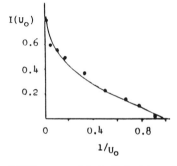

Figure 3. Plot of $I(U_0)$ versus; • Monte Carlo data, — formula in paper.

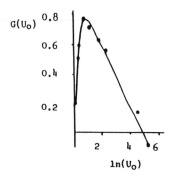

Figure 4. Plot of $G(U_0)$ versus ln U_0; • Monte Carlo data, — formula in paper.

with the particular experimental conditions employed, its general form does not. It is possible, therefore, to use a curve-fitting approach for determining $f(\chi)$ in which equations are developed to model experimentally established $\phi(\rho z)$ curves.

The procedure employed in the Love-Scott model is to describe the $\phi(\rho z)$ curve by two straight-line segments joining the points A, B and B, C, also plotted in figure 1. Point A represents the magnitude of x-ray generation at the sample surface, B is the height and position of the maximum, and C is close to the x-ray range; their coordinates are $[0, \phi(0)]$, $[\rho z_m, \phi(m)]$ and $(\rho z_r, 0)$, respectively. Using eq (1) it is relatively simple to obtain an analytical solution for $f(\chi)$ and this may be written as

$$f(\chi)=2\left[(\rho z_r-\rho z_m)\,(\rho z_m+h\,\rho z_r)\chi^2\right]^{-1}\left\{[-\exp\,(-\chi\rho z_m)+h\exp\,(-\chi\rho z_r)\right.$$

$$\left.+\chi(\rho z_r-\rho z_m)-h+1]+\left[\frac{\exp\,(-\chi\rho z_m)\,(\rho z_r-h\,\rho z_r)+h\,\rho z_r-\rho z_r}{\rho z_m}\right]\right\}\ ;$$

$h=\phi(m)/\phi(0)$.

Although at first sight the representation of $\phi(\rho z)$ by two straight lines may seem a very crude approximation, it has been shown [10] to be accurate to within 1 percent for $f(\chi)$ values ranging from 1 (no absorption) to 0.05 (95% absorption), i.e., the complete range of interest in practical microanalysis work. Thus any serious defect in the quadrilateral model will arise not from inadequacies in the quadrilateral shape itself but from deficiencies in the equation used to describe the parameters.

The method used to formulate the parameters ρz_m, ρz_r, and h was to utilize the recent tracer measurements and Monte-Carlo simulations of Sewell et al. [9]. Procedural details are described fully in a previous paper [10] and consequently only the final formulae will be given here:

$h = a_1 - a_2 \exp\,(-a_3 U_0^x)$,

where

$x = 1.225 - 1.25\eta$
$a_1 = 2.2 + 1.88 \times 10^{-3}\,Z$
$a_3 = 0.01 + 7.19 \times 10^{-3}\,Z$

and

$a_2 = (a_1 - 1)\,\exp\,(a_3)$.

To take advantage of the scaling properties of $\phi(\rho z)$ curves [11] and to constrain the quadrilateral model to work accurately for $f(\chi)$ values above 0.5, both ρz_m and ρz_r were expressed in terms of the mean depth of x-ray generation, $\overline{\rho z}$, which controls the magnitude of $f(\chi)$ when absorption is relatively small. The formulae derived were

$$\rho z_m = \overline{\rho z} \; [0.29 + (0.662 + 0.443 \; U_0^{0.2})Z^{-0.5}]$$

and

$$\overline{\rho z} = \frac{[\rho z_m^2 + h\rho z_r^2 + h\rho z_m \cdot \rho z_r]}{3(\rho z_m + h\rho z_r)}.$$

The general form of the equation for $\overline{\rho z}$ was developed using a combination of tracer measurements and Monte Carlo calculations, but the final expression was established by carrying out an optimization exercise utilizing EPMA measurements. The following formula[2] was obtained:

$$\overline{\rho z} = \frac{(1.1 \times 10^{-5} \, J^{1.1} E_0^{1.2} + 3 \times 10^{-6} \, J^{0.13} \, E_0^{1.75} \, E_0^{-0.0008 \eta U_0}) \ln \, U_0}{[(1.1 + 6.5\eta + 3.5J - 3\eta^{1.5}) \ln \, U_0 + 1 + 0.08/\eta] \, \Sigma \, c_i Z_i / A_i}.$$

The method we adopt for averaging ρz_m, ρz_r, h and $\overline{\rho z}$ for multi-element samples involves weight averaging the constituent terms Z, η, J, etc.

VII. The Database

To test the effectiveness of correction models a database of microanalysis measurements has been collected which totals, to date, 945 results. The data are the result of careful sifting through published information and are treated here under four separate listings.

The first set consists of binary alloys containing elements ranging from magnesium ($Z = 12$) to uranium ($Z = 92$) in the periodic table, the microanalysis data being referred to the pure element as the reference standard. The measurements may be regarded as being representative of much microanalysis work in that no single correction factor is wholly dominant; sometimes absorption is the largest, sometimes the atomic number correction and, very occasionally, the characteristic fluorescence factor. There are 554 results in this set and these have been listed in Sewell et al. [10], together with the experimental conditions.

The other three sets of data all refer to measurements of the ultra-light elements and, almost without exception, absorption is the largest of the correction factors. Clearly, these data will provide the most stringent test for any correction procedure. The first of the three ultra-light element sets consists of fluorine measurements on binary fluorides and oxygen measurements on binary and ternary oxides, the standards being magnesium fluoride and alumina, respectively. There are 94 results altogether and they are also listed in Sewell et al. [10]. Next, 117 carbon analyses on binary carbides with an iron carbide standard (Bastin and Heijligers [12]) are used for assessment and, finally, 180 recent boron measurements on binary borides using a boron standard (Bastin and Heijligers [13]).

VIII. Mass Absorption Coefficients

With regard to the first data set which consists of analysis of elements $Z \geqslant 12$ in binary alloys the following practice has been adopted. For x rays of energy above 2 keV the mass

---

[2] The expression for $\overline{\rho z}$ is slightly different from the version published in Sewell et al. [10] since we are now able to take advantage of some new EPMA measurements on ultra-light elements for optimization.

absorption coefficients of Heinrich [14] have been used via the algorithms of Springer and Nolan [15], i.e., for K radiation from elements above silicon ($Z = 14$) in the periodic table. For x rays of energy below 2 keV the values of Henke et al. [16] are preferred, our previous tests showing that their usage always gave the lower root mean square (RMS) errors, irrespective of which particular correction model was being assessed.

The situation regarding mass absorption coefficients for the ultra-light element data is less clear. Previously, we have used exclusively Henke's values (Henke et al. [16]) which do not differ substantially from the earlier data of Henke and Ebisu [17] deduced from x-ray attenuation measurements in gas mixtures (Henke et al. [18]). In the present assessment, however, we have used values from other sources as well and from a comparison of our results some comments will later be made concerning the reliability of the different sets of μ/ρ data. Hence we include the recent tabulation of Heinrich [19] which, to some extent, is based upon the data of Henke and Ebisu [17] but also includes theoretical predictions of Veigele [20] as well as a few isolated experimental measurements. Next we refer to some μ/ρ values given by Bastin and Heijligers as detailed earlier (Bastin and Heijligers [13, 21]). Interestingly, these workers appear to have used a method previously proposed by Love et al. [22], and subsequently to some extent criticized, in which the "thin film" model of Duncumb and Melford [23] was applied to microanalysis measurements to derive μ/ρ values for oxygen K radiation. Bastin's approach appears, however, not so straightforward since he seems to have carried out further, and to some extent arbitrary, adjustments based upon the manner in which his correction model behaved when using these μ/ρ values. Nevertheless, apart from values relating to x-ray line energies close to absorption edges, there is a measure of agreement between these and those of Henke and Ebisu [17].

IX.   Assessment of Love-Scott II Correction Model

The Love-Scott II correction model consists of the foregoing atomic number and absorption corrections combined with the characteristic fluorescence correction of Reed [24]; continuum fluorescence is ignored since previous experience suggests it is very small, in most cases less than 1 percent. We shall present the corrected data first in the form of

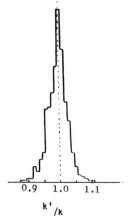

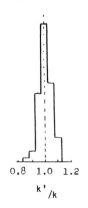

Figure 5. Histogram of corrected microanalysis data plotted as k'/k where k' is the x-ray intensity ratio predicted by the Love-Scott model and k is the measured value; data set of 554 heavier element systems.

Figure 6. As figure 5, but for data set of oxides and fluorides.

histograms, plotted with values of k'/k along the horizontal axis, where k is the measured intensity ratio and k' is the ratio predicted by the correction model from a knowledge of the specimen composition. Hence the closer k'/k is to unity, the better the model is working, with high values of k'/k indicating undercorrection and low values overcorrection. Thereafter RMS errors are quoted and these are referred with respect to the 1.0 position.

The correction model was applied first to heavier element data, i.e., the set of 554 measurements on binary systems, and the results are illustrated in figure 5. The first point to note is that the histogram is centered about 1.0 with very little bias and, secondly, that it has a fairly narrow base. Analysis of these data gave a mean value of 0.994 for k'/k and an RMS error of 3.1 percent. The actual errors attributable to the correction model itself will, of course, be less than this since inaccuracies in the measured data will undoubtedly be present due to beam instability and any imprecision in experimental settings, especially with data obtained on earlier instruments. If, for example, the RMS error in measured k values amounted to ~ 2 percent, the error associated with our model would be little more than 2 percent.

With regard to the ultra-light element data we treat separately the three sets of EPMA measurements referred to in section V, i.e., (a) binary carbides, (b) binary borides, and (c) with the fluorides and oxides together, the 25 measurements on fluorides hardly constituting sufficient information for a proper assessment to be made. The result of applying the mass absorption coefficients of Henke et al. [16] and Heinrich [19] to each set of measurements is given as well as the use of Bastin's values (Bastin and Heijligers [13, 21]) with the carbon and boron measurements.

The histogram (fig. 6) shows the result of applying, to the combined oxygen and fluorine measurements, the Love-Scott correction model with Heinrich's mass absorption coefficients. The distribution of data is symmetrical with a mean k'/k value of 1.003 and an RMS error of 4.73 percent (table 2). It was observed that the correction model appeared to work equally well on the oxide (4.74%) and fluoride (4.71%) data. However, the use of Henke's mass absorption coefficients produced much less satisfactory results giving an RMS error of 6.82 percent and a mean value of k'/k of 1.047.

Table 2. Assessment of Love-Scott model

Database	μ/ρ	% RMS error	mean (k'/k)
554 medium to heavy systems	Heinrich [14] plus Henke et al. [16] for radiation <2 keV	3.1	0.994
Oxides, Fluorides	Henke et al. [16]	6.8	1.047
	Heinrich [19]	4.7	1.003
Carbides	Bastin et al. [21]	3.8	0.997
	Henke et al. [16]	14.3	1.074
	Heinrich [19]	12.3	0.999
Borides	Bastin et al. [12]	9.0	1.043
	Henke et al. [16]	13.3	0.995
	Heinrich [19]	17.9	0.948
Borides (Ni & Co) systems removed	Bastin et al. [13]	5.3	1.017
	Henke et al. [16]	12.0	0.959
	Heinrich [19]	18.7	0.919

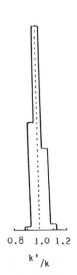

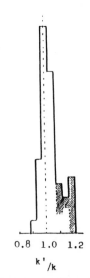

Figure 7. As figure 5, but for data set of carbides.

Figure 8. As figure 5, but for data set of borides; shaded area refers to borides of nickel and cobalt.

When tested on Bastin's carbon measurements (Bastin and Heijligers [12]) the correction model performed excellently if Bastin's own mass absorption coefficients (Bastin and Heijligers [21]) were employed. The histogram (fig. 7) shows very little bias with a mean k'/k value of 0.997 and an RMS error of only 3.8 percent (table 2); in fact, only 30 of the 117 results gave an error of greater than 7.5 percent. Interestingly, the use of either Henke's or Heinrich's mass absorption coefficients led to much larger RMS errors, with a tendency for k'/k values to be above unity.

Finally, the correction model was assessed on the boride measurements, using initially the mass absorption coefficients of Bastin. Results were somewhat disappointing, the RMS error being 9.0 percent and the mean (k'/k) equal to 1.043. Closer examination of the histogram (fig. 8), indicated that most of the results which produced the positive bias concerned borides of cobalt and nickel (shaded in the histogram). Bastin experienced a similar problem with these particular measurements due, he suggested, to an abnormally low emission of boron $K\alpha$ from these compounds. In fact, when these were removed from the data set (leaving 150 measurements) the RMS error is reduced substantially to 5.3 percent and the mean k'/k value to 1.017. Whilst a small degree of bias remains, it is now almost insignificant. The situation is quite different when using either Henke's or Heinrich's mass absorption coefficients and after omitting the cobalt and nickel borides from the assessment, the respective errors were 12 percent and 18.7 percent with mean k'/k values of 0.959 and 0.919 (table 2).

X. The Tilt Factor

So far the question of inclined specimens has not been discussed and clearly, in the search for a universal correction procedure it is important that this should be included. We have, in fact, addressed this issue in a recent paper [25] and for further details the reader is referred to this publication. In brief, the modifications involve introducing a trigonometrical factor, which includes the angle of incidence, to parameters such as $\overline{\rho z}$ as used in the quadrilateral absorption correction. The Monte Carlo calculations and tracer measurements on inclined specimens which were used are given in the above paper. As regards the atomic number correction, the necessary modification involves replacing η values for normal incidence with those for inclined incidence, viz, $\eta_\beta = 0.891 \, (\eta/0.891)^{\sin\beta}$, where β is the angle the incident electron beam makes with the specimen surface (see Darlington [26]).

It was shown that the programme worked well for aluminium $K\alpha$ measurements in $NiAl_3$ and $CuAl_2$ specimens and it now remains to be tested on a wider range of systems when published data are available.

XI. Conclusions

The new Love-Scott correction model incorporating the quadrilateral absorption correction has been assessed using a database of microanalysis measurements. As far as the medium to heavy element analyses are concerned the method works extremely well giving an RMS error close to 3 percent.

Ultra-light element performance is more difficult to assess because results are very dependent on the mass absorption coefficients used. Nevertheless RMS errors on the oxides and fluorides are less than 5 percent if Heinrich's latest coefficients are adopted and employment of Bastin's mass absorption coefficients on the carbides and borides (less the nickel and cobalt systems) produces RMS errors of 3.8 percent and 5.3 percent, respectively, with little if any evidence of bias.

One must also consider the extreme range of experimental conditions used when acquiring the ultra-light element data, ranging from accelerating voltages of 4 to 30 keV. For the very low accelerating voltages the measurements will be sensitive to the nature of the specimen surface whereas for the very high voltages the absorption correction will be extremely large and therefore more prone to error. If the accelerating voltages were restricted to the range 6 to 15 kV (typical values used in ultra-light element analyses) then RMS errors produced by the Love-Scott model would be reduced significantly.

In view of their impact on the results it is worthwhile commenting on the mass absorption coefficients for the ultra-light elements. Adoption of either Henke's or Heinrich's values tended to produce rather large errors when used with our correction model. Although agreement between all three sets of coefficients was reasonable for most elements, in some instances (particularly when close to an absorption edge) Heinrich's and Henke's values differed substantially from those of Bastin. Where such differences occurred it was the coefficients of Bastin that almost without exception produced much the superior results when used with our correction model and also, although not reported here, in tests we have performed using the PAP correction programme (Pouchou and Pichoir [27]). Indeed, with the greatly improved accuracy of the more recent correction programmes, the precision of the mass absorption coefficients themselves are often responsible for the larger errors experienced when carrying out quantitative soft x-ray analysis.

XII. Acknowledgments

To the SERC for their support.

XIII. References

[1] Scott, V. D. and Love, G., eds. (1983), Quantitative Electron Probe Microanalysis, Chichester: Ellis Horwood.

[2] Philibert, J. and Tixier, R. (1968), NBS Spec. Publ. 298, Heinrich, K. F. J., ed., 13-34.

[3] Bethe, H. A. and Ashkin, J. (1953), Experimental Nuclear Physics, John Wiley, New York.

[4] Love, G., Cox, M. G. C., and Scott, V. D. (1978), J. Phys. D: Appl. Phys. **11**, 7–21.

[5] Bethe, H. A. (1930), Ann. Phys. Leipz. **5**, 325–400.

[6] Duncumb, P. and Reed, S. J. B. (1968), NBS Spec. Publ. 298, Heinrich, K. F. J., ed., 133–154.

[7] Love, G., Cox, M. G. C., and Scott, V. D. (1977), J. Phys. D: Appl. Phys. **10**, 7–23.

[8] Fitting, H. J. (1975), J. Phys. D: Appl. Phys. **8**, 1481–6.

[9] Sewell, D. A., Love, G., and Scott, V. D. (1985), J. Phys. D: Appl. Phys. **18**, 1233–1243.

[10] Sewell, D. A., Love, G., and Scott, V. D. (1985), J. Phys. D: Appl. Phys. **18**, 1245–1268.

[11] Bishop, H. E. (1974), J. Phys. D: Appl. Phys. **7**, 2009–20; Darlington, E. H. (1975), J. Phys. D: Appl. Phys. **8**, 85–93.

[12] Bastin, G. F. and Heijligers, H. J. M. (1984), Quantitative Electron Probe Microanalysis of Carbon in Binary Carbides, University of Technology, Eindhoven Report.

[13] Bastin, G. F. and Heijligers, H. J. M. (1986), Quantitative Electron Probe Microanalysis of Boron in Binary Borides, University of Technology, Eindhoven Report.

[14] Heinrich, K. F. J. (1966), The Electron Microprobe, McKinley, T. D., Heinrich, K. F. J., and Wittry, D. B., eds., Wiley, New York, 296–377.

[15] Springer, G. and Nolan, B. (1976), Canad. J. of Spectrosc. **21**, 134–138.

[16] Henke, B. L., Lee, P., Tanaka, T. J., Shimabukuro, R. L., and Fujikawa, B. K. (1982), Atom. Data Nucl. Data Tables **27**, (New York: Academic Press).

[17] Henke, B. L. and Ebisu, E. S. (1974), Adv. in X-Ray Anal. **17**, 150–213.

[18] Henke, B. L., Elgin, R. L., Lent, R. E., and Ledingham, R. B. (1967), Norelco Reporter **14**, 112–131.

[19] Heinrich, K. F. J. (1987), X-Ray Optics and Microanalysis, Brown, J. D. and Packwood, R. H., eds., Ontario, University of Western Ontario, 67–119.

[20] Veigele, W. J. (1973), Atomic Data **5**, 51.

[21] Bastin, G. F. and Heijligers, H. J. M. (1986), X-Ray Spectrometry **15**, 143–150.

[22] Love, G., Cox, M. G. C., and Scott, V. D. (1974), J. Phys. D: Appl. Phys. **7**, 2131–2141.

[23] Duncumb, P. and Melford, D. A. (1966), X-Ray Optics and Microanalysis, Castaing, R., Deschamps, P. and Philibert, J., eds., Hermann, Paris, 240–253.

[24] Reed, S. J. B. (1965), Br. J. Appl. Phys. **16**, 913–926.

[25] Sewell, D. A., Love, G., and Scott, V. D. (1987), J. Phys. D: Appl. Phys. **20**, 1567–1573.

[26] Darlington, E. H. (1975), J. Phys. D: Appl. Phys. **8**, 85–93.

[27] Pouchou, J. L. and Pichoir, F. (1984), Recherche Aerospatiale, no. 3, 167–192.

QUANTITATIVE ANALYSIS OF HOMOGENEOUS OR STRATIFIED MICROVOLUMES APPLYING THE MODEL "PAP"

JEAN-LOUIS POUCHOU AND FRANÇOISE PICHOIR

Office National d'Etudes et de Recherches Aerospatiales
92320 Chatillon, France

I. Introduction

For about 20 years, quantitative analysis of homogeneous microvolumes has been performed with the aid of correction models which transform into mass concentrations C_A the ratio k_A between the emerging intensities from the specimen and a standard obtained for a characteristic line of element A:

$$k_A = (I_A)_{spc}/(I_A)_{std} \ .$$

The most frequently used correction procedure has been the ZAF procedure, which can be defined by the equation

$$k_A = C_A \cdot ZAF(C_A, C_B, \ldots) \ . \tag{1}$$

In this expression Z, A, and F represent, respectively, the effects of atomic number, of absorption and of fluorescence, which are calculated independently from each other on the base of the concentrations of all elements in the specimen.

While the atomic number correction may be difficult to obtain at very low operating voltages, it is the absorption correction that limits the conventional ZAF procedure most severely. It uses in effect the simplified Philibert model [1] which corresponds to a very crude representation of the depth distribution of primary x-ray generation $\phi(\rho z)$ by means of a linear combination of two exponentials (fig. 1). When this model was proposed, little information existed on the depth distribution, apart from some determinations by the tracer

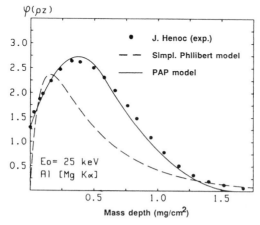

Figure 1. Depth distribution of Mg $K\alpha$ radiation in aluminum at 25 kV. Comparison of the simplified Philibert model (curve) and of the distribution obtained by the tracer method [72].

Electron Probe Quantitation, Edited by K.F.J. Heinrich and
D. E. Newbury, Plenum Press, New York, 1991

31

method by Castaing and Descamps [2]. The Philibert model, based on simple physical arguments, had the advantage of producing a simple formulation of the absorption correction factor $f(\chi)$, at a time when the electron probe microanalyzers were not in general outfitted with a computer. Furthermore, due to the lack of detecting instrumentation the problem of quantitative analysis of strongly absorbed radiation did not usually present itself.

At present the evolving requirements in material sciences include accurate routine analyses of light elements in an absorbing matrix such as aluminum in nickel or gallium arsenide. The development of new materials also made it necessary to analyze as quantitatively as possible the very light elements (boron, carbon, nitrogen, oxygen), the emissions of which are of low energy and hence always strongly, often very strongly, absorbed. At the same time there is an increasing need for characterizing compositionally and dimensionally superficial segregations such as thin (a few tens to hundreds of nm) films, as are prepared, for instance, in microelectronics, or those of a few μm in dimension which could be used as diffusion barriers in some composite high temperature materials.

The experimental conditions for such applications to the analysis in depth vary considerably with the problem. For very thin superficial segregations one would in general prefer an analytical line of low energy, and a low accelerating voltage for the incident electrons. For the dimensional analysis of thicker layers one may have to use the maximum available accelerating voltage so that a line of high energy (and hence low absorption) is excited. In even more complex cases, such as specimens with multiple layers or concentration gradients close to the surface, measurements must usually be made at various conveniently chosen tensions.

We believe that, rather than treating separately the improvement of corrections for homogeneous specimens and the characterization of specimens of composition variable in depth, it would be more logical and efficient to conceive a general and unique model for the calculation of emergent x rays. For this we have used the usual fundamental physical concepts applied in x-ray microanalysis and we apply to them as accurate a description of the depth distribution, $\phi(\rho z)$ as possible.

II. The PAP (Pouchou and Pichoir) Procedure

A. GENERAL PRINCIPLES

From the start [3,4], the PAP model was conceived not only to provide an operative model for correction, but also to be applicable to problems such as the determination of absorption coefficients [5], and especially to the analysis of stratified specimens [6,7]. For such applications the calculation of correction factors is not sufficient; we must be able to calculate precisely the emergent intensity for soft as well as for hard lines and for a wide range of excitation potentials. This requires a distribution of $\phi(\rho z)$ as realistic as possible. To obtain a reliable model it is desirable that the parameters of the shape of $\phi(\rho z)$ should not influence the value of the generated primary intensity, contrary to what happens with the modified Gaussian model of Brown and Packwood [8]. To avoid such an interference it is sufficient to use the distribution $\phi(\rho z)$ in conformity with the definition that Castaing has given for it [9]: $\phi(\rho z)$ is the ratio between the intensity of an elemental layer of mass thickness $d\rho z$ at mass depth ρz and that of an identical but unsupported layer.

Paradoxically, none of the other existing models satisfies that definition which is equivalent to saying that the integral F of $\phi(\rho z)$ is proportional to the number of primary ionizations generated by the incident electron on the level l of the atoms A:

$$n_A = C_A \cdot (N^0/A) \cdot Q_l^A(E_0) \cdot F \qquad (2)$$

where

$$F = \int_0^\infty \phi(\rho z) \cdot \mathrm{d}\rho z \, ,$$

N^0 is Avogadro's number and $Q_l^A(E_0)$ the ionization cross section at the level l of the atoms A for an electron incident at the energy E_0.

Except for a number of factors, the integral F of $\phi(\rho z)$ must provide the generated primary intensity and thus carry the effects of the atomic number. Hence, the area F is a fundamental parameter of the PAP model, and it ensures a proper consideration of the atomic number effects, whatever may be the detailed shape of the depth distribution $\phi(\rho z)$. The rest of the parameters of the model serves to specify the shape of the distribution, and their choice depends therefore on the mathematical description adopted for $\phi(\rho z)$. This mathematical description must be able to provide a distribution of quite variable aspects, depending on the atomic number of specimen (figs. 2 and 3), but it must also be such that the resulting system of equations should have an analytical solution when conditions are imposed to the area and to certain form parameters.

Therefore, the PAP procedure is neither like a ZAF structure with separate Z and A corrections, nor like a global structure such as the model of Brown and Packwood in which emerging intensity is obtained directly from a parametric description of $\phi(\rho z)$ without the appearance of the atomic number effects. The PAP procedure in fact agrees more with the fundamental principles of microanalysis, with a realistic description of $\phi(\rho z)$ hav-

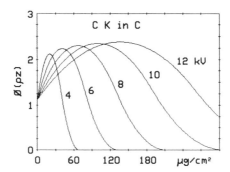

Figure 2. Depth distributions of the C K radiation in carbon from 4 to 12 kV (Parabolic PAP model).

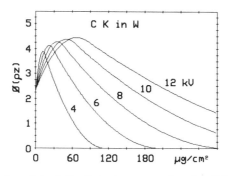

Figure 3. Depth distributions of the C K radiation in tungsten, from 4 to 12 kV (Parabolic PAP model).

ing an area in accordance with Castaing's definition. The PAP calculation is performed in two steps:

- calculation of the area of the distribution, equivalent in principle to that of an atomic number correction.

- direct calculation of the generated intensity, on the basis for the distribution $\phi(\rho z)$ which is defined by its area and by the parameters of form adapted to the selected mathematical representation.

The ensemble of the expressions and parameters used in the PAP procedure is not the simple result of an adjustment to an analytical data base, but it is derived from taking into account the following inputs of diverse nature:

- Distributions $\phi(\rho z)$ experimentally obtained by the tracer method,

- Measurements of the depth of penetration of electrons in function of their energy,

- Monte Carlo simulations,

- Analysis of binary specimens of known composition at a wide range of accelerating tensions,

- Analyses of stratified specimens (layers of known thickness and composition on known substrates) as a function of accelerating tension,

- Measurement at variable tension of the intensity emerging from known specimens.

B.   AREA PARAMETER (NUMBER OF PRIMARY IONIZATIONS)

The average number of primary ionizations due to an electron of initial energy E_0 at the level l of the atoms A is commonly expressed by means of a deceleration factor $1/S$ and a backscatter factor R:

$$n_A = C_A \cdot (N^0/A) \cdot (R/S) \quad \text{with} \quad 1/S = \int_{E_0}^{E_l} [Q_l^A (E)/(dE/d\rho s)] \cdot dE . \tag{3}$$

For the calculation of $1/S$ one should seek expressions for the ionization cross section $Q(E)$ and the deceleration $dE/d\rho s$ that agree well with the available experimental evidence, and, if possible, have a form such that they do not require a numerical integration.

Deceleration of Electrons

Commonly the law of Bethe [10] is used in the ZAF procedure to calculate the average energy loss:

$$dE/d\rho s = -(2\pi e^4 N^0/E) \cdot \sum_i (C_i Z_i/A_i) \cdot \ln (1.166 \, E/J_i) \tag{4}$$

(J_i is the mean ionization potential of each constituent). The above expression is known to be satisfactory for electrons of high energy (say, from 30 keV upwards), but it produces lower results than the measured penetrations at low energies [11]. It is also desirable to use an expression mathematically defined at very low energies, contrary to expression (4) in which the logarithmic factor can change sign. As other authors have done [12], it is possible to represent the energy-dependent term in eq (4) by $1/f(V)$, where $V=E/J$, and write the deceleration as follows:

$$dE/d\rho s = -(M/J) \cdot [1/f(V)] \tag{5}$$

with $M = \sum_i C_i \cdot Z_i/A_i$ and J defined by

$$\ln(J) = \sum_i [C_i (Z_i/A_i) \cdot \ln(J_i)]/M \ . \tag{6}$$

The expression we adopt for the mean ionization potential is that of Zeller [13] which was obtained by a fit to the results of slowing-down measurements on protons [14,15]:

$$J_i = 10^{-3} Z_i \cdot [10.04 + 8.25 \exp (-Z_i/11.22)] \quad (J_i \text{ in keV}) \ . \tag{7}$$

Having so defined J, we must find a semiempirical expression for $f(V)$ such that the algorithm (5) is equivalent to Bethe's law for 30–50 keV; i.e., giving in this domain a course that varies approximately as $E^{1.7}$ [16]. Experiments show that the power of E must decrease with the energy so as to reach 1.3 for electrons of an energy of a few keV [11,17,18,19]. Finally, a slowing-down model for an electron gas predicts at very low energies a deceleration proportional to the velocity of the particle [20].

These requirements have led us to adopt for $f(V)$ a sum of three powers of V, of the form:

$$f(V) = \sum_{k=1}^{3} D_k \cdot V^{P_k} \tag{8}$$

with $\quad D_1 = 6.6 \ 10^{-6} \qquad\qquad\qquad P_1 = 0.78$

$\qquad\quad D_2 = 1.12 \ 10^{-5} (1.35 - 0.45 \ J^2) \qquad P_2 = 0.1$

$\qquad\quad D_3 = 2.2 \ 10^{-6}/J \qquad\qquad\qquad P_3 = -(0.5 - 0.25 \ J)$

(J in keV; $dE/d\rho s$ in keV $\cdot$ g^{-1} cm^2) .

This expression, useful up to 50 keV, gives at high energy a deceleration similar to that of Bethe and at lower (1–10 keV) energies a weaker slowdown, and it follows a reasonable trend as the electron energy tends to zero (fig. 4). Figure 5 shows the evolution with energy of the range, calculated for diverse targets. The expression of the total trajectory R_0 (in g/cm^2) for an electron of energy E_0 can easily be deduced from eq (5):

$$R_0 = (1/M) \cdot \sum_{k=1}^{3} [J^{1-P_k} \cdot D_k E_0^{1+P_k}/(1+P_k)] \ . \tag{9}$$

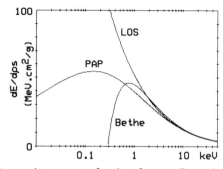

Figure 4. Deceleration of electrons in copper as a function of energy. Comparison of the expressions used in the PAP model and in the Love-Scott model with Bethe's law.

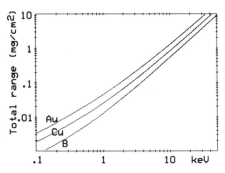

Figure 5. Variation with energy of the path of electrons in boron, copper, and gold (PAP model).

Ionization Cross Section

The only ZAF programs that require an expression for the ionization cross section are those using the rigorous model for atomic number correction by Philibert and Tixier [21]. The cross section used in this model, derived from the equations of Bethe, is proportional to $\ln(U)/U$, where U is the overvoltage E/E_l. This simple formulation gives a maximum cross section known to be too pronounced at $2 < U < 3$. Other theoretical or empirical expressions that were proposed are generally too mathematically complex to allow for an analytical calculation of the integral (3). A satisfactory way of varying the cross section with U is obtained with the expression proposed by Hutchins [23]:

$$Q_l^A(U) \propto \ln(U)/(U^m \cdot E_l^2) \tag{10}$$

where the power m, equal to 1 in the programs of type ZAF, is taken to be between 0.7 and 1, following Hutchins.

While the experimental evidence is quite inconsistent, a recent compilation of results [24] confirms a variation of the ionization cross sections within the above limits. These limits agree with the values of m we have adopted in the PAP model in order to satisfy our measurements of emerging intensity from solid targets at variable energy, as well as the measurements of relative intensity of thin surface films at varying tension:

for the K level: $m = 0.86 + 0.12 \exp[-(Z_A/5)^2]$
for the L levels: $m = 0.82$
for the M levels: $m = 0.78$

(Z_A is the atomic number of the excited element, E_l the excitation energy of the level considered, and $U = E/E_l$ the overvoltage.)

The expression (10) for the ionization cross section, associated with eq (5) for deceleration, permits the analytical calculation of the deceleration factor. When T_k is defined by $T_k = 1 + P_k - m$, we obtain:

$$1/S = [U_0/(V_0 \cdot M)] \cdot \sum_{k=1}^{3} D_k (V_0/U_0)^{P_k} \cdot [T_k U_0^{T_k} \cdot \ln(U_0) - U_0^{T_k} + 1]/T_k^2 . \tag{11}$$

Backscattering Losses

In the classical ZAF procedure the losses due to backscattering are expressed by a factor R smaller than 1, calculated by Duncumb and Reed [25] from energy distributions of backscattered electrons measured by Bishop [26]. In practice, one uses polynomial expressions of Z and $1/U$ [27].

In the PAP model we have preferred using an expression which refers explicitly to the basic physical parameter which is the backscatter coefficient η, so that in the future the method could be adapted easily to the case of analysis with oblique electron beam incidence. Although the expression of Love and Scott [12] uses η as the basic parameter, we did not employ it since it comports itself incorrectly at high overvoltages in heavy targets. (A new parametrization of R is proposed by Scott and Love in the present volume.) We have expressed the backscatter factor R in function of the mean backscatter coefficient $\bar{\eta}$ of the electrons, and of the mean reduced energy of the backscattered electron, $\bar{W}$:

$$R = 1-\bar{\eta} \cdot \bar{W} \cdot [1-G(U_0)] \ . \tag{12}$$

The expressions needed for the calculation are given in Appendix 1.

After calculating R/S, representing the primary intensity, and $Q_l^A(E_0)$, one determines the area F of the distribution $\phi(\rho z)$:

$$F = (R/S)Q_l^A(E_0) \ . \tag{13}$$

C. PARABOLIC REPRESENTATION OF THE DISTRIBUTION $\phi(\rho z)$

To satisfy the requirements exposed in paragraph A, the distribution $\phi(\rho z)$ was mathematically expressed by means of two parabolic branches, equal in value and slope at a certain mass depth level R_c (fig. 6). The conditions imposed on the distribution are:

- the integral must be equal to F
- at the surface, a certain value of $\phi(0)$ must be obtained
- the maximum must appear at a certain value R_m
- the function must vanish, with a horizontal tangent, at a predetermined range of ionization, R_x.

Therefore, the distribution is written as follows:

- from $\rho z = 0$ to $\rho z = R_c$ $\phi_1(\rho z) = A_1 \cdot (\rho z - R_m)^2 + B_1$ (14)
- from $\rho z = R_c$ to $\rho z = R_x$ $\phi_2(\rho z) = A_2 \cdot (\rho z - R_x)^2 \ .$ (15)

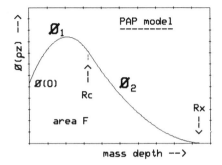

Figure 6. Schema of the representation of $\phi(\rho z)$ by two parabolic branches in the PAP model.

These conditions lead to a system of equations which has under the usual conditions an appropriate root R_c between R_m and R_x:

$$R_c = \frac{3}{2}\{[F-\phi(0)R_x/3]/\phi(0)-d^{1/2}/[\phi(0)\cdot(R_x-R_m)]\}$$

with

$$d = (R_x-R_m)\cdot[F-\phi(0)\cdot R_x/3]\cdot[(R_x-R_m)\cdot F-\phi(0)\cdot R_x(R_m+R_x/3)] \ . \tag{17}$$

Using this root and the form parameters of the distribution, we obtain the coefficients of the parabolic branches:

$$A_1 = \phi(0)/\{R_m\cdot[R_c-R_x\cdot(R_c/R_m-1)]\} \tag{18}$$
$$B_1 = \phi(0)-A_1\cdot R_m^2$$
$$A_2 = A_1\cdot(R_c-R_m)/(R_c-R_x) \ .$$

The expressions of the form parameters $\phi(0)$, R_x and R_m are given in Appendix 2.

The total emerging intensity is, except for proportionality factors identical for specimen and standard, the sum of two integrals:

$$I_A \propto C_A \int_0^{R_c} \phi_1(\rho z)\cdot\exp(-\chi\cdot\rho z)\cdot d\rho z + C_A \int_{R_c}^{R_x} \phi_2(\rho z)\cdot\exp(-\chi\cdot\rho z)\cdot d\rho z \ . \tag{19}$$

The ionization cross section $Q_I^A(E_0)$ is among the proportionality factors omitted in the above expression, and must be reintroduced if we are interested in the variation of intensity with the acceleration potential. The fluorescent yield, the weights of the line of interest and the instrumental factors (solid angle and detection efficiency, incident current) are other factors omitted in eq (19).

The simplicity of the mathematical formulation of $\phi(\rho z)$ allows an easy analytical calculation of the integrals of eq (19). Although only the emergent intensity is of practical interest one can, if desired, deduce the absorption correction factor, usually defined as $f(\chi)=F(\chi)/F(0)$, and which is written the PAP model as follows:

$$f(\chi) = [F_1(\chi)+F_2(\chi)]/F \tag{20}$$

where $F_1(\chi)$ and $F_2(\chi)$ are the integrals in expression (19), and F is the area of the distribution:

$$F_1(\chi) = (A_1/\chi)\cdot\{[(R_c-R_m)\cdot(R_x-R_c-2/\chi)-2/\chi^2]\cdot\exp(-\chi\cdot R_c)$$
$$-(R_c-R_m)\cdot R_x+R_m(R_c-2/\chi)+2/\chi^2\}$$
$$F_2(\chi) = (A_2/\chi)\cdot\{[(R_x-R_c)\cdot(R_x-R_c-2/\chi)-2/\chi^2]\cdot\exp(-\chi\cdot R_c)$$
$$-(2/\chi^2)\cdot\exp(-\chi\cdot R_x)\} \ . \tag{21}$$

With certain computers, at very low values of U_0 and χ it may be necessary to use a development of exponentials in order to avoid numerical roundoff errors. This problem exists in all models where integrations are performed between finite limits such as, for instance, the quadrilateral model of Love and Scott [29].

At very low overvoltages, rarely used in present practice, the problem of the splicing of the two parabolic branches of $\phi(\rho z)$ may not have a satisfying solution. The modifications to be applied in such case are indicated in Appendix 3.

The version of the PAP model presented above is that presently distributed by CAMECA.

D. SIMPLIFIED MODEL: XPP

The preceding absorption model corresponds to the needs for a general approach, adapted at the same time to the correction problems, to the determination of x-ray absorption coefficients and to the analysis of stratified specimens, for which the most accurate description of $\phi(\rho z)$ is required. But when we limit ourselves to the problem of corrections in quantitative analysis of homogeneous microvolumes, one can be satisfied with models for $\phi(\rho z)$ of a less rigorous fit but of a mathematical formulation conducive to more rapid calculations.

The best way to produce a mathematically simple model for $\phi(\rho z)$ is to use exponential functions integrated between zero and infinite. The experience of the exponential formulae produced by Philibert's model has shown that it is impossible to obtain for the general case correct depth distributions. The problem of such distributions is their tendency to concentrate the generated intensities at too shallow depths below the surface. Heinrich [34] has only partially attenuated this tendency when he recently proposed an approach equivalent to using as distribution in depth the product of an exponential and a linear function.

We have shown recently [35] that one can shift the maximum of the distributions to greater depth and obtain a better description of $\phi(\rho z)$ (fig. 7) with an expression of the form

$$\phi(\rho z) = A \cdot \exp(-a\rho z) + [B\rho z + \phi(0) - A] \cdot \exp(-b\rho z) \tag{22}$$

where the coefficients A, B, a, and b are deduced from the following characteristical parameters: the area F of the distribution, its value $\phi(0)$, its slope P at the surface, and the average depth of ionization R. One obtains:

$$F = \int_0^\infty \phi(\rho z) \cdot d(\rho z) = A/a + [\phi(0) - A]/b + B/b^2 \tag{23}$$

$$\bar{R} = (1/F) \int_0^\infty \rho z \cdot \phi(\rho z) \cdot d(\rho z) = A/a^2 + [\phi(0) - A]/b^2 + 2B/b^3 \tag{24}$$

and

$$P = B - aA - b[\phi(0) - A] \; . \tag{25}$$

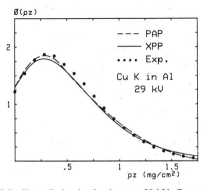

Figure 7. Depth distribution of Cu $K\alpha$ radiation in aluminum at 29 kV. Comparison of the PAP model (– – – –) and the XPP model (–––––) with a distribution obtained by the tracer method [2].

The emergent intensity is, omitting some factors identical for specimen and target:

with
$$I_A \propto C_A \cdot Q_l^A(E_0) \cdot F(\chi) \tag{26}$$

$$F(\chi) = A/(a+\chi) + [\phi(0)-A]/(b+\chi) + B/(b+\chi)^2 \ . \tag{27}$$

The absorption correction factor is:
$$f(\chi) = F(\chi)/F \ .$$

Having calculated the area F and the surface ionization $\phi(0)$ as was done in the parabolic PAP model, the average depth of ionization $\bar{R}$ is obtained by:

$$F/\bar{R} = 1 + \{X \cdot \ln[1 + Y \cdot (1-1/U_0^{0.42})]\}/\ln(1+Y) \tag{28}$$

with
$$X = 1 + 1.3 \ln (\bar{Z}_b) \qquad \text{and} \qquad Y = 0.2 + \bar{Z}_b/200 \ .$$

$\bar{Z}_b$ is the mean atomic number defined in Appendix 1 for the calculation of the backscattering factor.

The initial slope P can be related to the ratio $F/\bar{R}^2$ by:

$$P = g \cdot h^4 \cdot F/\bar{R}^2 \tag{29}$$

with
$$g = 0.22 \ln (4\bar{Z}_b) \cdot \{1 - 2 \exp[-\bar{Z}_b \cdot (U_0-1)/15]\}$$

and
$$h = 1 - 10[1 - 1/(1 + U_0/10)]/\bar{Z}_b^2 \ .$$

The correction procedure based on this simplified model (called XPP) has a structure similar to that of the usual PAP procedure, because of the condition of area imposed to $\phi(\rho z)$. This model has been recently adapted to analysis with oblique beam incidence which is a configuration of interest for scanning electron microscopes [77].

When a and b have very similar values, difficulties of roundoff may occur with some computers. The expression for b and the practical fashion of performing the calculations are described in Appendix 4.

E. CHOICE OF ABSORPTION COEFFICIENTS

The problem of modeling the mass absorption coefficients was recently discussed by Heinrich [36]. The new proposed model (MAC 30) has been shown to provide absorption coefficients which are much more satisfactory in a number of situations such as that of radiation emitted by light elements (aluminum, magnesium, silicon...) in elements of medium atomic number such as nickel or copper. There remain however cases, particularly among very light emitters, but also among the discontinuities M of heavy elements, where much work is still to be done. The electron probe microanalyzer is one of the instruments which lend themselves very well to this work [5].

The absorption coefficients associated with the PAP procedure are those of the model MAC 30 of Heinrich except for very light emitters ($Z < 10$); for these the coefficients are read from a table. Originally this table was constructed from the values proposed by Henke et al. [38], but it is continually reviewed as new data appear. Besides, in recent versions of programs a supplementary table is consulted when the radiation if of energy very close to an absorption discontinuity or for certain particular resonances (see Appendix 5).

F. ITERATIVE PROCEDURE

Analyses for boron in lanthanum hexaboride [37] have demonstrated that the classical methods of iterations resulted in divergence starting from the first cycle when very light elements were analyzed in heavy matrices which absorbed weakly compared with the pure standard [40]. The probability of finding such a situation is small at present but it may increase when analyses for beryllium become more current, due to the multilayer monochromators which are appearing in the market.

In such situations, which are infrequent, where the effects of atomic number and absorption act in the same direction to produce an intensity ratio larger than the concentration (and more so as the tension is raised), the curve $k = f(C)$ can have a maximum above one. Therefore two different concentrations can give the same measured intensity ratio (fig. 8), a case not considered in the conventional hyperbolic approximation. In such a case, Newton's method also poses some problems, since near the maximum $d(C/k)/dC$ is close to $1/k$ and divergence occurs.

The solution we have adopted is to test for each element before each iterative cycle the value of k/C. If this value is below or equal to 1, the classical hyperbolic approximation is used. If not, the hyperbolic approximation is replaced by parabolic approximation:

- When $k/C < 1$, $C/k = \alpha + (1-\alpha) \cdot C$ (hyperbolic approx.)
- When $k/C > 1$, $k/C = \alpha + (1-\alpha) \cdot C$ (parabolic approx.) .

This method not only resolves the particular problems cited above but also reduces the number of cycles needed for convergence in a number of cases where the atomic number effects are preponderant.

III. Evaluation of Performances of Models

A. ANALYSIS OF LIGHT TO HEAVY ELEMENTS

It would exceed the scope of this publication to give in detail the study of the many cases required to evaluate a given correction model, and especially to understand the rea-

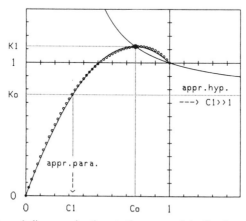

Figure 8. Parabolic and hyperbolic approximations to the output of the first iterative cycle. Case of boron in La B_6 at 30 kV and 40 degrees of x-ray emergence. Application of the PAP correction to the relative intensity k_0 measured by [37].

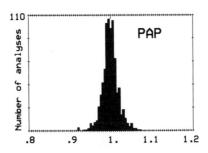

Figure 9. Error histogram of the ratio k_{calc}/k_{exp} obtained by the PAP procedure applied to 826 entries given in Appendix 6.

sons for its success or failure. A number of curves of relative intensity as a function of acceleration potential can be found in previous publications [3,40,41]. Here we will present an overall evaluation with the aid of error histograms derived from applying the models to an experimental database.

Clearly, the conclusions one may deduce from a comparison of results calculated on the basis of experimental data can depend strongly on the selection of the data. To set up a significant database is a fairly difficult operation which must be based on certain objective criteria.

For the analysis of light to heavy elements, the selected criteria for inclusion in the set were the following:

- binary conducting specimens of well-defined nominal compositions;
- sets of measurements at variable acceleration voltages, with reference to elemental standards;
- measurements with x-ray emergence angles of at least 40°, to limit the effect on the measurements of surface conditions (this requirement excludes most old measurements);
- measurements of relative intensities of at least 2% so as to limit the effects of errors or spread of results due to counting statistics, of background correction, and, lastly, of nominal composition;
- coherence between series of measures obtained on similar compositions analyzed under comparable conditions.

Following these criteria we have formed a database of 826 entries (see Appendix 6). The error histograms of the ratio k(calculated)/k (experimental) are presented in figures 9 and 10 for the PAP and ZAF procedures. Table 1 gives the average m and the standard deviation s for each model of reference.

The first conclusion from table 1 is that no model provides a deviation lower than 1–2% relative as long as pure elements are used as standards.

As far as individual trends are concerned we find that the simplified XPP gives results which are a little better than the PAP results because its parameters have in part been adjusted to the database of consideration.

Three methods give quite similar overall results: the procedure of Bastin [37] (which uses an optimized form of the Packwood-Brown [8] model), the quadrilateral model of Love and Scott [12,29] and the model of normalized distribution of Tanuma and Nagashima [42]. It should be remarked, however, that the version 861 of the Bastin model which gives good results concerning absorption gives poorer results concerning the atomic-number

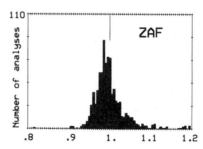

Figure 10. Error histogram of the ratio k_{calc}/k_{exp} obtained by applying the ZAF procedure to 826 entries given in Appendix 6.

correction. A recent unpublished version of Bastin aims to eliminate this deficiency by abandoning the combined $Z \cdot A$ approach in favor of a separate atomic-number effect treatment according to the equations of the PAP model.

The ZAF procedure gives a bias of 1–1.5% in the average on the atomic-number correction according to Philibert and Tixier [21] with an average ionization potential $J = 11.5 \cdot Z$. The expression for J of Ruste reduces this bias somewhat.

Concerning the absorption correction, the model proposed by Heinrich introduces a significant but insufficient improvement in the ZAF procedure; to the contrary, the modifications of Ruste and Zeller [43] produce in some cases a significant deterioration, specially in the analysis of light elements such as aluminum in targets of medium atomic number such as nickel and copper.

B. ANALYSIS OF VERY LIGHT ELEMENTS

It is quite difficult to obtain a significant database for the determination of very light elements. It is in first place desirable to eliminate data on insulating materials, even if metal-coated, in which the charge effects could be significant. But even for conductors, the measurement of soft radiation from elements of low atomic number is difficult for several reasons:

– low sensitivity of detection (low intensity, poor peak-to-background ratio),
– strong effects of lack of flatness, surface irregularity and contamination of both specimens and standards,

Table 1. Performances of eight correction procedures applied to light-to-heavy element analyses: average and standard deviation of the distribution of the ratio [k (calculated)/k (experimental)]. 826 analyses, μ/ρ MAC 30 Heinrich [36].

Procedure	826 Analyses		577 Analyses Effect of Z predominant		242 Analyses Absorption predominant	
	m	s	m	s	m	s
XPP	0.9997	1.79%	0.9983	1.58%	1.0023	2.17%
PAP	0.9982	1.91%	0.9975	1.60%	0.9993	2.50%
Love-Scott	0.9915	2.59%	0.9941	1.79%	0.9845	3.79%
Tanuma-Nagashima	0.9917	2.73%	0.9893	2.05%	0.9967	3.85%
Bastin 861	1.0090	2.92%	1.0138	2.84%	0.9976	2.83%
ZAF f(x) Heinrich	0.9850	2.74%	0.9881	2.01%	0.9765	3.84%
ZAF Philibert simpl.	0.9999	4.05%	0.9854	1.96%	1.0339	5.50%
ZAF Ruste-Zeller	1.0089	4.21%	0.9958	2.02%	1.0401	6.07%

- chemical binding effects which modify the position and spectral distribution of the characteristic emission bands. It is even possible that the chemical binding may affect the characteristic features of emission such as fluorescence yield or transition probabilities between levels. No model takes into account such modifications, which apparently occur in cases such as borides of cobalt and especially of nickel.
- polarization of the radiation emitted from anisotropic materials, which modifies the band shapes observed depending of the crystallographic orientation of the analyzed microvolume with respect to the spectrometer. The importance of these polarization effects depends on the Bragg angle with which the radiation has been analyzed.

The measurements of boron in binary borides by Bastin and Heijligers [37] seem to form a reliable database, well adapted to the evaluation of the correction models for very light elements (provided that the borides of cobalt and nickel are omitted for the reasons indicated above). Hence one has a set of 153 area k-ratios measured between 4 and 30 kV, with reference to a pure element. Table 2 shows the comparison of models, using average and standard deviation of the ratio [k(calculated)/k(experimental)] for this purpose.

Two sets of absorption coefficients were used in this comparison: those proposed by Bastin [37] for his model and those referred to in Appendix 5 by the PAP model. The absorption coefficients proposed by Bastin are essentially obtained by adjusting them to the results of analyses forming that database. For the PAP model, they result mainly from the use of emergent intensity measures at varying tension. The coefficients referred to in both sets differ but little [40]. Table 2 shows that the use of one or the other does not produce variation of more than 1% in Bastin's model. It is remarkable that two models of very different concepts and used in different manner should lead to very similar absorption coefficients. Besides it is known [31] that with respect to the values tabulated by Henke et al. [38] the coefficients of absorption adjusted to the Bastin model improve the performances of the ZAF, Ruste-Zeller or Love-Scott procedures.

It is useful to note that after this adjustment of absorption coefficients the PAP and BAS 861 models reflect correctly the changes of relative intensity in the entire domain of acceleration voltages, for all carbides and borides considered. With models of less realistic $\phi(\rho z)$ distributions it is more often impossible to obtain such an agreement: an adjusted absorption coefficient then produces agreement of calculation and experience only for a given composition and a single operating condition; i.e., for one tension and one take-off angle only.

Table 2. Performances of eight correction procedures applied to very light elements. Average and standard deviation of the ratio [k(calculated)/k(experimental)]. 153 analyses of binary borides; standard is pure boron. μ/ρ according to Bastin (B) or Pouchou and Pichoir (P)

Procedure	μ/ρ	k(calculated)/k(experimental)	
		Ave.	Standard Deviation
PAP	P	1.0075	3.95
XPP	P	1.0106	4.76
Bastin 861	B	1.0080	5.17
Bastin 861	P	1.0162	6.25
ZAF Ruste-Zeller	P	1.0139	9.59
Love-Scott quadrilateral	P	1.0148	12.24
ZAF + $f(\chi)$ Heinrich	P	1.0657	14.25
Tanuma-Nagashima	P	1.1018	17.51
ZAF Philibert simplified	P	1.1433	19.72

The main message of table 2 is that the results corresponding to different models are significantly different, and that the best models at present allow, in the particularly difficult situation of the analysis of a very light element with respect to an elementary standard, obtaining standard deviations in the order of 4%; i.e., a precision similar to that of the usual ZAF procedure applied to elements other than the very light ones.

One can also observe that the simplified procedure XPP conserves the satisfactory performance of the model PAP although the database considered here, which uses a very light standard, is particularly unfavorable.

The model Bastin 861 gives for very light elements results similar to those of XPP. The models of Love-Scott and of Ruste-Zeller produce results of significantly higher dispersion, although their average remains fairly satisfactory.

The absorption correction proposed by Heinrich is significantly better than the Philibert model usually employed in ZAF programs, but cannot always eliminate completely the biases characteristic of exponential models. Finally, the model of Tanuma and Nagashima which was established for the analysis of oxygen is conducive to error when applied to the analysis of a very light element using an elementary standard.

IV.   Analysis of Surface Segregations

A.   GENERALITIES

Under the usual working conditions x-ray microanalysis permits the investigation of materials to a depth in the order of 1 mg/cm^2, i.e., in the order of 1 μm for material of average density. However, the maximum depth of excitation can in fact vary by more than two orders of magnitude with the acceleration energy of the incident electrons (fig. 11). It follows that in the area of analysis of segregations close to the surface, one should in general select the optimum operating conditions for the problem of interest; this may imply using lines of low energy at very low tension or, to the contrary, very high tension. This explains the big effort made to enable the PAP model to give realistic distributions in depth including in extreme situations.

At the time of the first tentatives of using the electron probe for superficial analysis it was hardly possible to vary the operating conditions since the only known depth distributions $\phi(\rho z)$ were those of Castaing and Descamps [2] at 29 kV. Sweeney et al. [44] were the first trying to use depth distributions for thicknesses of various metals evaporated on glass, in a depth range up to 300 nm. Cockett and Davis [45] also used the distributions to show, by measuring the emissions from both the layer and the substrate, that the segrega-

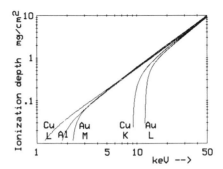

Figure 11.   Variation with electron energy of the depth range of excitation of K lines in aluminum, of K and L lines in copper, and of L and M lines in gold (PAP model).

Table 3. Count rates and line-to-background ratios for the lines Cu $K\alpha$ and Cu $L\alpha$ of pure copper under working conditions giving 20, 50, or 100 nm for limit in depth of ionization. Microprobe Camebax, $I = 100$ nA.

Line		Ultimate ionization depth		
		20 nm	50 nm	100 nm
Cu $K\alpha$	Acc. volt.	–	9.3 kV	9.7 kV
	Peak (c/s)	–	60 c/s	250 c/s
	Peak/Bkgd	–	4	13
Cu $L\alpha$	Acc. volt.	1.6 kV	2.6 kV	4.2 kV
	Peak (c/s)	350 c/s	1600 c/s	4000 c/s
	Peak/Bkgd	100	130	100

tions of a few tens of nm could be detected at the surface and that thicknesses up to 2 μm could be measured. Shortly afterwards, Philibert [46] saw the possibilities of application of quantitative surface analysis with his analytical expression for $\phi(\rho z)$. More recently, Reuter [30] proposed several modifications of the Philibert model in order to analyze semiquantitatively very thin films on substrates. At the same time, Bishop and Poole [47] published an interesting paper explaining the basic principles of surface analysis and proposing an approximative graphic method for evaluating the thickness of superficial films with the hypothesis that the distribution of emission was the same in film and substrate, and assuming that the absorption was moderate. Other approaches, including those of Hutchins [48] and Colby [49], who take into account the effects of backscatter by the substrate, have been reviewed by Heinrich [50].

The major problem was that of knowing the depth distribution of emission in the target; therefore, Monte Carlo simulations were widely used. The usual simple calculations, choosing for electron diffusion the screened Rutherford cross sections, were proven to be insufficiently accurate for truly quantitative results, particularly at low tensions (i.e., below 10 kV), and for heavy elements. More complex and larger programs, particularly those using the scattering cross sections of Mott, were used in order to improve the results [51,52].

The Monte Carlo techniques require a huge effort if it is attempted to apply them to real problems of superficial segregation. In our opinion they are of no interest except for establishing the general trends of various parameters needed for later use in an analytical model of rapid and easy application.

B. DEPTH RESOLUTION AND SENSITIVITY

When reviewing instruments for surface analysis, the electron microprobe is usually not mentioned, probably because its depth resolution is poor in comparison with electron spectrometric methods (ESCA, AES) or with secondary ion mass spectrometry (SIMS). Nonetheless, the electron probe microanalyzer permits the investigation of shallow depths of materials, as demonstrated in figure 11: one can, for instance, excite the line Cu $L\alpha$ in copper ($E_c = 0.933$ keV) over a depth not exceeding about 20 nm, operating at a tension of 1.6 kV. At 2.6 kV, the maximum depth of excitation of the Cu $L3$ subshell is extended to about 50 nm. As shown in table 3, these experimental conditions produce count rates and line-to-background ratios amply sufficient for the detection of very thin surface segregations; from results obtained at 1.6 kV one can easily obtain a limit of detection of copper at the surface below 0.1 μg/cm^2 (i.e., about 0.1 nm), for a counting time of 100 s and a beam

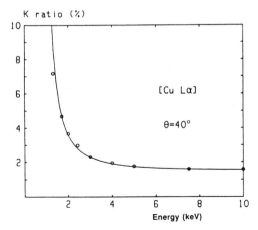

Figure 12.   Variation of the relative intensity of Cu $L\alpha$ at low energies, evidencing a segregation in copper of 0.36 μg/cm^2 at the surface of an electrolytically polished aluminum alloy type 7010 (1.6% copper).

current of 100 nA. Figure 12 shows that, in effect, it is easy to detect by measuring at low energy a surface segregation of 0.36 μg/cm^2 of copper at the surface of an aluminum alloy of low copper concentration and electrolytically polished. Such examples show the limits of x-ray microanalysis: the method provides a sensitivity of interest for surface layers, with the possibility of quantifying the mass of the element segregated on the surface, but it does not have enough depth resolution for determining the chemical nature of so thin a segregation.

Table 3 shows that, unlike the lines of low energy, the copper K lines, poorly excited, would not provide useful signals at the same time as a good depth resolution.

Figure 11 shows that contrary to the case of analyzing very thin films, for measuring the thickness of heavy coatings (several mg/cm^2) it is preferable to use the highest tensions delivered by the instrument. If there is a choice it will in general be preferable for such applications to use lines of high energy, hence less absorbed.

In the measurement of both thin and heavy coatings, the measurement of the radiation from the substrate often provides information useful for quantitation, and should not be neglected.

C.   BASIC PRINCIPLES

Case of a Thick Surface Layer

When the surface layer to be analyzed is sufficiently thick (typically more than 0.1 mg/cm^2) one can usually reduce the electron energy so that the excitation volumes for all used analytical lines are entirely included in the layer. Under such conditions and if the substrate does not emit an appreciable amount of fluorescent radiation, the analysis of the layer can be performed with the usual correction procedures applicable to homogeneous volumes. Obviously such an approach does not tell us the thickness of the surface layer. To determine it one must increase the electron energy until, at least for one of the lines used, the depth of ionization exceeds the layer thickness. Whatever its thickness, the layer of reference can then by considered thin, in the sense that its mass thickness μ_f is below the depth range of ionization R_x.

Case of a Thin Surface Layer

If the layer is thin with respect to R_x, the characteristic emergent intensity from the element A can be written, ignoring a few proportional factors such as the ionization cross section $Q_l^A(E_0)$:

$$I_{film} \propto C_A \int_0^{\mu_f} \phi_A(\rho z) \cdot \exp(-\chi_A \cdot \rho z) \cdot d\rho z \ . \tag{30}$$

The equation as written implies that the depth distribution $\phi(\rho z)$ of the radiation of A conforms, as in the PAP model, with the definition of Castaing, i.e., that the integral of $\phi_A(\rho z)$ between zero and μ_f represents the primary intensity generated in the layer.

All proportionality factors omitted in eq (30) are eliminated when one considers the relative intensity k_A, taking a massive standard as reference. For an elementary standard we obtain:

$$k_A = I_{film}/I_{std} = C_A \int_0^{\mu_f} \phi_A(\rho z)\exp(-\chi_A \cdot \rho z) \cdot d(\rho z) / \int_0^\infty \phi^*_A(\rho z)\exp(-\chi^*_A \cdot \rho z) \cdot d\rho z \ . \tag{31}$$

In the general case, the distribution $\phi_A(\rho z)$ in the film is different from that in the standard, $\phi^*_A(\rho z)$. The two distributions are only identical when the stopping power and backscatter are the same for the two targets. In the same manner, the absorption factors χ_A and χ^*_A for film and standard are, generally, different.

It should be noted that in the preceding expressions the intensity emitted by a film depends both of the mass concentration C_A and the mass thickness μ_f. This means that the emitted intensity depends mainly of the mass of atoms A in the excited volume (the same as in a homogeneous volume). In the case of layered specimens, the problem consists of separating the contributions of thickness and of concentration. If the element A is considered in isolation, this separation is possible due to the fact that $\phi_A(\rho z)$ varies with the acceleration tension. Usually, the determination of both the film thickness and its concentration in element A requires two measurements at different energies:

- one measurement at high tension such that the range of ionization R_x be clearly larger than μ_f, which will mainly reflect the total mass segregated at the surface,
- one at lower tension, which is therefore more sensitive to the composition of the film as R_x becomes closer to μ_f.

These two measurements would be in effect indispensable in the case when element *A* only is analyzed in a layer *A*-*C* on the surface of a substrate *B*. But frequently in practice one tension is sufficient to determine thickness and composition at the same time. To understand this, one can consider that in the classical analysis of a homogeneous volume containing N elements it is sufficient to analyze N-1 elements so as to obtain by a ZAF (or PAP) method for one element a result by difference. To add the measurement of the last element is somewhat of a safety measure, since one can then verify that the sum of concentrations obtained is close to one.

If for a homogeneous volume the analysis of N-1 elements at one tension suffices to obtain a composition, clearly for a film at the surface of a known substrate the analysis at one tension only of N elements in the film provides the information for both the thickness and the composition. For this, it suffices that in at least one of the elements the emergent intensity be sensitive to the film thickness; hence the acceleration tension must be such that the corresponding ionization range R_x is larger than the thickness μ_f, for an analytical line that is moderately absorbed at depth μ_f. This practice has been followed to automatically

Table 4. Application of the PAP model to the automatic thickness and composition determination of ternary Cu-Pd-Au films on an SiO_2 substrate. Experimental data for 20 kV and an emergence angle of 52.5° [52]

#	Element	Nominal from NBS (Nucl. backscatt.) C	Nominal from NBS (Nucl. backscatt.) $\mu g/cm^2$	Experimental (EPMA) K-ratio	Experimental (EPMA) Line	Calculated (PAP model) C	Calculated (PAP model) $\mu g/cm^2$
	Cu	0.319		0.191	K	0.317	
1	Pd	0.347	282	0.157	L	0.350	278
	Au	0.334		0.157	M	0.333	
	Cu	0.607		0.301	K	0.605	
2	Pd	0.198	235	0.0766	L	0.204	229
	Au	0.195		0.0718	M	0.193	

handle in a program using the PAP model the measurements of Murata et al. [52] presented in table 4.

Another way of characterizing a surface film with the least number of measurements is to analyze N-1 elements of the film, and one element of the substrate which serves to obtain the thickness.

If it is desired to have one more degree of safety, and to allow for the possibility to eventually eliminate the forcing the sum of concentrations to one, one must analyze N elements of the layer plus at least one element of the substrate.

Case of a Substrate

The intensity of a characteristic line of element B generated in the substrate and emerging from the specimen is written:

$$I_{sub} = C_B \exp(-\chi_B^f \cdot \mu_f) \cdot \int_{\mu_f}^{\infty} \phi_B(\rho z) \cdot \exp[-\chi_B \cdot (\rho z - \mu_f)] \cdot d\rho z \qquad (32)$$

χ_B corresponds to the absorption coefficient of radiation of B in the substrate itself, and χ_B^f to its absorption coefficient in the coating material. Equation (32) can also be written in the form:

$$I_{sub} = C_B \exp[-\mu_f \cdot (\chi_B^f - \chi_B)] \cdot \int_{\mu_f}^{\infty} \phi_B(\rho z) \cdot \exp(-\chi_B \cdot \rho z) \cdot d\rho z \qquad (33)$$

General Case of an Imbedded Layer

In a more general way, the emerging intensity generated by the atoms A of an imbedded layer, indexed i, and located between thicknesses μ_i and μ_{i+1} (fig. 13) can be written:

$$I_{lay} = C_A^i T_A^i \cdot \int_{\mu_i}^{\mu_{i+1}} \phi_A(\rho z) \exp(-\chi_A^i \cdot \rho z) \cdot d\rho z \qquad (34)$$

$T_A^i = \prod_{j=1}^{i-1} \exp[-\Delta\mu_j(\chi_A^j - \chi_A^i)]$ takes into account the absorption by the $i-1$ layers of thickness $\Delta\mu_j$ which cover the layer i (provided that $i > 1$).

When the parabolic PAP model is used as distribution in depth of the radiation, one can associate with each of the parabolic branches an index k (equal to 1 to 2), and define a

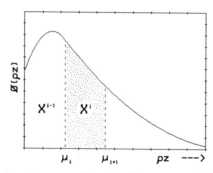

Figure 13. Intensity generated in a layer between mass depths μ_i and μ_{i+1}.

function $H_k(\rho z, \chi)$ representing an integral of the product $\phi_k(\rho z) \cdot \exp(-\chi \cdot \rho z)$:

$$H_k(\rho z, \chi) = -(A_k/\chi) \cdot \exp(-\chi \cdot \rho z) \cdot [(\rho z - R_k)^2 + 2(\rho z - R_k)/\chi + 2/\chi^2 + B_k/A_k] \ . \quad (35)$$

A_k and B_k are the coefficients of the parabolic branch of reference (see expressions (14) and (15)), and R_k is the depth at which the extreme of branch k is situated ($R_k = R_m$ for the first branch, $R_k = R_x$ and $B_k = 0$ for the second branch).

With this definition of H_k, the emergent intensity from an imbedded layer is simply:

$$I_{lay} = C_A^i \cdot T_A^i \cdot [H_k(\mu_{i+1}, \chi_A^i) - H_k(\mu_i, \chi_A^i)] \ . \quad (36)$$

The structure of this expression is analogous to that one can obtain with Packwood and Brown's [53] modified Gaussian model in which the complementary error function plays the same role as the above mentioned function H_k.

When the limits μ_i and μ_{i+1} of the layer are on different sides from the depth R_c where the parabolic branches ϕ_1 and ϕ_2 meet, the calculation must be performed in two parts, from μ_1 to R_c for $k = 1$, and from R_c to μ_{i+1} for $k = 2$:

$$I_{lay} = C_A^i \cdot T_A^i [H_1(R_c, \chi_A^i) - H_1(\mu_i, \chi_A^i) + H_2(\mu_{i+1}, \chi_A^i) - H_2(R_c, \chi_A^i)] \ . \quad (37)$$

When this is applied to the particular case of a massive semi-infinite specimen, this equation reduces to

$$I_{bulk} = C_A[H_1(R_c, \chi_A) - H_1(0, \chi_A) + H_2(R_x, \chi_A) - H_2(R_c, \chi_A)] \ . \quad (38)$$

D. PARAMETRIZATION OF THE DISTRIBUTION $\phi(\rho z)$

We recall that the parabolic model PAP for the analysis of homogeneous volumes is based on four parameters:

- the area F under the distribution curve $\phi(\rho z)$, representative of the generated intensity; i.e., the atomic-number effects;
- the ionization at the surface, $\phi(0)$;
- the range of ionization, R_x;
- the depth R_m of the maximum of $\phi(\rho z)$.

These parameters depend on the energy of the incident electrons, the characteristic line chosen, and the composition of the target, which strongly affects each of the above parameters: for instance, for a line of low energy and medium acceleration tension, R_x only increases about 20% when we pass from a light target to a heavy target, while at the same time $\phi(0)$ about doubles, R_m diminishes about 60%, and the integral F increases about 50%.

In stratified specimens the situation is complicated by the fact that each of the above parameters depends in a specific way on the position in depth of the regions of low and high average atomic number. However, the weak dependence of the range R_x on the nature of the specimen is the property which greatly simplifies the analysis of heterogeneous specimens; even if the compositional nature of the segregation is unknown one can always make a reasonable estimate of its mass thickness, for instance in observing the potential of appearance of a substrate radiation. On the other hand the weak dependence of R_x on the composition favors the convergence in iterative schemes for the simultaneous determination of thickness and composition.

Simple Approximations

When the surface layer or layers and the substrate have very similar atomic numbers one can assume that the material is homogeneous as far as deceleration and backscattering are concerned. In such case the depth distribution function $\phi(\rho z)$ is essentially identical to that of a massive homogeneous target of the same atomic number. It was mainly the measures at varying tension made with this approximation on stratified specimens of diverse thickness (Al/Si, Cu/Ni, Pd/Mo, Au/Ta, Au/Pt) that permitted our establishing the depth distributions of the PAP model [4,6]. Figure 14 is an example of the variation with tension of the relative intensities of Pd $L\alpha$ and Mo $L\alpha$ from a film of 80 nm of palladium on a substrate of molybdenum, and it indicates that the accuracy and sensitivity regarding the mass segregated on the surface are a few percent.

When the mass thickness μ_f of the surface layer is only a little shallower than the ionization range R_x one can postulate that the distribution $\phi(\rho z)$ is very close to that of a massive homogeneous specimen of the same composition as the layer. If, on the contrary, the film thickness is small compared to R_x, the effect of a difference in atomic number of layer and substrate on the electron penetration can be neglected so that the function $\phi(\rho z)$

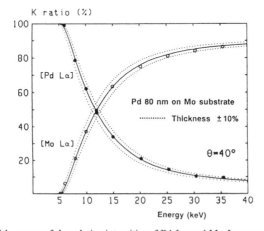

Figure 14. Variation with energy of the relative intensities of Pd $L\alpha$ and Mo $L\alpha$ emerging at 40° from a layer of 80 nm of palladium on top of a molybdenum substrate. The dotted curves give the result of the PAP model when the nominal thickness is changed by 10%.

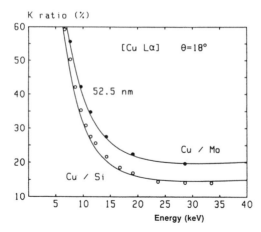

Figure 15. Influence of the substrate (silicon or molybdenum) on the $L\alpha$ emission of a superficial layer of copper. Experimental (OOOO); PAP model: (-----).

of the target is practically identical to that of the substrate. In practice, when the substrate is known this approximation is often used since the treatment is then based on a distribution of $\phi(\rho z)$ independent of the nature of the segregation. Willich [54] for instance has treated the problem of analysis of very thin layers ($<250 \mu g/cm^2$) by operating at tensions above 30 kV, which permit using this approximation. The drawback of this approach is that the intensities emitted by the layer are fairly weak compared with the continuum and therefore difficult to measure with precision, and that the calculated intensities depend mainly on the surface ionization $\phi(0)$ which is at high overvoltages a quite poorly defined parameter.

More Refined Approximations

When one wishes to obtain a general approach useful for the analysis of very thin, intermediate and thick layers, the effects of differences in atomic number between the different layers of the specimen must be taken into account. In the case of a layer on a substrate one can see easily that a heavy substrate generates an increase in electron

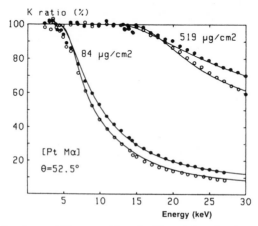

Figure 16. Influence of the substrate (silicon or gold) on the $M\alpha$ emission of platinum surface films. Experimental (OOOO) from [19]; PAP model: (-----).

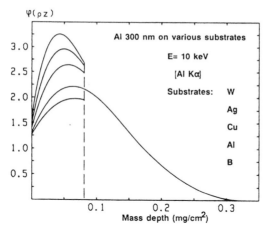

Figure 17. Change with the nature of the substrate of the depth distribution of Al $K\alpha$ in a layer of a aluminum of 300 nm at 10 kV.

backscatter which results in additional ionizations in the surface layer. On figures 15 and 16 one sees that the nature of the substrate produces relative variations of several tenths of the emission. At the limit, for a very thin film and high overvoltage, one must expect that the emission should vary with the atomic number of the substrate in the same fashion as the surface ionization $\phi(0)$ (fig. 24); in other words, more or less by a factor of two when one passes from a very light substrate to a very heavy one.

In general, when the substrate becomes heavier the emission distribution $\phi(\rho z)$ tends towards that of a heavy material. This is to say that at the same time that $\phi(0)$ increases, the maximum of the distribution $\phi(\rho z)$ also increases in height and approaches the surface (fig. 17). The distribution $\phi(\rho z)$ suffers a similar deformation when the thickness of a light film on a heavy substrate diminishes (fig. 18).

Obviously the electrons which can influence the value of surface ionization and of the backscatter factor R cannot come from deeper than a maximum depth shown by Monte Carlo calculations to be in the order of half the ionization range R_x. It is also clear that the

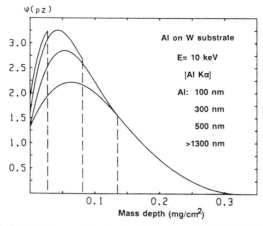

Figure 18. Change with layer thickness of the depth distribution at 10 kV of Al $K\alpha$ radiation in a layer of aluminum on a tungsten substrate.

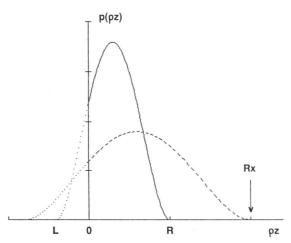

Figure 19. Weighting rules used in the PAP model for the calculation of surface ionization and the backscatter factor (continuous curve) and for calculating the ionization range (dotted curve).

height and the position R_m of the maximum $\phi(\rho z)$ are the result of ionizations by electrons which have penetrated a little deeper in the material (between $R_x/2$ and R_x). Finally, the value of R_x, though not very variable, is affected by the composition of the target between the surface and R_x. The values of R_x in a target which is inhomogeneous with respect to depth must thus be obtained by an iterative procedure.

Weighting Laws Used in the PAP Model

To obtain the depth distributions $\phi(\rho z)$ used in the PAP model for stratified specimens one calculates the four basic parameters as if they were those of fictitious homogeneous specimens, each different for each of the parameters of reference. These fictitious materials are defined by the laws of weighting of the form:

$$p(\rho z) = N \cdot (\rho z - L)^2 \cdot (\rho z - R)^2 \tag{39}$$

L and R are the double roots of $p(\rho z)$ of the left and right sides (fig. 19), and N ensures the normalization of the area of $p(\rho z)$ between 0 and R.

To calculate the range of ionization R_x one must start from a first approximation; this does not require an accurate knowledge of the target since R_x is not strongly sensitive to the composition of the target. After this the law of weighting $\phi(\rho z)$ is applied to the concentrations of the stratified specimen, with $R = R_x$ and $L = -0.4\ R$. After normalization one obtains in this way a set of fictitious concentrations which permit the calculation of a new value of R_x by the formulae given in Appendix 2 for homogeneous specimens. In practice this loop can be stopped after the value of R_x has varied less than 1% from the previous value.

A weighting which gives more importance to the regions near the surface ($R = R_x/2$ and $L = -0.4\ R$) is used to obtain a set of fictitious concentrations and hence a mean atomic number $\bar{Z}_b$ appropriate for calculating the backscatter coefficient $\bar{\eta}$, the backscatter factor $\bar{R}$, and the surface ionization $\phi(0)$, with the aid of the formulae of Appendices 1 and 2.

A weighting law using the roots $R = 0.7\ R_x$ and $L = -0.6\ R$ leads to the mean atomic number $\bar{Z}$ used in the calculation of the term $1/S$ of the generated intensity [eq (11)].

Table 5. Relative intensity calculated for 5 kV and a take-off angle of 40° (with pure elemental standards) for six materials from boron to gold deposited on SiO_2 in three thicknesses (10, 20, and 40 nm).

Metal	Line	Ec (keV)	Density	Relative intensity for a film thickness of		
				10 nm	20 nm	40 nm
B	K	0.192	2.34	2.8%	5.9%	12.8%
Al	$K\alpha$	1.559	2.70	4.3%	9.1%	19.9%
Cr	$L\alpha$	0.574	7.19	9.0%	21.0%	48.7%
Cu	$L\alpha$	0.933	8.96	11.7%	27.8%	59.6%
Pd	$L\alpha$	3.173	12.0	27.4%	53.3%	88.4%
Au	$M\alpha$	2.123	19.3	32.6%	64.4%	96.8%

Finally, to calculate the position of the maximum of $\phi(\rho z)$ one introduces in the equation relating R_m to R_x (see Appendix 2) an atomic number equal to the mean between $\bar{Z}$ and $\bar{Z}_b$, previously defined.

This procedure which at this point is generally satisfactory is only an approximation which may vary slightly as new experimental evidence is gained in the future.

V. Examples of Application to Specimens Inhomogeneous in Depth

A. DETERMINATION OF THICKNESS OF SURFACE LAYERS

The determination of the mass thickness μ_f of a layer of known (or supposedly known) composition is a common practical problem very simple to handle with the electron probe microanalyzer. The measurement of only one analytical line at only one tension is sufficient in principle provided that the associated depth of ionization R_x is larger than μ_f and that the radiation generated at depth μ_f can emerge without excessive absorption.

This type of problem justifies a procedure with automatic iteration. Table 5 permits the determination of some orders of magnitude: it gives for some pure materials the relative intensity corresponding at 5 kV to films of 10, 20, and 40 nm on the substrate SiO_2.

With the multilayer monochromators presently offered on the market one can obtain good conditions for very light elements. Figure 20 shows that a signal of oxygen from a layer of natural oxide of mechanically polished aluminum specimen can be easily detected. For this specimen the relative intensity of O K measured at 5 kV with respect to a standard

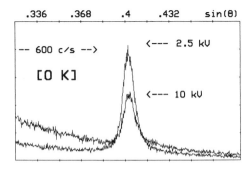

Figure 20. Oxygen spectra corresponding to a natural film of alumina at 2.5 and 10 kV, obtained with a tungsten-silicon multilayer monochromator ($2d = 5.9$ nm). Beam current: 100 nA; 1 second/channel, 512 channels. Pulse height setting: $E = 1$ V, $\Delta E = 1.5$ V.

of $Y_3Fe_5O_{12}$ ($\simeq 26\%$ of oxygen) is equal to 5%, which with the PAP model corresponds to 2.1 $\mu g/cm^2$ of alumina or 6 nm with an assumed density of 3.5. The comparison of this thickness and the spectra of figure 20 shows that even for very light elements the sensitivity to surface segregation is excellent and can reach a fraction of a $\mu g/cm^2$.

The program PAPTH2 prepared for the problems of automatic thickness determination also considers the case of two layers superposed on the surface. Because one given element may be present in several strata, the program requires measurement of relative intensities at least two tensions, except when the thickness of one of the layers is already known.

B. SIMULTANEOUS DETERMINATION OF THICKNESS AND COMPOSITION

As demonstrated previously (table 4), both thickness and composition of a layer on a known substrate can be obtained by an iterative procedure of the measured relative intensities for each of the elements present, with only one operative setting. However, one should always be careful in applying this type of approach, which assumes that the film is uniform in composition with respect to depth and that the interface film-substrate is perfect; i.e., there is no oxidation or diffusion at the interface.

Usually we prefer performing measurements at different tensions, so as to verify whether the hypothesis of a unique layer is appropriate and the composition of the layer can be considered uniform. If this structure is verified to be complex, then the only other general method of characterization is one that requires measuring at several tensions and trying to reproduce the results by formulating reasonable hypotheses regarding the structure of the specimen (a gradient can be described by an assembly of layers which vary progressively in composition). For such an approach it is practically indispensable to have a graphics terminal on which the measured relative intensities and the curves calculated in function of tension can be visually compared for each hypothesis. We have already presented several examples of this type of reconstructive analysis: for a layer of aluminum on magnesium before and after diffusion, for a double layer Al/Cu on aluminum before and after diffusion [6], for layers of Ga-As on glass [7], and more recently for antireflex deposits consisting of four layers (oxide and fluoride) which in total have a thickness in the order of 300 nm, and have an additional carbon coating to ensure electrical conduction [55]. Another application for which the reconstruction "by eye" proved rapid and efficient is the determination in alloys Fe-Gd-Tb which on the surface form a natural oxide film of about 10 nm, from the amount of oxygen incorporated in the material itself during the fabrication. Supposing that the nature of the oxide is known, and basing ourselves on two measurements, at 5 and 10 kV, of the relative intensity of O K [56], we can, after considering some hypotheses, determine the oxide thickness and the level of oxygen incorporated which varies according to the conditions of preparation between a fraction of percent and several percent. The results we obtained are in very good agreement with those that can be obtained by calibration methods [57] which are much more cumbersome to perform.

C. SPECIAL APPLICATIONS

We have very recently [5] published examples of the application of the PAP procedure to substrates of molybdenum and of mercury sulfide covered by a layer of gold, in order to obtain the coefficients of absorption of S K, Mo L, and Hg M lines close to the gold M edges. This simple and efficient method has already permitted eliminating the ambiguities concerning the absorption coefficient of Si $K\alpha$ in tantalum [7].

VI. Fluorescence (Secondary) Emission

For the classical analysis of homogeneous specimens the contribution of fluorescence by characteristic lines is usually determined with sufficient accuracy by the simple model of Reed [58], derived from the equations of Castaing [9]. Fluorescence by the continuum is usually neglected, except in the program COR of Henoc et al. [59].

In the case of stratified specimens the treatment of secondary emission is more complex since besides the excitation by the continuum which cannot always be neglected one must take into account the many possibilities of excitation by characteristic lines: excitation of an element of a layer by another of the same layer, excitation of an element of the layer by one of another layer and/or the substrate, and excitation of a substrate element by an element of layer.

Some authors have treated specific cases in which a simplification can be introduced in the calculation: for instance, very thin films excited by a characteristic line of the substrate [60], and the case of a substrate excited by the characteristic emission from a layer of thickness superior to the depth of excitation by the electrons [61]. Other authors have attacked the problem differently by applying numerical integration procedures to diverse configurations [62]. Though we get to rather complicated formulae we have adopted for our programs with variable tension a general analytical approach of fluorescence by characteristic and continuous radiation. This approach leads to calculation times compatible with the use of interactive software for stratified specimens. The principle of the calculation and the main expressions obtained are summarized in Appendices 7 and 8.

From a practical viewpoint we believe that the most unpleasant consequence of the fluorescent effects arises in the case of heavy layers which excite a substrate element not present in the layer. This situation is illustrated in figure 21, for a layer of 1.22 μm of nickel on a substrate of iron. It is seen that at tensions up to 20 kV the substrate is not excited by the incident electrons, and that its emission of Fe $K\alpha$ which reaches a relative intensity of 10% is entirely due to secondary emission, mainly that induced by the K lines of nickel.

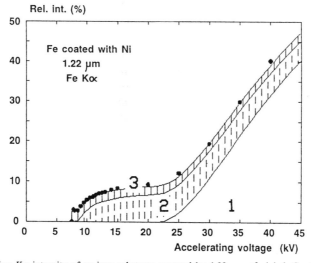

Figure 21. Relative $K\alpha$ intensity of an iron substrate covered by 1.22 μm of nickel. Contributions of primary emission (1), of fluorescence due to the Ni K lines (2) and fluorescence due to the continuum (3).

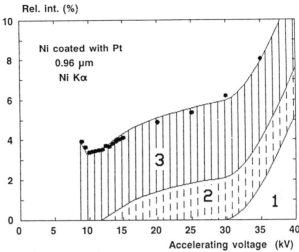

Figure 22. Relative $K\alpha$ intensity of a nickel substrate covered by 0.96 μm of platinum. Contributions of primary emission (1), of fluorescence due to the Pt L lines (2) and to fluorescence excited by the continuum (3).

In the case of figure 22 (0.96 μm of platinum on a nickel substrate), the continuum produced in the platinum layer is more intense, and the fluorescence of the substrate produced by it prevails over the excitation by the Pt L lines.

These two examples present a typical situation in which one could be tempted to analyze the layer at a medium tension (e.g., 15 kV), considering that its thickness suffices for approximating the case of a homogeneous semi-infinite specimen. Such an approach would from the qualitative view point conduce to a serious error: to conclude that the layer contains an element that in reality only exists in the substrate.

VII. Conclusions

We believe that the quantitation procedures by x-ray microanalysis which have appeared in the few last years have produced an observable and significant improvement of the analytical results for homogeneous specimens, especially for cases of strong absorption (light and very light elements). The best performing models now offer for the very light elements correction calculations of the same order of accuracy as those of ZAF for the heavier elements.

Among these models, the PAP model was conceived for wider applications than correction calculations only, particularly for the quantitative study of specimens which are inhomogeneous with respect to depth. Its possibilities in this domain derive from the use of realistic distributions in depth of the emission used, obtained with respect of the basic physical concepts by adjusting the laws and parameters to numerous experimental results including measurements on synthetic stratified specimens.

As a whole, x-ray microanalysis has found until now very little use in problems of surface analysis. It is true that its depth resolution is not sufficient for competing with true surface techniques. But it could certainly complement them effectively in many applications, where one not only tries to characterize the surface but also the more or less profound regions close to it. For such applications x-ray analysis should in the future be used more frequently since it uses an equipment available in many laboratories, and because it adds to the local and non- (or almost so) destructive character of the analysis a sensitivity of interest for surface segregations, as well as authentic quantitative possibilities.

Appendix 1

Backscatter Loss Factor R

$$R = 1 - \bar{\eta} \cdot \bar{W} \cdot [1 - G(U_0)] \ .$$

The Monte Carlo simulations show that the relative mean energy of the backscattered electrons depends but little on the initial energy. Under normal beam incidence it can be related to the average backscatter coefficient as follows:

$$\bar{W} = \bar{E}_r/E_0 = 0.595 + \bar{\eta}/3.7 + \bar{\eta}^{4.55} \ .$$

The slight variation of the backscatter coefficient with energy has been neglected. The coefficient is expressed as a function of the mean atomic number of the target as follows:

$$\bar{\eta} = 1.75 \ 10^{-3} \bar{Z}_b + 0.37 \ [1 - \exp(-0.015 \ \bar{Z}_b^{1.3})] \ .$$

In a nonelementary target, $\bar{Z}_b$ is defined by the following averaging rule:

$$\bar{Z}_b = (\Sigma \ C_i \cdot Z_i^{1/2})^2 \ .$$

The term $G(U_0)$ is extracted from the theoretical model of Coulon and Zeller [28]:

$$G(U_0) = [U_0 - 1 - (1 - 1/U_0^{1+q})/(1+q)]/[(2+q) \cdot J(U_0)]$$

where

$$J(U_0) = 1 + U_0 \cdot [\ln(U_0) - 1] \quad \text{and} \quad q = (2\bar{W} - 1)/(1 - \bar{W}) \ .$$

This expression results from an energy distribution for backscattered electrons of the form:

$$(1/\eta) \cdot (d\eta/dW) = (q+1) \cdot W^q \ .$$

The fairly coarse approximation implicit in this expression for light targets has little influence on the final results, since for them the backscatter coefficient is small.

The values of R calculated as above are compared to those of Duncumb on figure 23.

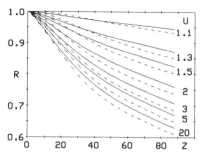

Figure 23. Variation of the backscatter factor R with atomic number and overvoltage. PAP model (——) and Duncumb's expression (– – –).

Appendix 2

Form Parameters for $\phi(\rho z)$ in the Parabolic PAP Model

Surface Ionization:

The surface ionization $\phi(0)$ is expressed as a function of the overvoltage U_0 and the mean backscatter coefficient $\bar{\eta}$ (see Appendix 1):

$$\phi(0) = 1 + 3.3(1 - 1/U_0^r) \cdot \bar{\eta}^{1 \cdot 2} \qquad \text{with: } r = 2 - 2.3\bar{\eta} \ .$$

This expression is, in its general form, reminiscent of the formula of Reuter [30] except that it tends to 1 as U_0 tends to 1.

Figure 24 shows that it implies a larger range of variation with overvoltage for heavy targets, as indicated by our measurements on thin layers, as well as other recent observations [31,32]. We note in passing that the formulation of Love and Scott [33], which predicts too early a saturation with overvoltage does not agree with these experimental results.

Range of Ionization:

The range of ionization of the energy level E_l, R_x, is a fraction $Q \cdot D$ of the total trajectory of the electrons between E_0 and E_l, R_0:

$$R_x = Q \cdot D \cdot R_0$$

where Q is a function of overvoltage and atomic number, and D a factor which only plays a role at very weak overvoltages.

As the overvoltage tends to infinity R_x becomes equal to one penetration. If furthermore the atomic number approaches zero, R_x becomes equal to a full trajectory. Based on these principles, and taking into account the trends revealed by Monte Carlo calculations, we established the following expressions:

$$Q = Q_0 + (1 - Q_0) \cdot \exp[-(U_0 - 1)/b]$$

with

$$b = 40/\bar{Z} \text{ and } \bar{Z} = \sum_i C_i \cdot Z_i$$

$$Q_0 = 1 - 0.535 \exp[-(21/\bar{Z}_n)^{1 \cdot 2}] - 2.5 \ 10^{-4}(\bar{Z}_n/20)^{3 \cdot 5}$$

with

$$\ln(\bar{Z}_n) = \sum C_i \cdot \ln(Z_i) \ .$$

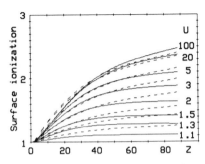

Figure 24. Variation of the surface ionization $\phi(0)$ with the atomic number and the overvoltage. PAP model (——) and Reuter's expression (- - -).

The factor D is introduced to account, in a purely empirical and approximative fashion, for the statistical dispersion of the trajectories at low overvoltages:

$$D = 1 + 1/U_0^h \qquad \text{with } h = \bar{Z}^{0.45} .$$

The total trajectory R_0 between the energies E_0 and E_l is calculated from the rule of deceleration defined in section B:

$$R_0 = (1/M) \sum_{k=1}^{3} [J^{1-P_k} \cdot D_k \cdot (E_0^{1+P_k} - E_l^{1+P_k})/(1+P_k)] .$$

Depth of the Maximum of the Distribution:

The depth R_m at which the maximum of $\phi(\rho z)$ is located has been related to R_x through the results obtained by simulation:

$$R_m = G1(\bar{Z}) \cdot G2(U_0) \cdot G3(U_0, \bar{Z}) \cdot R_x .$$

The preponderant factor of this expression is the function of $\bar{Z}$:

$$G1 = 0.11 + 0.41 \exp[-(\bar{Z}/12.75)^{0.75}] .$$

The other two terms express small deviations from the tendency expressed by $G1$:

$$G2 = 1 - \exp[-(U_0 - 1)^{0.35}/1.19]$$
$$G2 = 1 - \exp[-(U_0 - 0.5) \cdot \bar{Z}^{0.4}/4] .$$

Appendix 3

Modifications of the Parabolic PAP Model at Very Small Overvoltages

At very low values of U_0 where the uncertainties of the parameters are the most important, the form conditions imposed upon $\phi(\rho z)$ may be incompatible with the area conditions so that the relation (16) will have no satisfactory solution and the discriminant d is negative.

In such case an additional degree of freedom can be introduced in the distribution by suppressing the parametric relation between R_m and R_x (see Appendix 2), and imposing a value of zero upon d. We thus obtain:

$$R_m = R_x \cdot [F - \phi(0) \cdot R_x/3]/[F + \phi(0) \cdot R_x]$$

so that

$$R_c = 3 R_m \cdot [F + \phi(0) \cdot R_x]/[2\phi(0) \cdot R_x] .$$

If U_0 is indeed very close to one, it can also happen that the value of R_m is unacceptable ($R_m < 0$ or $R_m > R_x$). This means that $\phi(\rho z)$ does no longer have a maximum. It is then necessary to describe the distribution in an other way, such as a simple decreasing exponential

$$\phi(\rho z) = \phi(0) \cdot \exp(-s \cdot \rho z)$$

in which the coefficient s derived from the area condition must not be smaller than a minimum value s_{min} such that $\phi(R_x)$ is negligible. For instance with $s_{min} = 4.5$ one obtains: $\phi(R_x) = \phi(0)/100$. This representation of $\phi(\rho z)$ is permissible when $F < \phi(0)/s_{min}$. The factor $F(\chi)$ is then simply equal to $\phi(0) \cdot F/[\phi(0) + \chi \cdot F]$.

If the above inequality is not satisfied, $\phi(\rho z)$ can be described by the sum of an exponential and a parabola:

$$\phi(\rho z) = \phi_e(\rho z) + \phi_p(\rho z) .$$

The conditions imposed at the surface and at the range R_x imply that

$$\phi_p(\rho z) = [\phi_p(0)/R_x^2] \cdot (\rho z - R_x)^2$$

and

$$\phi_e(\rho z) = [\phi(0) - \phi_p(0)] \cdot \exp(-s_{min} \cdot \rho z) .$$

If the area condition is imposed on the distribution, we obtain:

$$\phi_p(0) = 3 [F \cdot s_{min} - \phi(0)]/(R_x \cdot s_{min} - 3)$$

and the factor $F(\chi)$ is:

$$F(\chi) = \{(2/\chi^3) \cdot [1 - \exp(-\chi R_x)] - 2R_x/\chi^2 + R_x^2/\chi\} \cdot [\phi_p(0)/R_x^2] + [\phi(0) - \phi_p(0)]/(s_{min} + \chi).$$

Appendix 4

Practical Computation Scheme for the Simplified Model XPP

– Calculate the area F of the distribution $\phi(\rho z)$ (see paragraph II.B).

– Calculate the surface ionization $\phi(0)$ (see Appendix 2).

– Calculate the mean depth of ionization $\bar{R}$ [eq (28)].

If $F/\bar{R} < \phi(0)$, impose the condition $\bar{R} = F/\phi(0)$.

– Calculate the initial slope P [eq (29)].

If necessary, limit the value $g \cdot h^4$ to the value $0.9 \, b \cdot \bar{R}^2 \cdot [b - 2 \cdot \phi(0)/F]$.

– Calculate $b = \sqrt{2} \cdot [1 + \sqrt{1 - \bar{R} \cdot \phi(0)/F}]/\bar{R}$

– Calculate $a = \{P + b \cdot [2\phi(0) - b \cdot F]\}/[b \cdot F \cdot (2 - b \cdot \bar{R}) - \phi(0)]$

– Define: $\epsilon = (a - b)/b$.

If necessary, impose on ϵ a minimum absolute value (e.g., 10^{-6}), and then assume that $a = b \cdot (1 + \epsilon)$.

– Calculate $B = [b^2 \cdot F(1 + \epsilon) - P - \phi(0) \cdot b \cdot (2 + \epsilon)]/\epsilon$
    and $A = [B/b + \phi(0) - b \cdot F] \cdot (1 + \epsilon)/\epsilon$

– Calculate the emergent intensity by relation (26), preferably expressing $F(\chi)$ as follows:

$$F(\chi) = \{\phi(0) + B/(b + \chi) - A \cdot b \cdot \epsilon/[b \cdot (1 + \epsilon) + \chi]\}/(b + \chi) .$$

Appendix 5

Absorption Coefficients Associated With the PAP Procedure

A number of absorption coefficients for very light elements could be determined (particularly from the measurements of Bastin and Heijligers [31,37]) and have replaced the values of Henke [38]:

	Emitter		
Absorber	Boron	Carbon	Nitrogen
B	3500	39000	15800
C	6750	2170	
N	11000		1640
O	16500		
Al	64000		13800
Si	80000	35000	17000
Ti	15000	8100	4270
V	18000	8850	4950
Cr	20700	10700	5650
Fe	27800	13500	7190
Co	(32000)		
Ni	(37000)		
Zr	4400	25000	24000
Nb	4500	24000	25000
Mo	4600	20500	25800
La	2500		
Hf		18000	14000
Ta	23000	17000	15500
W	21000	18000	
U	7400		

A second table is used for the emitters which are not very light elements. This table contains a number of absorption coefficients corresponding to situations where the line is close to an absorption edge, or to particular resonance situations which are ignored in the model MAC 30.

It cannot be precluded that these absorption coefficients may depend on the state of chemical bonding of the emitter or absorber. This seems to occur in the case of self-absorption of the line Ni $L\alpha$ in the alloys Ni-Al, depending on the level of addition of the aluminum atoms [39]. This is the reason for which in the table below we include the nature of the specimen.

Emitter	Line	Absorber	μ/ρ	Peculiarity		Specimen
Si	$K\alpha$	Ta	1490	Proximity	Ta $M5$	Ta$_2$O$_5$ film on Si
S	$K\alpha$	Au	2200	"	Au $M4$	Au film on HgS
Cu	$L\beta$	Cu	6750	"	Cu $L2$	Cu (pure)
As	$L\alpha$	Ga	7000	"	Ga $L1$	Ga-As (bulk)
Mo	$L\alpha$	Au	2200	"	Au $M4$	Au film on Mo
Gd	$M\beta$	Gd	4700	"	Gd $M4,M5$	Gd (pure)
Hf	$M\beta$	Hf	3000	"	Hf $M4,M5$	Hf (pure)
Ta	$M\beta$	Ta	2500	"	Ta $M4,M5$	Ta (pure)
W	$M\beta$	W	2080	"	W $M4,M5$	W (pure)
Au	$M\beta$	Pt	2550	"	Pt $M4$	Au-Pt-Pd alloy
Hg	$M\beta$	Au	2170	"	Au $M4$	Au film on HgS
Sc	$L\alpha$	Sc	4750	Resonance		Sc (pure)
Ti	$L\alpha$	Ti	4550	"		Ti (pure)
V	$L\alpha$	V	4370	"		V (pure)
Cr	$L\alpha$	Cr	3850	"		Cr (pure)
Mn	$L\alpha$	Mn	3340	"		Mn (pure)
Fe	$L\alpha$	Fe	3350	"		Fe (pure)
Co	$L\alpha$	Co	3260	"		Co (pure)
Ni	$L\alpha$	Ni	3560	"		Ni (pure)
Cu	$L\alpha$	Cu	1755	"		Cu (pure)

Appendix 6

Data Basis of 826 Entries Used for the Performance Tests of Models

#13 to #208	#475 to #619	#772 to #826	Bastin et al. [31,37,63]
#209 to #224	#620 to #716	#737 to #771	Pouchou and Pichoir [3,40,41]
#268 to #392	#717 to #738		Heinrich et al. [64]
#225 to #267			Willich [65]
#393 to #430			Christ et al. [66]
#431 to #466			Colby [67]
#7 to #12	#467 to #474		Springer [68]
#1 to #6			Shimizu et al. [69]

Atomic number effect predominant:

#4 to #208	#225 to #287	#332 to #376	#431 to #474	#484 to #619
#671 to #704	#710 to #712	#781 to #826		

Absorption effect predominant:

#1 to #3	#209 to #224	#288 to #331	#377 to #430	#475 to #483
#620 to #670	#705 to #708	#713 to #764	#772 to #780	

1 - Analysed element (A)	5 - Weight fraction of A
2 - Line	6 - K-ratio of A
3 - Companion (B)	7 - X-ray take-off angle
4 - Accelerating voltage	

	1	2	3	4	5	6	7		1	2	3	4	5	6	7
1	Al	Ka	Fe	20.0	.2410	.1240	52.5	26	Al	Ka	B	12.0	.1722	.1503	40.0
2	Al	Ka	Fe	25.0	.2410	.0980	52.5	27	Al	Ka	B	15.0	.1722	.1504	40.0
3	Al	Ka	Fe	30.0	.2410	.0830	52.5	28	Al	Ka	B	20.0	.1722	.1495	40.0
4	Fe	Ka	Al	20.0	.7590	.7360	52.5	29	Al	Ka	B	25.0	.1722	.1490	40.0
5	Fe	Ka	Al	25.0	.7590	.7420	52.5	30	Al	Ka	B	30.0	.1722	.1455	40.0
6	Fe	Ka	Al	30.0	.7590	.7480	52.5	31	Si	Ka	B	4.0	.4780	.4465	40.0
7	Fe	Ka	S	10.0	.4660	.4060	75.0	32	Si	Ka	B	6.0	.4780	.4528	40.0
8	Fe	Ka	S	12.0	.4660	.4210	75.0	33	Si	Ka	B	8.0	.4780	.4580	40.0
9	Fe	Ka	S	15.0	.4660	.4250	75.0	34	Si	Ka	B	10.0	.4780	.4622	40.0
10	Fe	Ka	S	20.0	.4660	.4250	75.0	35	Si	Ka	B	12.0	.4780	.4658	40.0
11	Fe	Ka	S	25.0	.4660	.4220	75.0	36	Si	Ka	B	15.0	.4780	.4696	40.0
12	Fe	Ka	S	30.0	.4660	.4190	75.0	37	Si	Ka	B	20.0	.4780	.4738	40.0
13	Al	Ka	B	4.0	.5551	.5150	40.0	38	Si	Ka	B	25.0	.4780	.4758	40.0
14	Al	Ka	B	6.0	.5551	.5178	40.0	39	Si	Ka	B	30.0	.4780	.4760	40.0
15	Al	Ka	B	8.0	.5551	.5188	40.0	40	Si	Ka	B	4.0	.3143	.2900	40.0
16	Al	Ka	B	10.0	.5551	.5200	40.0	41	Si	Ka	B	6.0	.3143	.2940	40.0
17	Al	Ka	B	12.0	.5551	.5198	40.0	42	Si	Ka	B	8.0	.3143	.2978	40.0
18	Al	Ka	B	15.0	.5551	.5170	40.0	43	Si	Ka	B	10.0	.3143	.3002	40.0
19	Al	Ka	B	20.0	.5551	.5130	40.0	44	Si	Ka	B	12.0	.3143	.3036	40.0
20	Al	Ka	B	25.0	.5551	.5080	40.0	45	Si	Ka	B	15.0	.3143	.3072	40.0
21	Al	Ka	B	30.0	.5551	.5030	40.0	46	Si	Ka	B	20.0	.3143	.3120	40.0
22	Al	Ka	B	4.0	.1722	.1500	40.0	47	Si	Ka	B	25.0	.3143	.3140	40.0
23	Al	Ka	B	6.0	.1722	.1502	40.0	48	Si	Ka	B	30.0	.3143	.3156	40.0
24	Al	Ka	B	8.0	.1722	.1503	40.0	49	Ti	Ka	B	6.0	.8322	.7922	40.0
25	Al	Ka	B	10.0	.1722	.1503	40.0	50	Ti	Ka	B	8.0	.8322	.7937	40.0

51	Ti	Ka	B	10.0	.8322	.7942	40.0	116	Ni	Ka	B	20.0	.9422	.9358	40.0
52	Ti	Ka	B	12.0	.8322	.7960	40.0	117	Ni	Ka	B	25.0	.9422	.9380	40.0
53	Ti	Ka	B	15.0	.8322	.7978	40.0	118	Ni	Ka	B	30.0	.9422	.9382	40.0
54	Ti	Ka	B	20.0	.8322	.8012	40.0	119	Ni	Ka	B	10.0	.9157	.8858	40.0
55	Ti	Ka	B	25.0	.8322	.8042	40.0	120	Ni	Ka	B	12.0	.9157	.8899	40.0
56	Ti	Ka	B	30.0	.8322	.8077	40.0	121	Ni	Ka	B	15.0	.9157	.8950	40.0
57	Ti	Ka	B	6.0	.6993	.6330	40.0	122	Ni	Ka	B	20.0	.9157	.9005	40.0
58	Ti	Ka	B	8.0	.6993	.6400	40.0	123	Ni	Ka	B	25.0	.9157	.9040	40.0
59	Ti	Ka	B	10.0	.6993	.6462	40.0	124	Ni	Ka	B	30.0	.9157	.9060	40.0
60	Ti	Ka	B	12.0	.6993	.6513	40.0	125	Ni	Ka	B	10.0	.8560	.8083	40.0
61	Ti	Ka	B	15.0	.6993	.6578	40.0	126	Ni	Ka	B	12.0	.8560	.8180	40.0
62	Ti	Ka	B	20.0	.6993	.6662	40.0	127	Ni	Ka	B	15.0	.8560	.8283	40.0
63	Ti	Ka	B	25.0	.6993	.6738	40.0	128	Ni	Ka	B	20.0	.8560	.8388	40.0
64	Ti	Ka	B	30.0	.6993	.6782	40.0	129	Ni	Ka	B	25.0	.8560	.8438	40.0
65	V	Ka	B	6.0	.7160	.6378	40.0	130	Ni	Ka	B	30.0	.8560	.8450	40.0
66	V	Ka	B	8.0	.7160	.6500	40.0	131	Zr	La	B	4.0	.8211	.7540	40.0
67	V	Ka	B	10.0	.7160	.6602	40.0	132	Zr	La	B	6.0	.8211	.7662	40.0
68	V	Ka	B	12.0	.7160	.6681	40.0	133	Zr	La	B	8.0	.8211	.7770	40.0
69	V	Ka	B	15.0	.7160	.6770	40.0	134	Zr	La	B	10.0	.8211	.7855	40.0
70	V	Ka	B	20.0	.7160	.6878	40.0	135	Zr	La	B	12.0	.8211	.7930	40.0
71	V	Ka	B	25.0	.7160	.6939	40.0	136	Zr	La	B	15.0	.8211	.8024	40.0
72	V	Ka	B	30.0	.7160	.6977	40.0	137	Zr	La	B	20.0	.8211	.8140	40.0
73	Cr	Ka	B	8.0	.8320	.7758	40.0	138	Zr	La	B	25.0	.8211	.8225	40.0
74	Cr	Ka	B	10.0	.8320	.7840	40.0	139	Zr	La	B	30.0	.8211	.8290	40.0
75	Cr	Ka	B	12.0	.8320	.7905	40.0	140	Nb	La	B	4.0	.8958	.8588	40.0
76	Cr	Ka	B	15.0	.8320	.7981	40.0	141	Nb	La	B	6.0	.8958	.8680	40.0
77	Cr	Ka	B	20.0	.8320	.8070	40.0	142	Nb	La	B	8.0	.8958	.8758	40.0
78	Cr	Ka	B	25.0	.8320	.8122	40.0	143	Nb	La	B	10.0	.8958	.8820	40.0
79	Cr	Ka	B	30.0	.8320	.8156	40.0	144	Nb	La	B	12.0	.8958	.8878	40.0
80	Cr	Ka	B	8.0	.7269	.6230	40.0	145	Nb	La	B	15.0	.8958	.8942	40.0
81	Cr	Ka	B	10.0	.7269	.6374	40.0	146	Nb	La	B	20.0	.8958	.9036	40.0
82	Cr	Ka	B	12.0	.7269	.6480	40.0	147	Nb	La	B	25.0	.8958	.9100	40.0
83	Cr	Ka	B	15.0	.7269	.6603	40.0	148	Nb	La	B	30.0	.8958	.9157	40.0
84	Cr	Ka	B	20.0	.7269	.6738	40.0	149	Nb	La	B	4.0	.8285	.7502	40.0
85	Cr	Ka	B	25.0	.7269	.6810	40.0	150	Nb	La	B	6.0	.8285	.7640	40.0
86	Cr	Ka	B	30.0	.7269	.6850	40.0	151	Nb	La	B	8.0	.8285	.7760	40.0
87	Fe	Ka	B	8.0	.9118	.8680	40.0	152	Nb	La	B	10.0	.8285	.7862	40.0
88	Fe	Ka	B	10.0	.9118	.8730	40.0	153	Nb	La	B	12.0	.8285	.7950	40.0
89	Fe	Ka	B	12.0	.9118	.8764	40.0	154	Nb	La	B	15.0	.8285	.8060	40.0
90	Fe	Ka	B	15.0	.9118	.8821	40.0	155	Nb	La	B	20.0	.8285	.8210	40.0
91	Fe	Ka	B	20.0	.9118	.8888	40.0	156	Nb	La	B	25.0	.8285	.8338	40.0
92	Fe	Ka	B	25.0	.9118	.8940	40.0	157	Nb	La	B	30.0	.8285	.8442	40.0
93	Fe	Ka	B	30.0	.9118	.8970	40.0	158	Mo	La	B	4.0	.8987	.8401	40.0
94	Fe	Ka	B	8.0	.8378	.7823	40.0	159	Mo	La	B	6.0	.8987	.8500	40.0
95	Fe	Ka	B	10.0	.8378	.7922	40.0	160	Mo	La	B	8.0	.8987	.8580	40.0
96	Fe	Ka	B	12.0	.8378	.8002	40.0	161	Mo	La	B	10.0	.8987	.8660	40.0
97	Fe	Ka	B	15.0	.8378	.8098	40.0	162	Mo	La	B	12.0	.8987	.8731	40.0
98	Fe	Ka	B	20.0	.8378	.8203	40.0	163	Mo	La	B	15.0	.8987	.8828	40.0
99	Fe	Ka	B	25.0	.8378	.8280	40.0	164	Mo	La	B	20.0	.8987	.8978	40.0
100	Fe	Ka	B	30.0	.8378	.8330	40.0	165	Mo	La	B	25.0	.8987	.9100	40.0
101	Co	Ka	B	10.0	.9160	.8660	40.0	166	Mo	La	B	30.0	.8987	.9202	40.0
102	Co	Ka	B	12.0	.9160	.8832	40.0	167	Ta	Ma	B	4.0	.9436	.9160	40.0
103	Co	Ka	B	15.0	.9160	.8980	40.0	168	Ta	Ma	B	6.0	.9436	.9208	40.0
104	Co	Ka	B	20.0	.9160	.9101	40.0	169	Ta	Ma	B	8.0	.9436	.9258	40.0
105	Co	Ka	B	25.0	.9160	.9120	40.0	170	Ta	Ma	B	10.0	.9436	.9295	40.0
106	Co	Ka	B	30.0	.9160	.9078	40.0	171	Ta	Ma	B	12.0	.9436	.9324	40.0
107	Co	Ka	B	10.0	.8450	.7823	40.0	172	Ta	Ma	B	15.0	.9436	.9365	40.0
108	Co	Ka	B	12.0	.8450	.7939	40.0	173	Ta	Ma	B	20.0	.9436	.9430	40.0
109	Co	Ka	B	15.0	.8450	.8058	40.0	174	Ta	Ma	B	25.0	.9436	.9462	40.0
110	Co	Ka	B	20.0	.8450	.8180	40.0	175	Ta	Ma	B	30.0	.9436	.9482	40.0
111	Co	Ka	B	25.0	.8450	.8244	40.0	176	Ta	La	B	12.0	.9436	.9158	40.0
112	Co	Ka	B	30.0	.8450	.8278	40.0	177	Ta	La	B	15.0	.9436	.9193	40.0
113	Ni	Ka	B	10.0	.9422	.9238	40.0	178	Ta	La	B	20.0	.9436	.9230	40.0
114	Ni	Ka	B	12.0	.9422	.9262	40.0	179	Ta	La	B	25.0	.9436	.9258	40.0
115	Ni	Ka	B	15.0	.9422	.9303	40.0	180	Ta	La	B	30.0	.9436	.9260	40.0

181	Ta	Ma	B	4.0	.9077	.8478	40.0	246	Cu	Ka	Au	25.0	.7990	.8270	40.0
182	Ta	Ma	B	6.0	.9077	.8560	40.0	247	Cu	Ka	Au	20.0	.7990	.8350	40.0
183	Ta	Ma	B	8.0	.9077	.8625	40.0	248	Cu	Ka	Au	16.0	.7990	.8450	40.0
184	Ta	Ma	B	10.0	.9077	.8690	40.0	249	Au	Ma	Cu	14.0	.8020	.7420	40.0
185	Ta	Ma	B	12.0	.9077	.8740	40.0	250	Au	Ma	Cu	12.0	.8020	.7450	40.0
186	Ta	Ma	B	15.0	.9077	.8803	40.0	251	Au	Ma	Cu	10.0	.8020	.7560	40.0
187	Ta	Ma	B	20.0	.9077	.8895	40.0	252	Cu	Ka	Au	14.0	.1980	.2570	40.0
188	Ta	Ma	B	25.0	.9077	.8960	40.0	253	Cu	Ka	Au	12.0	.1980	.2640	40.0
189	Ta	Ma	B	30.0	.9077	.9002	40.0	254	Au	Ma	Cu	14.0	.6040	.5110	40.0
190	Ta	La	B	12.0	.9077	.8230	40.0	255	Au	Ma	Cu	12.0	.6040	.5140	40.0
191	Ta	La	B	15.0	.9077	.8365	40.0	256	Au	Ma	Cu	10.0	.6040	.5230	40.0
192	Ta	La	B	20.0	.9077	.8525	40.0	257	Cu	Ka	Au	14.0	.3960	.4760	40.0
193	Ta	La	B	25.0	.9077	.8622	40.0	258	Cu	Ka	Au	12.0	.3960	.4940	40.0
194	Ta	La	B	30.0	.9077	.8690	40.0	259	Au	Ma	Cu	14.0	.4010	.3140	40.0
195	W	Ma	B	4.0	.9445	.8850	40.0	260	Au	Ma	Cu	12.0	.4010	.3160	40.0
196	W	Ma	B	6.0	.9445	.8950	40.0	261	Au	Ma	Cu	12.0	.4010	.3220	40.0
197	W	Ma	B	8.0	.9445	.9038	40.0	262	Cu	Ka	Au	14.0	.5990	.6750	40.0
198	W	Ma	B	10.0	.9445	.9105	40.0	263	Cu	Ka	Au	12.0	.5990	.6940	40.0
199	W	Ma	B	12.0	.9445	.9178	40.0	264	Au	Ma	Cu	14.0	.2010	.1430	40.0
200	W	Ma	B	15.0	.9445	.9258	40.0	265	Au	Ma	Cu	12.0	.2010	.1470	40.0
201	W	Ma	B	20.0	.9445	.9375	40.0	266	Cu	Ka	Au	14.0	.7990	.8450	40.0
202	W	Ma	B	25.0	.9445	.9463	40.0	267	Cu	Ka	Au	12.0	.7990	.8530	40.0
203	W	Ma	B	30.0	.9445	.9530	40.0	268	Au	La	Ag	15.0	.2243	.2030	52.5
204	W	La	B	12.0	.9445	.8918	40.0	269	Au	La	Ag	20.0	.2243	.2030	52.5
205	W	La	B	15.0	.9445	.8990	40.0	270	Au	La	Ag	30.0	.2243	.2110	52.5
206	W	La	B	20.0	.9445	.9103	40.0	271	Au	La	Ag	40.0	.2243	.2160	52.5
207	W	La	B	25.0	.9445	.9198	40.0	272	Au	La	Ag	48.5	.2243	.2140	52.5
208	W	La	B	30.0	.9445	.9264	40.0	273	Au	La	Ag	15.0	.4003	.3680	52.5
209	Au	Ma	Cu	5.2	.5080	.4259	40.0	274	Au	La	Ag	20.0	.4003	.3660	52.5
210	Au	Ma	Cu	10.4	.5080	.4203	40.0	275	Au	La	Ag	30.0	.4003	.3760	52.5
211	Au	Ma	Cu	15.7	.5080	.4034	40.0	276	Au	La	Ag	40.0	.4003	.3840	52.5
212	Au	Ma	Cu	20.8	.5080	.3865	40.0	277	Au	La	Ag	48.5	.4003	.3850	52.5
213	Au	Ma	Cu	26.1	.5080	.3668	40.0	278	Au	La	Ag	15.0	.6005	.5660	52.5
214	Au	Ma	Cu	31.5	.5080	.3499	40.0	279	Au	La	Ag	20.0	.6005	.5730	52.5
215	Au	Ma	Cu	36.6	.5080	.3405	40.0	280	Au	La	Ag	30.0	.6005	.5800	52.5
216	Au	Ma	Cu	39.7	.5080	.3349	40.0	281	Au	La	Ag	40.0	.6005	.5840	52.5
217	Au	Ma	Cu	5.2	.3000	.2317	40.0	282	Au	La	Ag	48.5	.6005	.5840	52.5
218	Au	Ma	Cu	10.4	.3000	.2317	40.0	283	Au	La	Ag	15.0	.8005	.7890	52.5
219	Au	Ma	Cu	15.7	.3000	.2186	40.0	284	Au	La	Ag	20.0	.8005	.7830	52.5
220	Au	Ma	Cu	20.8	.3000	.2054	40.0	285	Au	La	Ag	30.0	.8005	.7830	52.5
221	Au	Ma	Cu	26.1	.3000	.1886	40.0	286	Au	La	Ag	40.0	.8005	.7900	52.5
222	Au	Ma	Cu	31.5	.3000	.1773	40.0	287	Au	La	Ag	48.5	.8005	.7880	52.5
223	Au	Ma	Cu	36.6	.3000	.1717	40.0	288	Au	Ma	Ag	10.0	.2243	.2010	52.5
224	Au	Ma	Cu	39.7	.3000	.1689	40.0	289	Au	Ma	Ag	20.0	.2243	.1970	52.5
225	Au	La	Cu	25.0	.8020	.7620	40.0	290	Au	Ma	Ag	30.0	.2243	.1920	52.5
226	Au	La	Cu	20.0	.8020	.7610	40.0	291	Au	Ma	Ag	40.0	.2243	.1850	52.5
227	Au	La	Cu	16.0	.8020	.7510	40.0	292	Au	Ma	Ag	48.5	.2243	.1810	52.5
228	Cu	Ka	Au	25.0	.1980	.2350	40.0	293	Au	Ma	Ag	10.0	.4003	.3620	52.5
229	Cu	Ka	Au	20.0	.1980	.2450	40.0	294	Au	Ma	Ag	20.0	.4003	.3640	52.5
230	Cu	Ka	Au	16.0	.1980	.2520	40.0	295	Au	Ma	Ag	30.0	.4003	.3510	52.5
231	Au	La	Cu	25.0	.6040	.5370	40.0	296	Au	Ma	Ag	40.0	.4003	.3440	52.5
232	Au	La	Cu	20.0	.6040	.5300	40.0	297	Au	Ma	Ag	48.5	.4003	.3370	52.5
233	Au	La	Cu	16.0	.6040	.5180	40.0	298	Au	Ma	Ag	10.0	.6005	.5590	52.5
234	Cu	Ka	Au	25.0	.3960	.4470	40.0	299	Au	Ma	Ag	20.0	.6005	.5510	52.5
235	Cu	Ka	Au	20.0	.3960	.4630	40.0	300	Au	Ma	Ag	30.0	.6005	.5470	52.5
236	Cu	Ka	Au	16.0	.3960	.4720	40.0	301	Au	Ma	Ag	40.0	.6005	.5410	52.5
237	Au	La	Cu	25.0	.4010	.3380	40.0	302	Au	Ma	Ag	48.5	.6005	.5370	52.5
238	Au	La	Cu	20.0	.4010	.3340	40.0	303	Au	Ma	Ag	10.0	.8005	.7710	52.5
239	Au	La	Cu	16.0	.4010	.3230	40.0	304	Au	Ma	Ag	20.0	.8005	.7640	52.5
240	Cu	Ka	Au	25.0	.5990	.6430	40.0	305	Au	Ma	Ag	30.0	.8005	.7660	52.5
241	Cu	Ka	Au	20.0	.5990	.6550	40.0	306	Au	Ma	Ag	40.0	.8005	.7610	52.5
242	Cu	Ka	Au	16.0	.5990	.6700	40.0	307	Au	Ma	Ag	48.5	.8005	.7550	52.5
243	Au	La	Cu	25.0	.2010	.1550	40.0	308	Ag	La	Au	5.0	.1996	.2230	52.5
244	Au	La	Cu	20.0	.2010	.1530	40.0	309	Ag	La	Au	10.0	.1996	.1910	52.5
245	Au	La	Cu	16.0	.2010	.1470	40.0	310	Ag	La	Au	20.0	.1996	.1440	52.5

311	Ag	La	Au	30.0	.1996	.1090	52.5
312	Ag	La	Au	40.0	.1996	.0887	52.5
313	Ag	La	Au	48.5	.1996	.0795	52.5
314	Ag	La	Au	5.0	.3992	.4360	52.5
315	Ag	La	Au	10.0	.3992	.3860	52.5
316	Ag	La	Au	20.0	.3992	.3090	52.5
317	Ag	La	Au	30.0	.3992	.2450	52.5
318	Ag	La	Au	40.0	.3992	.2080	52.5
319	Ag	La	Au	48.5	.3992	.1810	52.5
320	Ag	La	Au	5.0	.5993	.6240	52.5
321	Ag	La	Au	10.0	.5993	.5840	52.5
322	Ag	La	Au	20.0	.5993	.5040	52.5
323	Ag	La	Au	30.0	.5993	.4260	52.5
324	Ag	La	Au	40.0	.5993	.3750	52.5
325	Ag	La	Au	48.5	.5993	.3390	52.5
326	Ag	La	Au	5.0	.7758	.8070	52.5
327	Ag	La	Au	10.0	.7758	.7640	52.5
328	Ag	La	Au	20.0	.7758	.7060	52.5
329	Ag	La	Au	30.0	.7758	.6360	52.5
330	Ag	La	Au	40.0	.7758	.5860	52.5
331	Ag	La	Au	48.5	.7758	.5420	52.5
332	Cu	Ka	Au	15.0	.1983	.2540	52.5
333	Cu	Ka	Au	20.0	.1983	.2470	52.5
334	Cu	Ka	Au	25.0	.1983	.2400	52.5
335	Cu	Ka	Au	30.0	.1983	.2350	52.5
336	Cu	Ka	Au	40.0	.1983	.2180	52.5
337	Cu	Ka	Au	48.5	.1983	.2180	52.5
338	Cu	Ka	Au	15.0	.3964	.4800	52.5
339	Cu	Ka	Au	20.0	.3964	.4640	52.5
340	Cu	Ka	Au	25.0	.3964	.4530	52.5
341	Cu	Ka	Au	30.0	.3964	.4430	52.5
342	Cu	Ka	Au	40.0	.3964	.4160	52.5
343	Cu	Ka	Au	48.5	.3964	.4140	52.5
344	Cu	Ka	Au	15.0	.5992	.6760	52.5
345	Cu	Ka	Au	20.0	.5992	.6630	52.5
346	Cu	Ka	Au	25.0	.5992	.6510	52.5
347	Cu	Ka	Au	30.0	.5992	.6440	52.5
348	Cu	Ka	Au	40.0	.5992	.6160	52.5
349	Cu	Ka	Au	48.5	.5992	.6040	52.5
350	Cu	Ka	Au	15.0	.7985	.8510	52.5
351	Cu	Ka	Au	20.0	.7985	.8410	52.5
352	Cu	Ka	Au	25.0	.7985	.8340	52.5
353	Cu	Ka	Au	30.0	.7985	.8260	52.5
354	Cu	Ka	Au	40.0	.7985	.8150	52.5
355	Cu	Ka	Au	48.5	.7985	.8000	52.5
356	Au	La	Cu	15.0	.2012	.1450	52.5
357	Au	La	Cu	20.0	.2012	.1580	52.5
358	Au	La	Cu	25.0	.2012	.1540	52.5
359	Au	La	Cu	30.0	.2012	.1570	52.5
360	Au	La	Cu	48.5	.2012	.1520	52.5
361	Au	La	Cu	15.0	.4010	.3120	52.5
362	Au	La	Cu	20.0	.4010	.3330	52.5
363	Au	La	Cu	25.0	.4010	.3310	52.5
364	Au	La	Cu	30.0	.4010	.3310	52.5
365	Au	La	Cu	48.5	.4010	.3260	52.5
366	Au	La	Cu	15.0	.6036	.5110	52.5
367	Au	La	Cu	20.0	.6036	.5310	52.5
368	Au	La	Cu	25.0	.6036	.5290	52.5
369	Au	La	Cu	30.0	.6036	.5330	52.5
370	Au	La	Cu	48.5	.6036	.5210	52.5
371	Au	La	Cu	15.0	.8015	.7400	52.5
372	Au	La	Cu	20.0	.8015	.7520	52.5
373	Au	La	Cu	25.0	.8015	.7450	52.5
374	Au	La	Cu	30.0	.8015	.7540	52.5
375	Au	La	Cu	35.0	.8015	.7580	52.5
376	Au	La	Cu	40.0	.8015	.7510	52.5
377	Au	Ma	Cu	10.0	.2012	.1540	52.5
378	Au	Ma	Cu	20.0	.2012	.1400	52.5
379	Au	Ma	Cu	30.0	.2012	.1220	52.5
380	Au	Ma	Cu	40.0	.2012	.1100	52.5
381	Au	Ma	Cu	10.0	.4010	.3320	52.5
382	Au	Ma	Cu	20.0	.4010	.3040	52.5
383	Au	Ma	Cu	30.0	.4010	.2750	52.5
384	Au	Ma	Cu	40.0	.4010	.2530	52.5
385	Au	Ma	Cu	10.0	.6036	.5310	52.5
386	Au	Ma	Cu	20.0	.6036	.5000	52.5
387	Au	Ma	Cu	30.0	.6036	.4620	52.5
388	Au	Ma	Cu	40.0	.6036	.4280	52.5
389	Au	Ma	Cu	10.0	.8015	.7490	52.5
390	Au	Ma	Cu	20.0	.8015	.7280	52.5
391	Au	Ma	Cu	30.0	.8015	.7040	52.5
392	Au	Ma	Cu	40.0	.8015	.6750	52.5
393	Sb	La	Bi	10.0	.6255	.6255	52.5
394	Sb	La	Bi	15.0	.6255	.5908	52.5
395	Sb	La	Bi	20.0	.6255	.5567	52.5
396	Sb	La	Bi	25.0	.6255	.5268	52.5
397	Sb	La	Bi	30.0	.6255	.4990	52.5
398	Sb	La	Bi	10.0	.5344	.5344	52.5
399	Sb	La	Bi	15.0	.5344	.4980	52.5
400	Sb	La	Bi	20.0	.5344	.4632	52.5
401	Sb	La	Bi	25.0	.5344	.4335	52.5
402	Sb	La	Bi	30.0	.5344	.4063	52.5
403	Sb	La	Bi	10.0	.4602	.4602	52.5
404	Sb	La	Bi	15.0	.4602	.4242	52.5
405	Sb	La	Bi	20.0	.4602	.3906	52.5
406	Sb	La	Bi	25.0	.4602	.3624	52.5
407	Sb	La	Bi	30.0	.4602	.3370	52.5
408	Sb	La	Bi	10.0	.3130	.3130	52.5
409	Sb	La	Bi	15.0	.3130	.2825	52.5
410	Sb	La	Bi	20.0	.3130	.2552	52.5
411	Sb	La	Bi	25.0	.3130	.2330	52.5
412	Sb	La	Bi	30.0	.3130	.2136	52.5
413	Sb	La	Bi	10.0	.1664	.1664	52.5
414	Sb	La	Bi	15.0	.1664	.1471	52.5
415	Sb	La	Bi	20.0	.1664	.1305	52.5
416	Sb	La	Bi	25.0	.1664	.1174	52.5
417	Sb	La	Bi	30.0	.1664	.1064	52.5
418	Sb	La	Bi	10.0	.1058	.1058	52.5
419	Sb	La	Bi	15.0	.1058	.0928	52.5
420	Sb	La	Bi	20.0	.1058	.0817	52.5
421	Sb	La	Bi	25.0	.1058	.0731	52.5
422	Sb	La	Bi	30.0	.1058	.0659	52.5
423	Sb	La	Bi	10.0	.0502	.0502	52.5
424	Sb	La	Bi	15.0	.0502	.0437	52.5
425	Sb	La	Bi	20.0	.0502	.0382	52.5
426	Sb	La	Bi	25.0	.0502	.0340	52.5
427	Sb	La	Bi	30.0	.0502	.0306	52.5
428	Sb	La	Bi	10.0	.0277	.0277	52.5
429	Sb	La	Bi	15.0	.0277	.0240	52.5
430	Sb	La	Bi	20.0	.0277	.0210	52.5
431	U	Mb	C	10.0	.9084	.8493	52.5
432	U	Mb	C	15.0	.9084	.8588	52.5
433	U	Mb	C	20.0	.9084	.8648	52.5
434	U	Mb	C	25.0	.9084	.8709	52.5
435	U	Mb	C	30.0	.9084	.8747	52.5
436	U	Mb	C	35.0	.9084	.8746	52.5
437	U	Mb	N	10.0	.9444	.9061	52.5
438	U	Mb	N	15.0	.9444	.9129	52.5
439	U	Mb	N	20.0	.9444	.9179	52.5
440	U	Mb	N	25.0	.9444	.9218	52.5

441	U	Mb	N	30.0	.9444	.9243	52.5	506	V	Ka	C	20.0	.8400	.8220	40.0
442	U	Mb	N	35.0	.9444	.9259	52.5	507	V	Ka	C	25.0	.8400	.8265	40.0
443	U	Mb	Si	10.0	.9622	.9395	52.5	508	V	Ka	C	30.0	.8400	.8295	40.0
444	U	Mb	Si	15.0	.9622	.9433	52.5	509	Cr	Ka	C	8.0	.9432	.9252	40.0
445	U	Mb	Si	20.0	.9622	.9471	52.5	510	Cr	Ka	C	10.0	.9432	.9257	40.0
446	U	Mb	Si	25.0	.9622	.9493	52.5	511	Cr	Ka	C	12.0	.9432	.9259	40.0
447	U	Mb	Si	30.0	.9622	.9507	52.5	512	Cr	Ka	C	15.0	.9432	.9261	40.0
448	U	Mb	Si	35.0	.9622	.9521	52.5	513	Cr	Ka	C	20.0	.9432	.9265	40.0
449	U	Mb	P	10.0	.8849	.8242	52.5	514	Cr	Ka	C	25.0	.9432	.9271	40.0
450	U	Mb	P	15.0	.8849	.8349	52.5	515	Cr	Ka	C	30.0	.9432	.9278	40.0
451	U	Mb	P	20.0	.8849	.8432	52.5	516	Cr	Ka	C	8.0	.9090	.8790	40.0
452	U	Mb	P	25.0	.8849	.8506	52.5	517	Cr	Ka	C	10.0	.9090	.8819	40.0
453	U	Mb	P	30.0	.8849	.8544	52.5	518	Cr	Ka	C	12.0	.9090	.8842	40.0
454	U	Mb	P	35.0	.8849	.8592	52.5	519	Cr	Ka	C	15.0	.9090	.8875	40.0
455	U	Mb	S	10.0	.8813	.8191	52.5	520	Cr	Ka	C	20.0	.9090	.8918	40.0
456	U	Mb	S	15.0	.8813	.8291	52.5	521	Cr	Ka	C	25.0	.9090	.8950	40.0
457	U	Mb	S	20.0	.8813	.8394	52.5	522	Cr	Ka	C	30.0	.9090	.8972	40.0
458	U	Mb	S	25.0	.8813	.8452	52.5	523	Cr	Ka	C	8.0	.8666	.8310	40.0
459	U	Mb	S	30.0	.8813	.8500	52.5	524	Cr	Ka	C	10.0	.8666	.8322	40.0
460	U	Mb	S	35.0	.8813	.8530	52.5	525	Cr	Ka	C	12.0	.8666	.8358	40.0
461	U	Mb	Fe	10.0	.9624	.9436	52.5	526	Cr	Ka	C	15.0	.8666	.8385	40.0
462	U	Mb	Fe	15.0	.9624	.9478	52.5	527	Cr	Ka	C	20.0	.8666	.8422	40.0
463	U	Mb	Fe	20.0	.9624	.9506	52.5	528	Cr	Ka	C	25.0	.8666	.8455	40.0
464	U	Mb	Fe	25.0	.9624	.9524	52.5	529	Cr	Ka	C	30.0	.8666	.8478	40.0
465	U	Mb	Fe	30.0	.9624	.9538	52.5	530	Fe	Ka	C	8.0	.9333	.9172	40.0
466	U	Mb	Fe	35.0	.9624	.9549	52.5	531	Fe	Ka	C	10.0	.9333	.9194	40.0
467	Pb	La	S	20.0	.8660	.8070	75.0	532	Fe	Ka	C	12.0	.9333	.9215	40.0
468	Pb	La	S	25.0	.8660	.8090	75.0	533	Fe	Ka	C	15.0	.9333	.9239	40.0
469	Pb	La	S	30.0	.8660	.8130	75.0	534	Fe	Ka	C	20.0	.9333	.9270	40.0
470	Pb	La	S	35.0	.8660	.8180	75.0	535	Fe	Ka	C	25.0	.9333	.9290	40.0
471	Pb	La	S	40.0	.8660	.8260	75.0	536	Fe	Ka	C	30.0	.9333	.9304	40.0
472	Zn	Ka	S	20.0	.6710	.6200	75.0	537	Zr	La	C	4.0	.9145	.8981	40.0
473	Zn	Ka	S	25.0	.6710	.6260	75.0	538	Zr	La	C	6.0	.9145	.9011	40.0
474	Zn	Ka	S	30.0	.6710	.6280	75.0	539	Zr	La	C	8.0	.9145	.9033	40.0
475	B	Ka	C	4.0	.7981	.7768	40.0	540	Zr	La	C	10.0	.9145	.9055	40.0
476	B	Ka	C	6.0	.7981	.7754	40.0	541	Zr	La	C	12.0	.9145	.9070	40.0
477	B	Ka	C	8.0	.7981	.7471	40.0	542	Zr	La	C	15.0	.9145	.9088	40.0
478	B	Ka	C	10.0	.7981	.7309	40.0	543	Zr	La	C	20.0	.9145	.9095	40.0
479	B	Ka	C	12.0	.7981	.7132	40.0	544	Zr	La	C	25.0	.9145	.9100	40.0
480	B	Ka	C	15.0	.7981	.6853	40.0	545	Zr	La	C	30.0	.9145	.9110	40.0
481	B	Ka	C	20.0	.7981	.6544	40.0	546	Nb	La	C	4.0	.9145	.8750	40.0
482	B	Ka	C	25.0	.7981	.6516	40.0	547	Nb	La	C	6.0	.9145	.8822	40.0
483	B	Ka	C	30.0	.7981	.6574	40.0	548	Nb	La	C	8.0	.9145	.8880	40.0
484	Si	Ka	C	4.0	.7005	.6800	40.0	549	Nb	La	C	10.0	.9145	.8930	40.0
485	Si	Ka	C	6.0	.7005	.6798	40.0	550	Nb	La	C	12.0	.9145	.8970	40.0
486	Si	Ka	C	8.0	.7005	.6790	40.0	551	Nb	La	C	15.0	.9145	.9022	40.0
487	Si	Ka	C	10.0	.7005	.6787	40.0	552	Nb	La	C	20.0	.9145	.9088	40.0
488	Si	Ka	C	12.0	.7005	.6780	40.0	553	Nb	La	C	25.0	.9145	.9135	40.0
489	Si	Ka	C	15.0	.7005	.6762	40.0	554	Nb	La	C	30.0	.9145	.9175	40.0
490	Si	Ka	C	20.0	.7005	.6730	40.0	555	Mo	La	C	4.0	.9442	.9060	40.0
491	Si	Ka	C	25.0	.7005	.6682	40.0	556	Mo	La	C	6.0	.9442	.9128	40.0
492	Si	Ka	C	30.0	.7005	.6630	40.0	557	Mo	La	C	8.0	.9442	.9182	40.0
493	Ti	Ka	C	6.0	.8160	.7832	40.0	558	Mo	La	C	10.0	.9442	.9225	40.0
494	Ti	Ka	C	8.0	.8160	.7862	40.0	559	Mo	La	C	12.0	.9442	.9264	40.0
495	Ti	Ka	C	10.0	.8160	.7888	40.0	560	Mo	La	C	15.0	.9442	.9312	40.0
496	Ti	Ka	C	12.0	.8160	.7910	40.0	561	Mo	La	C	20.0	.9442	.9373	40.0
497	Ti	Ka	C	15.0	.8160	.7938	40.0	562	Mo	La	C	25.0	.9442	.9418	40.0
498	Ti	Ka	C	20.0	.8160	.7980	40.0	563	Mo	La	C	30.0	.9442	.9458	40.0
499	Ti	Ka	C	25.0	.8160	.8012	40.0	564	Ta	Ma	C	4.0	.9400	.8765	40.0
500	Ti	Ka	C	30.0	.8160	.8033	40.0	565	Ta	Ma	C	6.0	.9400	.8815	40.0
501	V	Ka	C	6.0	.8400	.7952	40.0	566	Ta	Ma	C	8.0	.9400	.8862	40.0
502	V	Ka	C	8.0	.8400	.8013	40.0	567	Ta	Ma	C	10.0	.9400	.8905	40.0
503	V	Ka	C	10.0	.8400	.8065	40.0	568	Ta	Ma	C	12.0	.9400	.8943	40.0
504	V	Ka	C	12.0	.8400	.8111	40.0	569	Ta	Ma	C	15.0	.9400	.9000	40.0
505	V	Ka	C	15.0	.8400	.8162	40.0	570	Ta	Ma	C	20.0	.9400	.9088	40.0

571	Ta	Ma	C	25.0	.9400	.9162	40.0	636	Al	Ka	Ni	10.5	.1250	.0848	40.0

571	Ta	Ma	C	25.0	.9400	.9162	40.0
572	Ta	Ma	C	30.0	.9400	.9220	40.0
573	Ta	La	C	12.0	.9400	.8820	40.0
574	Ta	La	C	15.0	.9400	.8886	40.0
575	Ta	La	C	20.0	.9400	.8980	40.0
576	Ta	La	C	25.0	.9400	.9060	40.0
577	Ta	La	C	30.0	.9400	.9130	40.0
578	Ta	Ma	C	4.0	.9700	.9342	40.0
579	Ta	Ma	C	6.0	.9700	.9390	40.0
580	Ta	Ma	C	8.0	.9700	.9428	40.0
581	Ta	Ma	C	10.0	.9700	.9462	40.0
582	Ta	Ma	C	12.0	.9700	.9490	40.0
583	Ta	Ma	C	15.0	.9700	.9540	40.0
584	Ta	Ma	C	20.0	.9700	.9602	40.0
585	Ta	Ma	C	25.0	.9700	.9660	40.0
586	Ta	Ma	C	30.0	.9700	.9702	40.0
587	Ta	La	C	12.0	.9700	.9348	40.0
588	Ta	La	C	15.0	.9700	.9406	40.0
589	Ta	La	C	20.0	.9700	.9453	40.0
590	Ta	La	C	25.0	.9700	.9503	40.0
591	Ta	La	C	30.0	.9700	.9580	40.0
592	W	Ma	C	4.0	.9387	.8560	40.0
593	W	Ma	C	6.0	.9387	.8712	40.0
594	W	Ma	C	8.0	.9387	.8823	40.0
595	W	Ma	C	10.0	.9387	.8910	40.0
596	W	Ma	C	12.0	.9387	.8987	40.0
597	W	Ma	C	15.0	.9387	.9081	40.0
598	W	Ma	C	20.0	.9387	.9212	40.0
599	W	Ma	C	25.0	.9387	.9328	40.0
600	W	Ma	C	30.0	.9387	.9428	40.0
601	W	La	C	12.0	.9387	.8558	40.0
602	W	La	C	15.0	.9387	.8782	40.0
603	W	La	C	20.0	.9387	.8975	40.0
604	W	La	C	25.0	.9387	.9093	40.0
605	W	La	C	30.0	.9387	.9178	40.0
606	W	Ma	C	4.0	.9700	.8865	40.0
607	W	Ma	C	6.0	.9700	.9060	40.0
608	W	Ma	C	8.0	.9700	.9178	40.0
609	W	Ma	C	10.0	.9700	.9272	40.0
610	W	Ma	C	12.0	.9700	.9350	40.0
611	W	Ma	C	15.0	.9700	.9443	40.0
612	W	Ma	C	20.0	.9700	.9568	40.0
613	W	Ma	C	25.0	.9700	.9668	40.0
614	W	Ma	C	30.0	.9700	.9756	40.0
615	W	La	C	12.0	.9700	.9153	40.0
616	W	La	C	15.0	.9700	.9290	40.0
617	W	La	C	20.0	.9700	.9432	40.0
618	W	La	C	25.0	.9700	.9522	40.0
619	W	La	C	30.0	.9700	.9580	40.0
620	Al	Ka	Ti	7.0	.3980	.3853	40.0
621	Al	Ka	Ti	10.5	.3980	.3543	40.0
622	Al	Ka	Ti	15.7	.3980	.3008	40.0
623	Al	Ka	Ti	20.8	.3980	.2585	40.0
624	Al	Ka	Ti	26.0	.3980	.2162	40.0
625	Al	Ka	Ti	31.3	.3980	.1870	40.0
626	Al	Ka	Ti	36.4	.3980	.1579	40.0
627	Al	Ka	Ti	40.0	.3980	.1429	40.0
628	Al	Ka	Ni	5.3	.3090	.2957	40.0
629	Al	Ka	Ni	10.5	.3090	.2184	40.0
630	Al	Ka	Ni	15.9	.3090	.1576	40.0
631	Al	Ka	Ni	21.2	.3090	.1141	40.0
632	Al	Ka	Ni	26.6	.3090	.0878	40.0
633	Al	Ka	Ni	31.9	.3090	.0683	40.0
634	Al	Ka	Ni	37.2	.3090	.0570	40.0
635	Al	Ka	Ni	5.3	.1250	.1246	40.0

636	Al	Ka	Ni	10.5	.1250	.0848	40.0
637	Al	Ka	Ni	15.9	.1250	.0585	40.0
638	Al	Ka	Ni	21.2	.1250	.0405	40.0
639	Al	Ka	Ni	26.6	.1250	.0307	40.0
640	Al	Ka	Ni	31.9	.1250	.0233	40.0
641	Al	Ka	Ni	5.3	.0490	.0473	40.0
642	Al	Ka	Ni	10.5	.0490	.0315	40.0
643	Al	Ka	Ni	15.9	.0490	.0210	40.0
644	Cu	La	Ni	2.5	.5650	.5135	40.0
645	Cu	La	Ni	5.0	.5650	.4487	40.0
646	Cu	La	Ni	7.6	.5650	.3876	40.0
647	Cu	La	Ni	10.2	.5650	.3303	40.0
648	Cu	La	Ni	12.6	.5650	.2823	40.0
649	Cu	La	Ni	15.3	.5650	.2419	40.0
650	Cu	La	Ni	20.5	.5650	.1930	40.0
651	Cu	La	Ni	25.6	.5650	.1648	40.0
652	Cu	La	Ni	30.8	.5650	.1489	40.0
653	Cu	La	Ni	35.5	.5650	.1395	40.0
654	Cu	La	Ni	40.0	.5650	.1310	40.0
655	Cu	La	Au	5.2	.7000	.7143	40.0
656	Cu	La	Au	10.4	.7000	.6259	40.0
657	Cu	La	Au	15.7	.7000	.5489	40.0
658	Cu	La	Au	20.8	.7000	.4981	40.0
659	Cu	La	Au	26.1	.7000	.4662	40.0
660	Cu	La	Au	31.3	.7000	.4398	40.0
661	Cu	La	Au	36.5	.7000	.4229	40.0
662	Cu	La	Au	39.5	.7000	.4173	40.0
663	Cu	La	Au	5.2	.4920	.5019	40.0
664	Cu	La	Au	10.4	.4920	.4098	40.0
665	Cu	La	Au	15.7	.4920	.3383	40.0
666	Cu	La	Au	20.8	.4920	.2932	40.0
667	Cu	La	Au	26.1	.4920	.2632	40.0
668	Cu	La	Au	31.3	.4920	.2444	40.0
669	Cu	La	Au	36.5	.4920	.2368	40.0
670	Cu	La	Au	39.5	.4920	.2331	40.0
671	Ti	Ka	Al	15.7	.6020	.5789	40.0
672	Ti	Ka	Al	20.9	.6020	.5605	40.0
673	Ti	Ka	Al	26.1	.6020	.5580	40.0
674	Ti	Ka	Al	31.3	.6020	.5435	40.0
675	Ti	Ka	Al	36.5	.6020	.5437	40.0
676	Ti	Ka	Al	40.0	.6020	.5380	40.0
677	Au	La	Cu	13.0	.3000	.2319	40.0
678	Au	La	Cu	15.7	.3000	.2278	40.0
679	Au	La	Cu	20.9	.3000	.2365	40.0
680	Au	La	Cu	26.1	.3000	.2402	40.0
681	Au	La	Cu	31.3	.3000	.2413	40.0
682	Au	La	Cu	36.5	.3000	.2381	40.0
683	Au	La	Cu	39.5	.3000	.2377	40.0
684	Au	La	Cu	13.0	.5080	.4174	40.0
685	Au	La	Cu	15.7	.5080	.4142	40.0
686	Au	La	Cu	20.9	.5080	.4285	40.0
687	Au	La	Cu	26.1	.5080	.4348	40.0
688	Au	La	Cu	31.3	.5080	.4337	40.0
689	Au	La	Cu	36.5	.5080	.4282	40.0
690	Au	La	Cu	39.5	.5080	.4279	40.0
691	Cu	Ka	Au	13.0	.7000	.7758	40.0
692	Cu	Ka	Au	15.7	.7000	.7583	40.0
693	Cu	Ka	Au	20.9	.7000	.7468	40.0
694	Cu	Ka	Au	26.1	.7000	.7387	40.0
695	Cu	Ka	Au	31.3	.7000	.7270	40.0
696	Cu	Ka	Au	36.5	.7000	.7223	40.0
697	Cu	Ka	Au	39.5	.7000	.7161	40.0
698	Cu	Ka	Au	13.0	.4920	.5850	40.0
699	Cu	Ka	Au	15.7	.4920	.5697	40.0
700	Cu	Ka	Au	20.9	.4920	.5534	40.0

No.	Element	Line	Std	kV	Pure	Meas	Take-off
701	Cu	Ka	Au	26.1	.4920	.5403	40.0
702	Cu	Ka	Au	31.3	.4920	.5298	40.0
703	Cu	Ka	Au	36.5	.4920	.5229	40.0
704	Cu	Ka	Au	39.5	.4920	.5160	40.0
705	Al	Ka	Cu	15.9	.9445	.8801	40.0
706	Al	Ka	Cu	21.2	.9445	.8441	40.0
707	Al	Ka	Cu	31.8	.9445	.7637	40.0
708	Al	Ka	Cu	39.9	.9445	.7065	40.0
709	Cu	Ka	Al	15.9	.0555	.0469	40.0
710	Cu	Ka	Al	21.2	.0555	.0479	40.0
711	Cu	Ka	Al	31.8	.0555	.0487	40.0
712	Cu	Ka	Al	39.3	.0555	.0483	40.0
713	Cu	La	Al	15.9	.0555	.0515	40.0
714	Cu	La	Al	21.2	.0555	.0523	40.0
715	Cu	La	Al	31.8	.0555	.0514	40.0
716	Cu	La	Al	39.9	.0555	.0509	40.0
717	Ag	Lb	Au	15.0	.1996	.1670	52.5
718	Ag	Lb	Au	20.0	.1996	.1460	52.5
719	Ag	Lb	Au	30.0	.1996	.1120	52.5
720	Ag	Lb	Au	40.0	.1996	.0898	52.5
721	Ag	Lb	Au	48.5	.1996	.0720	52.5
722	Ag	Lb	Au	15.0	.3992	.3490	52.5
723	Ag	Lb	Au	20.0	.3992	.3120	52.5
724	Ag	Lb	Au	30.0	.3992	.2530	52.5
725	Ag	Lb	Au	40.0	.3992	.2090	52.5
726	Ag	Lb	Au	48.5	.3992	.1750	52.5
727	Ag	Lb	Au	15.0	.5993	.5480	52.5
728	Ag	Lb	Au	20.0	.5993	.5060	52.5
729	Ag	Lb	Au	30.0	.5993	.4360	52.5
730	Ag	Lb	Au	40.0	.5993	.3680	52.5
731	Ag	Lb	Au	48.5	.5993	.3400	52.5
732	Ag	Lb	Au	15.0	.7758	.7420	52.5
733	Ag	Lb	Au	20.0	.7758	.7000	52.5
734	Ag	Lb	Au	30.0	.7758	.6450	52.5
735	Ag	Lb	Au	40.0	.7758	.5740	52.5
736	Ag	Lb	Au	48.5	.7758	.5420	52.5
737	Pd	La	Au	5.2	.3590	.3921	40.0
738	Pd	La	Au	10.3	.3590	.3342	40.0
739	Pd	La	Au	15.5	.3590	.2876	40.0
740	Pd	La	Au	20.7	.3590	.2445	40.0
741	Pd	La	Au	25.8	.3590	.2154	40.0
742	Pd	La	Au	31.0	.3590	.1898	40.0
743	Pd	La	Au	36.2	.3590	.1742	40.0
744	Au	Ma	Pd	5.2	.6410	.5786	40.0
745	Au	Ma	Pd	10.3	.6410	.5936	40.0
746	Au	Ma	Pd	15.5	.6410	.6005	40.0
747	Au	Ma	Pd	20.7	.6410	.6033	40.0
748	Au	Ma	Pd	25.8	.6410	.5962	40.0
749	Au	Ma	Pd	31.0	.6410	.5928	40.0
750	Au	Ma	Pd	36.2	.6410	.5883	40.0
751	Cu	Ka	Ni	10.1	.5650	.5580	40.0
752	Cu	Ka	Ni	12.5	.5650	.5547	40.0
753	Cu	Ka	Ni	15.3	.5650	.5556	40.0
754	Cu	Ka	Ni	20.4	.5650	.5574	40.0
755	Cu	Ka	Ni	25.6	.5650	.5610	40.0
756	Cu	Ka	Ni	30.7	.5650	.5587	40.0
757	Cu	Ka	Ni	40.0	.5650	.5557	40.0
758	Cu	Kb	Ni	10.2	.5650	.5494	40.0
759	Cu	Kb	Ni	12.5	.5650	.5418	40.0
760	Cu	Kb	Ni	15.3	.5650	.5334	40.0
761	Cu	Kb	Ni	20.5	.5650	.5308	40.0
762	Cu	Kb	Ni	25.6	.5650	.5143	40.0
763	Cu	Kb	Ni	30.7	.5650	.5014	40.0
764	Cu	Kb	Ni	40.0	.5650	.4770	40.0
765	Ni	Kb	Cu	10.2	.4350	.4473	40.0
766	Ni	Kb	Cu	12.5	.4350	.4509	40.0
767	Ni	Kb	Cu	15.3	.4350	.4504	40.0
768	Ni	Kb	Cu	20.5	.4350	.4387	40.0
769	Ni	Kb	Cu	25.6	.4350	.4454	40.0
770	Ni	Kb	Cu	30.7	.4350	.4457	40.0
771	Ni	Kb	Cu	40.0	.4350	.4460	40.0
772	B	Ka	N	4.0	.4355	.3747	40.0
773	B	Ka	N	6.0	.4355	.3317	40.0
774	B	Ka	N	8.0	.4355	.2856	40.0
775	B	Ka	N	10.0	.4355	.2539	40.0
776	B	Ka	N	12.0	.4355	.2316	40.0
777	B	Ka	N	15.0	.4355	.2049	40.0
778	B	Ka	N	20.0	.4355	.1823	40.0
779	B	Ka	N	25.0	.4355	.1800	40.0
780	B	Ka	N	30.0	.4355	.1834	40.0
781	Si	Ka	N	4.0	.6006	.5640	40.0
782	Si	Ka	N	6.0	.6006	.5655	40.0
783	Si	Ka	N	8.0	.6006	.5660	40.0
784	Si	Ka	N	10.0	.6006	.5640	40.0
785	Si	Ka	N	12.0	.6006	.5625	40.0
786	Si	Ka	N	15.0	.6006	.5570	40.0
787	Si	Ka	N	20.0	.6006	.5450	40.0
788	Cr	Ka	N	8.0	.8882	.8518	40.0
789	Cr	Ka	N	10.0	.8882	.8595	40.0
790	Cr	Ka	N	12.0	.8882	.8635	40.0
791	Cr	Ka	N	15.0	.8882	.8680	40.0
792	Cr	Ka	N	20.0	.8882	.8740	40.0
793	Cr	Ka	N	25.0	.8882	.8783	40.0
794	Cr	Ka	N	30.0	.8882	.8805	40.0
795	Cr	Ka	N	8.0	.7878	.7290	40.0
796	Cr	Ka	N	10.0	.7878	.7365	40.0
797	Cr	Ka	N	12.0	.7878	.7433	40.0
798	Cr	Ka	N	15.0	.7878	.7505	40.0
799	Cr	Ka	N	20.0	.7878	.7595	40.0
800	Cr	Ka	N	25.0	.7878	.7650	40.0
801	Cr	Ka	N	30.0	.7878	.7685	40.0
802	Fe	Ka	N	8.0	.9440	.9118	40.0
803	Fe	Ka	N	10.0	.9440	.9140	40.0
804	Fe	Ka	N	12.0	.9440	.9173	40.0
805	Fe	Ka	N	15.0	.9440	.9202	40.0
806	Fe	Ka	N	20.0	.9440	.9240	40.0
807	Fe	Ka	N	25.0	.9440	.9260	40.0
808	Fe	Ka	N	30.0	.9440	.9270	40.0
809	Nb	La	N	4.0	.9299	.9042	40.0
810	Nb	La	N	6.0	.9299	.9068	40.0
811	Nb	La	N	8.0	.9299	.9097	40.0
812	Nb	La	N	10.0	.9299	.9120	40.0
813	Nb	La	N	12.0	.9299	.9130	40.0
814	Nb	La	N	15.0	.9299	.9140	40.0
815	Nb	La	N	20.0	.9299	.9158	40.0
816	Nb	La	N	25.0	.9299	.9162	40.0
817	Nb	La	N	30.0	.9299	.9165	40.0
818	Mo	La	N	4.0	.9420	.8964	40.0
819	Mo	La	N	6.0	.9420	.9023	40.0
820	Mo	La	N	8.0	.9420	.9078	40.0
821	Mo	La	N	10.0	.9420	.9122	40.0
822	Mo	La	N	12.0	.9420	.9163	40.0
823	Mo	La	N	15.0	.9420	.9220	40.0
824	Mo	La	N	20.0	.9420	.9268	40.0
825	Mo	La	N	25.0	.9420	.9360	40.0
826	Mo	La	N	30.0	.9420	.9399	40.0

Appendix 7

Fluorescence Excited by Characteristic Lines in Stratified Specimens

To simplify calculations, the depth distribution of the exciting radiation is expressed as the sum of two exponentials:

$$\phi(t) = \sum_{j=1}^{2} A_j \cdot \exp(-a_j \cdot t) \qquad (t = \text{mass thickness}) .$$

The coefficients of this expression are obtained by imposing on the distribution, as was done in the model XPP, the area F and the mean ionization depth R, and defining $A_2 = \phi(0) - A_1$ and $a_2 = 1.3\,b$ (see Appendix 4).

Contrary to the case of homogeneous semi-infinite specimens, an element can receive secondary (fluorescence) excitation without being excited by primary electrons. Hence one cannot define the fluorescence as a correction relative to the primary intensity, but as a supplementary emission, the intensity of which can eventually be expressed by reference to a standard.

Contrary to the classic approximation, the distribution $\phi(\rho z)$ of the radiation of exciting element B has not been assimilated to that of the excited element A. Besides, the area F of $\phi(\rho z)$ represents in the PAP model the primary radiation. Therefore, if one uses for simplicity the approximation of Green and Cosslett [70] to relate the primary intensity of elements A and B, as does Reed [58], then the ratio I_B/I_A of Reed must be multiplied by the ratio F_A/F_B of the areas of the distributions of A and B.

Hence, the fluorescent intensities expressed below will be related to the intensity of a pure standard defined by eq (19) by means of the factor:

$$K_{AB} = 0.5\, P_{ij} \cdot C_A \cdot C_B \cdot \mu_B^A \cdot (r_A - 1)/r_A \cdot W_B \cdot (A/B) \cdot [(U_B - 1)/(U_A - 1)]^{1.67} \cdot (F_A/F_B) .$$

For this factor it is assumed that a complete series of lines of B excites the level of reference of A. If this is not the case, the relative weight of exciting lines in their series must also be introduced.

Notation common to all configurations:

T_k = mass thickness of the layer indexed k

M_k = mass absorption coefficient of exciter B in the layer k

χ_k = factor χ of absorption of radiation of A in the layer k

x = total mass thickness above the exciting layer

TT = $\exp(-\sum_e \chi_e \cdot T_e)$ = transmission of radiation of A across the layers covering the excited layer.

H_j = $A_j \cdot \exp(-a_j \cdot x) \qquad (j = 1 \text{ or } 2)$

Fluorescence inside a layer:

$$I_{\text{fluo}} = (K_{AB} \cdot TT/\chi_A) \cdot \sum_{j=1}^{2} H_j \cdot \{(I1 + I2 \cdot \chi_A/a_j)/(\chi_A + a_j) + I3/a_j$$
$$+ (I4 + I5 \cdot \chi_A/a_j) \cdot \exp[-(\chi_A + a_j)T_A]/(\chi_A + a_j)\}$$

with $\quad I1 = \ln(1 + \chi_A/M_A) - Ei[-(M_A + \chi_A)T_A]$

$\quad I2 = \ln(1 + a_j/M_A) - Ei[-(M_A + a_j)T_A]$

$\quad I3 = Ei(-M_A T_A) \cdot [\exp(-\chi_A T_A) + \exp(-a_j T_A)]$

$$I4 = \ln |1-\chi_A/M_A| - Ei[-(M_A-\chi_A)T_A]$$

$$I5 = \ln |1-a_j/M_A| - Ei[-(M_A-a_j)T_A]$$

and $\quad Ei[-\nu] = C + \ln |\nu| + \sum_{n=1}^{\infty} (-\nu)^n/(n \cdot n!)$

E_i = exponential integral

C = Euler's constant = 0.5772157

When T_A tends to infinity (case of a substrate), the expression reduces to

$$I_{fluo} = K_{AB} \cdot TT/\chi_A \cdot \sum_{j=1}^{2} H_j \cdot [\ln (1+\chi_A/M_A) + \ln (1+a_j/M_A) \cdot \chi_A/a_j]/(\chi_A+a_j) .$$

This expression has the same terms as that of Reed for a homogeneous material ($TT = 1$, $H_j = A_j$).

When T_A tends to zero (very thin layer), we obtain:

$$I_{fluo} = K_{AB} \cdot TT \cdot \left(\sum_{j=1}^{2} H_j \right) \cdot T_A^2 \cdot \{3/2 - [C + \ln (M_A \cdot T_A)]\} .$$

Except for some factors of proportionality, one arrives at the expression established for thin self-supporting films [71].

Fluorescence of a layer A excited by a layer B:
Specific notations:

$$D_A = D_I + M_A T_A \qquad D_B = D_I + M_B T_B \qquad D_T = D_B + M_A T_A$$

$D_I = \Sigma M_i T_i$ corresponds to the absorption of B by the strata between layers A and B.

Exciting layer above the excited layer ($s = 1$):

$$I_{fluo} = K_{AB} \cdot TT/\chi_A \cdot \sum_{j=1}^{2} H_j[I1 \cdot \exp(-a_j D_B/M_B) - I2 \cdot \exp(-\chi_A T_A - a_j D_T/M_B)]$$

with $\quad I1 = G_j^s(D_I) - G_j^s(D_B) \qquad$ and $\qquad I2 = G_j^s(D_A) - G_j^s(D_T) .$

Exciting layer below the excited layer ($s = -1$):

$$I_{fluo} = K_{AB} \cdot TT/\chi_A \cdot \sum_{j=1}^{2} H_j[I1 \cdot \exp(a_j D_A/M_B) - I2 \cdot \exp(-\chi_A T_A + a_j D_I/M_B)]$$

with $\quad I1 = G_j^s(D_T) - G_j^s(D_A) \qquad$ and $\qquad I2 = G_j^s(D_B) - G_j^s(D_I)$

$$G_j^s(D) = [G1+G2] \cdot \exp (s \cdot a_j \cdot D/M_B) + G3$$

with $\quad G1 = Ei(-D)/a_j$

$\qquad G2 = -\exp(s \cdot \chi_A \cdot D/M_A) \cdot Ei[-(1+s \cdot \chi_A/M_A) \cdot D]/(a_j+\chi_A M_B/M_A)$

$\qquad G3 = -Ei[-(1-s \cdot a_j/M_B) \cdot D] \cdot [1/a_j - 1/(a_j+\chi_A M_B/M_A)] .$

- Case of very thin superficial layer excited by the substrate:
  The previous expression is reduced to

$$I_{fluo} = K_{AB} \cdot T_A \cdot \sum_{j=1}^{2} (A_j/a_j) \cdot \ln (1+a_j/M_B) .$$

One arrives with this expression at the same results as Cox et al. [60].

Appendix 8

Fluorescence by the Continuous Radiation in Stratified Specimens

– Assumption: The exciting radiation is assumed to be generated at a point at the surface.

– Intensity of continuous radiation:

To take into account the massive appearance of the continuum when $E = E_0$, we have included in the spectral distribution of Small et al. [73] a supplementary term which only depends on the mean atomic number, as proposed by Kulenkampff [74]:

$$I(E) = q \cdot \exp(B) \cdot [\bar{Z} \cdot (E_0/E - 1)]^M + C \cdot \bar{Z}^2 \qquad \text{with} \qquad B = -3.22 \ 10^{-2}E_0 + 5.8$$

$$M = 5.99 \ 10^{-3} \ E_0 + 1.05 \qquad \text{and} \qquad C = 6 \ 10^{-10} \ .$$

The constant q, adjusted to 10^{-8} (keV)$^{-1}$ gives a continuous intensity in photons/electron.

In contrast to excitation by characteristic lines, some constants are not eliminated when the fluorescence intensity is related to that of a standard. Therefore one uses for the intensity emerging from an elementary standard the complete expression (except the efficiency of the detector system):

$$I_{std} = (N^0/A) \cdot W_A \cdot w_i^A \cdot Q_i^A(E_0) \cdot I_A$$

where I_A is defined in agreement with eq (19), W_A is the fluorescent yield, w_i^A the weight of the line in its series, and

$$Q_i^A(E_0) = 10^{-20} \cdot b_l \cdot z_{nl} \cdot \ln (U_0)/[(U_0)^m \cdot (E_l)^2] \qquad (E_l \text{ in keV}) \ .$$

For the K levels the coefficient b_l has been set to $b = 3.8m$ [m defined after eq (10)] so as to obtain a value of $2.8 \ 10^{-20}$cm^{2}keV^2 for the maximum of the product $Q_k^A \cdot (E_k)^2$ near $U = 3$ [22]. The coefficient to use at the L levels is less certain [75], but should be close to $5.7m$. $\bar{Z}_{nl}$ is the number of electrons at the excited level (2 for K, 4 for L$_3$). The correlation of the intensity of fluorescence with that of the standard is achieved by the factor:

$$K_F = 0.5 \cdot W_A \cdot C_A \cdot w_i^A \cdot TT \ .$$

– Specific notations:

$\mu^A(E)$ = mass absorption coefficient of radiation of energy E for element A

$M_k{}^E$ = mass absorption coefficient of radiation of energy E for the layer k

$$D_I^E = \sum_i M_i^E \cdot T_i \ ; \qquad D_A^E = D_I^E + M_A^E \cdot T_A$$

– The result of integration over the angle and the depth is $J(E)/\chi_A$, with

$$J(E) = -Ei[-D_I^F] + \exp(\chi_A \cdot D_I^F/M_A^E) \cdot Ei[-(1 + \chi_A/M_A^E) \cdot D_I^F]$$

$$+ \exp(-\chi_A \cdot T_A) \cdot Ei[-D_A^E] - \exp(\chi_A \cdot D_I^F/M_A^E) \cdot Ei[-(1 + \chi_A/M_A^E) \cdot D_A^E] \ .$$

The two last terms disappear when T_A tends to infinity (case of a substrate or a homogeneous material), and one returns to a classical expression [76].

–   The intensity of fluorescence of A is obtained by adding the contributions of the continuum between the various absorption edges of the material which the exciting radiation traverses:

$$I_{fluo} = (K_F/\chi_A) \cdot \sum_{j=1}^{n} (r_A - 1)/r_A \cdot \int_{E_j}^{E_{j+1}} Y(E) \cdot dE , \quad \text{with } Y(E) = \mu^A(E) \cdot I(E) \cdot J(E) .$$

In view of the way $Y(E)$ varies with E one can apply Simpson's approximation in good conditions:

$$\int_{E_j}^{E_{j+1}} Y(E) \cdot dE = \{Y[E_j] + Y[E_{j+1}]\} + 4Y[(E_j + E_{j+1})/2] \cdot (E_{j+1} - E_j)/6$$

–   Note: In the case of excitation by lines as well as by the continuum one must be alert to numerical roundoff errors when the arguments of exponentials or exponential integrals are very large or very small.

ADDENDUM

An error was introduced in the expression of primary intensity by the PAP model in the editing of the publication of J. L. Pouchou and F. Pichoir in the Proc. of the 11th Int. Congr. on X-Ray Optics and Microanalysis, London, Canada 1986, U. Western Ontario, 249 (1987).

The correct expression for the slowing-down term is given here in eq (11).

VIII.   References

[1]   Philibert, J., Métaux, 465, 157 (1964).
[2]   Castaing, R. and Descamps, J., J. Physique Radium 16, 304 (1955).
[3]   Pouchou, J. L. and Pichoir, F., 10th Int. Conf. X-Ray Optics and Microanalysis, Toulouse 1983, J. de Physique 45, C2-17 (1984).
[4]   Pouchou, J. L. and Pichoir, R., 10th Int. Conf. X-Ray Optics and Microanalysis, Toulouse 1983, J. de Physique 45, C2-47 (1984).
[5]   Pouchou, J. L. and Pichoir F., Microbeam Analysis, San Francisco Press, to be published (1988).
[6]   Pouchou, J. L. and Pichoir F., La Recherche Aérospatiale 5, 47 (1984).
[7]   Pouchou, J. L. and Pichoir, F., J. Microsc. Spectrosc. Electron. 10, 279 (1985).
[8]   Packwood, R. H. and Brown, J. D., X-Ray Spectr. 10, 138 (1981).
[9]   Castaing, R., Thesis, University of Paris 1951, Publication O.N.E.R.A. No. 55 (1952).
[10]   Bethe, H. A., Ann. Phys. (Leipzig) 5, 325 (1930).
[11]   Cosslett, V. E. and Thomas, R. N., Brit. J. Appl. Phys. 15, 1283 (1964).
[12]   Love, G., Cox, M. G. C., and Scott, V. D., J. Phys. D: Appl. Phys. 11, 7 (1978).
[13]   Zeller, C., cited by Ruste, J., and Gantois, M., J. Phys. D: Appl. Phys. 8, 872 (1975).
[14]   Bakker, C. J. and Segre, E., Phys. Rev. 81, 489 (1951).
[15]   Sachs, D. C. and Richardson, J. R., Phys. Rev. 89, 1163 (1953).
[16]   Castaing, R., Adv. Electron. Electron Phys., Academic Press, New York, 317 (1960).
[17]   Young, J. R., J. Appl. Phys. 27, 1 (1955).
[18]   Fitting, H. J., Phys. Stat. Solidi A 26, 525 (1974).
[19]   Reuter, W., Kuptsis, J. D., Lurio, A., and Kyser, D. F., J. Phys. D: Appl. Phys. 11, 2633 (1978).
[20]   Lindhard, J. and Winther, A., Mat. Fys. Medd. Dan. Vid. Selsk. 34, 4 (1964).
[21]   Philibert, J. and Tixier, R., Brit. J. Appl. Phys. 2, 1, 685 (1968).
[22]   Powell, C. J., in NBS Spec. Publ. 460, 97 (1976).
[23]   Hutchins, G. A., cited by [22].
[24]   Abe, H., Murata, K., Cvikevich, S., and Kuptsis, J. D., Microbeam Analysis 1985, San Francisco Press, 85 (1985).
[25]   Duncumb, P. and Reed, S. J. B., in NBS Spec. Publ. 298, 133 (1968).
[26]   Bishop, H. E., in X-Ray Optics and Microanalysis, Hermann, Paris, 153 (1966).
[27]   Heinrich, K. F. J., Electron Beam X-Ray Analysis, Van Nostrand Reinhold, New York, 249 (1981).
[28]   Coulon, J. and Zeller, C., C. R. Acad. Sci. Paris, 276 B, 215 (1973).

[29] Sewell, D. A., Love, G., and Scott, V. D., J. Phys. D: Appl. Phys. **18**, 1245 (1985).
[30] Reuter, W., Proc. 6th Int. Conf. on X-Ray Optics and Microanalysis 1971, University of Tokyo Press, 121 (1972).
[31] Bastin, G. F. and Heijligers, H. J. M., Report THE Eindhoven, ISBN 90-6819-002-4 (1984).
[32] Rehbach, W. and Karduck, P., Proc. 11th Int. Congr. on X-Ray Optics and Microanalysis, University of Western Ontario 1986, 244 (1987).
[33] Love, G., Cox, M. G., and Scott, V. D., J. Phys. D: Appl. Phys. **11**, 23 (1978).
[34] Heinrich, K. F. J., Microbeam Analysis, San Francisco Press, 79 (1985).
[35] Pouchou, J. L. and Pichoir, F., Microbeam Analysis, San Francisco Press, to be published (1988).
[36] Heinrich, K. J. F., Proc. 11th Int. Congr. on X-Ray Optics and Microanalysis, University of Western Ontario 1986, 67 (1987).
[37] Bastin, G. F. and Heijligers, H. J. M., Report THE Eindhoven, ISBN 90-6819-006-7 (1986).
[38] Henke, B. L., Lee, P., Tanaka, T. J., Shimabukuro, R. L., and Fujikawa, B. K., Atomic Data and Nuclear Data Tables **27**, 1 (1982).
[39] Pouchou, J. L. and Pichoir, F., J. Microsc. Spectrosc. Electron. **10**, 291 (1985).
[40] Pouchou, J. L. and Pichoir, F., J. Microsc. Spectrosc. Electron. **11**, 229 (1986).
[41] Pouchou, J. L. and Pichoir, F., La Recherche Aérospatiale **3**, 13 (1984).
[42] Tanuma, S. and Nagashima, K., Mikrochimica Acta **3**, 265 (1984).
[43] Ruste, J. and Zeller, C., C. R. Acad. Sci. Paris, **284** Ser. B, 507 (1977).
[44] Sweeney, W. E., Seebold, R. E., and Birks, L. S., J. Appl. Phys. **31**, 1061 (1960).
[45] Cockett, G. H. and Davis, C. D., Brit. J. Appl. Phys. **14**, 813 (1963).
[46] Philibert, J. and Penot, D., in X-Ray Optics and Microanalysis, Hermann, Paris, 365 (1966).
[47] Bishop, H. E. and Poole, D. M., J. Phys. D: Appl. Phys. **6**, 1142 (1973).
[48] Hutchins, G. A., in The Electron Microprobe, Wiley and Sons, New York, 390 (1966).
[49] Colby, J. W., Adv. in X-Ray Analysis **11**, 287 (1968).
[50] Heinrich, K. F. J., Electron Beam X-Ray Analysis, Van Nostrand Reinhold, New York, 1981, 249.
[51] Murata, K., Kotera, M., and Nagami, K., J. Appl. Phys. **54**, 1110 (1983).
[52] Murata, K., Cvikevich, S., and Kuptsis, J. D., 10th Int. Conf. X-Ray Optics and Microanalysis, Toulouse 1983, J. de Physique **45**, C2-13 (1984).
[53] Packwood, R., Microbeam Analysis, 268 (1986).
[54] Willich, P., J. Microsc. Spectrosc. Electron. **10**, 269 (1986).
[55] Pouchou, J. L., in Quantitative Microanalysis—Advanced Topics, Turku 1988, Heikinheimo, E., ed., TKK Helsinki (1988).
[56] Willich, P., personal communication (1986).
[57] Willich, P., Obertop, D., and Tolle, H. J., X-Ray Spectrometry **14**, 84 (1985).
[58] Reed, S. J. B., Brit. J. Appl. Phys. **16**, 913 (1965).
[59] Henoc, J., Heinrich, K. F. J., and Myklebust, R. L., NBS Tech. Note 769 (1973).
[60] Cox, M. G. C., Love, G., and Scott, V. D., J. Phys. D: Appl. Phys. **12**, 1441 (1979).
[61] Daguet, C. and Henoc, J., J. Microsc. Spectrosc. Electron. **6**, 77 (1981).
[62] Armstrong, J. T. and Buseck, P. R., Microbeam Analysis, 213 (1984).
[63] Bastin, G. F., personal communication (1987).
[64] Heinrich, K. F. J., Myklebust, R. C., Rasberry, S. D., and Michaelis, R. E., NBS Spec. Publ. 260-28 (1971).
[65] Willich, P., cited by Bastin, G. F., Heijligers, H. J. M., and Van Loo, F. J. J., Scanning **8**, 45 (1986).
[66] Christ, B., Delgart, G., and Stegmann, R., Phys. Stat. Solidii **A71**, 463 (1982).
[67] Colby, J. W., Adv. in X-Ray Analysis **8**, 352 (1965).
[68] Springer, G., X-Ray Optics and Microanalysis, 296 (1966).
[69] Shimizu, R., Nishigori, N., and Murata, K., Proc. 6th Int. Conf. on X-Ray Optics and Microanalysis 1971, University of Tokyo Press, 95 (1972).
[70] Green, M. and Cosslett, V. E., Proc. Phys. Soc. **78**, 1206 (1961).
[71] Reed, S. J. B., in Quantitative Electron-Probe Microanalysis, V. D. Scott and G. Love, Ellis Horwood Ltd, 215 (1983).
[72] Castaing, R. and Henoc, J., in X-Ray Optics and Microanalysis, Hermann, Paris, 120 (1966).
[73] Small, J. A., Leigh, S. D., Newbury, D. E., and Myklebust, R. L., J. Appl. Phys. **61**, 459 (1987).
[74] Kulenkampff, H., cited by Compton, A. H., and Allison, S. K., in X-Rays in Theory and Experiment, Van Nostrand, 106 (1935).
[75] Powell, C. J., in Electron Beam Interactions with Solids, SEM Inc., AMF O'Hare, 19 (1984).
[76] Springer, G., Proc. 6th Int. Conf. on X-Ray Optics and Microanalysis 1971, University of Tokyo Press, 141 (1972).
[77] Pouchou, J. L., Pichoir, F., and Boivin, D., Proc. of 12th Int. Congr. on X-Ray Optics and Microanalysis 1989, Krakow, to be published.

$\phi(\rho z)$ EQUATIONS FOR QUANTITATIVE ANALYSIS

J. D. BROWN

Materials Engineering Department and
The Centre for Interdisciplinary Studies in Chemical Physics
The University of Western Ontario
London, Ontario, Canada N6A 5B9

I. Introduction

The idea of $\phi(\rho z)$ equations and the depth distribution of x-ray intensities as a basis for the quantitative correction of measured x-ray intensities goes back to the origins of electron probe microanalysis. $\phi(\rho z)$ is the characteristic x-ray intensity generated in a thin layer $d\rho z$ at depth ρz in the specimen relative to intensity generated in an identical layer $d\rho z$, isolated in space. Castaing [1] first suggested that you could write the measured k-ratio in terms of $\phi(\rho z)$ equations as:

$$k_A = \frac{I_s^M}{I_A^M} = \frac{\int_0^\infty \phi_s(\rho z)\, e^{-\mu_A^s \rho z \csc\Psi}\, d\rho z}{\int_0^\infty \phi_A(\rho z)\, e^{-\mu_A^A \rho z \csc\Psi}\, d\rho z} \tag{1}$$

where subscripts s and A refer to specimen and pure element A, respectively, μ is the mass absorption coefficient, with subscript referring to characteristic line and superscript the absorber. ρz represents the mass depth in the specimen. Ψ is the x-ray take-off angle. Castaing (with Descamps) [2] also demonstrated that it was possible to measure $\phi(\rho z)$ curves using a sandwich sample technique (fig. 1). The advantage of the $\phi(\rho z)$ equation is its relative simplicity in concept and the fact that the major corrections of absorption, atomic number and characteristic fluorescence can be explicitly written. The equation for the fraction of x rays which escape from the specimen, $f(\chi)$, the absorption correction is:

$$f_A(\chi_A) = \frac{\int_0^\infty \phi_s(\rho z)\, e^{-\mu_A^s \rho z \csc\Psi}\, d\rho z}{\int_0^\infty \phi_s(\rho z)\, d\rho z} \tag{2}$$

where μ_A^s is the mass absorption coefficient for the characteristic x rays of element A in the specimen. The combined factor $\mu \csc\Psi$ is the so-called absorption parameter χ. The numerator is simply the number of x rays which escape from the specimen while the denominator represents the total number of x rays generated in the specimen.

In a similar manner, the atomic number effect can be stated as the area under the $\phi(\rho z)$ curve which represents the total number of x rays generated in the specimen per unit concentration per unit electron flux. Again, in terms of $\phi(\rho z)$ curves, the atomic number effect is given as:

$$\frac{I_s^g}{I_A^g} = \frac{\int_0^\infty \phi_s(\rho z)\, d\rho z}{\int_0^\infty \phi_A(\rho z)\, d\rho z} \ . \tag{3}$$

Electron Probe Quantitation, Edited by K.F.J. Heinrich and
D. E. Newbury, Plenum Press, New York, 1991

Again the numerator, I_s^g, represents the total number of x rays generated per unit concentration in the specimen while the denominator, I_A^g, is the number for the reference standard which is usually a pure element.

Normally, the absorption and atomic number corrections are combined in the $\phi(\rho z)$ approach since in eq (1), both of these effects are included.

For the characteristic fluorescence effect, an accurate equation can be derived based on $\phi(\rho z)$ curves [3] which requires no simplifying assumptions concerning the primary x-ray distributions. The resulting equations must be numerically integrated, a task easily performed by even a microcomputer.

Because of the paucity of $\phi(\rho z)$ data and the lack of easy computing power in the sixties, simplifications to the correction schemes were sought which led to the ZAF methods. In these methods the three major corrections to the x-ray data are separately modelled using appropriate simplifications. Much effort was then expended in trying to measure absorption correction $[f(\chi)]$ curves or the atomic number effect and to derive expressions for the fluorescence effect. In many of these models very simplified views of the depth distribution of x-ray production were used. Fortunately, the shape of the absorption correction curves is rather insensitive to the shape of the $\phi(\rho z)$ curve and fluorescent radiation is generated at depths which are large compared to the primary distribution. As a consequence, even crude approximations to the $\phi(\rho z)$ distribution have given acceptable accuracy at least in limited applications.

II. Measurements of $\phi(\rho z)$ Curves

The method which was proposed by Castaing and Descamps [2] to measure $\phi(\rho z)$ curves consisted of a thin layer of element A deposited on a polished block of element B then buried by successively thicker layers of matrix element B (fig. 1). The intensity of the characteristic radiation of the tracer element A is measured as a function of the depth of that tracer layer below the surface of the specimen. To obtain the generated x-ray intensity in the specimen a correction then has to be made for absorption in the overlying film. This correction can be explicitly stated as:

$$I_A^M (\rho z) = I_A^g (\rho z) \exp\left(-\mu_A^B \rho z \csc\Psi\right). \tag{4}$$

The accuracy of the x-ray generation curves depends not only on the measurements of x-ray intensity from the thin film but also on the accurate knowledge of the overlying film thickness and the mass absorption coefficients.

To make the measured curves independent of spectrometer efficiency and tracer layer thickness, Castaing and Descamps proposed that measurements be made relative to an isolated tracer layer, i.e., a tracer layer of the same thickness as in the sandwich specimen but without backing, i.e., a thin layer in space. In practice, the thin layers were deposited on a thin carbon layer such as is used in electron microscopy as a specimen support.

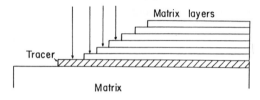

Figure 1.   Sandwich sample to measure $\phi(\rho z)$ curves.

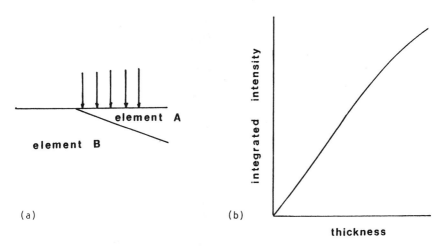

Figure 2. (a) Wedge sample; (b) Integral of $\phi(\rho z)$ curve.

The first such curves were measured for a Zn tracer in Cu, a Bi tracer in Au, and a Cu tracer in Al. For the former two pairs of elements, because the tracer atomic number was essentially the same as the matrix and no characteristic fluorescence of the tracer element occurred, the $\phi(\rho z)$ curves were deemed correct as x-ray generation curves for Au and Cu, but since the tracer atomic number of Cu is quite different than Al, the curve for Al was deemed to be in error. Castaing and Henoc [4] later measured a $\phi(\rho z)$ curve for a Mg tracer in Al which was deemed to be the proper Al curve. Later Brown and Parobek [5] pointed out that by using various tracer elements in the same matrix, the atomic number effect could be evaluated.

An alternate method to measure $\phi(\rho z)$ curves has been proposed and used by Buchner and Pitsch [6]. The specimen consists of a wedge, two elements which are brought together in close contact, perhaps by electroplating, then sectioned at a shallow angle (fig. 2). Characteristic x-ray intensity is measured as a function of distance from the surface junction, hence as a function of thickness of the top element. Again, a correction for x-ray absorption in the top layer has to be made to get the integral of the $\phi(\rho z)$ curve. Differentiation of this curve yields the actual $\phi(\rho z)$ curve. The objection that this method can be used to give undistorted $\phi(\rho z)$ curves only if the pairs of elements are adjacent in the periodic table can be overcome if an alloy containing say 1 wt% of a third element is used in the top layer and measurements are made from this third element [7].

During the past three decades considerable effort has been made to increase the number of measured $\phi(\rho z)$ curves so that generalized expressions could be obtained for $\phi(\rho z)$ on which to base quantitative analysis. The most recent measurements are those of Rehbach and Karduck [8] who have measured $\phi(\rho z)$ for very low energy x-ray lines. In these measurements, the mass absorption coefficients and knowledge of layer thickness are especially important since the correction for absorption is large for the low energy x-ray lines.

III. $\phi(\rho z)$ Equations

The first attempts to accurately model $\phi(\rho z)$ curves as opposed to using an approximation of acceptable accuracy for a specific correction can be traced to Criss [9], who used a multi-term polynomial exponential series. Wittry [10] proposed a Gaussian curve which was symmetric about the maximum value of the $\phi(\rho z)$ curve and later Kyser [11] developed

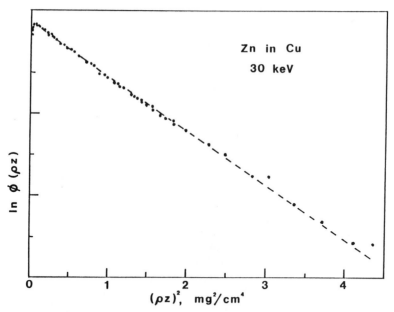

Figure 3. $\phi(\rho z)$ curve plotted to show Gaussian dependence.

a nonsymmetric modification. The problem in these attempts was that no method was available to generalize the curves to an arbitrary matrix, characteristic x-ray line or electron energy. The first equation which could be used to predict the shape of any $\phi(\rho z)$ curve for any matrix or characteristic line in the energy range from 20–30 keV was due to Parobek and Brown [12]. This equation took the form:

$$\phi(\rho z) = D \cdot K \cdot n \left[K \left(\rho z - \rho z_0 \right) \right]^{n-1} e^{-K(\rho z - \rho z_0)^n}.$$

The parameters, D, K, n, and z_0 were obtained by fitting the equation to approximately 80 measured $\phi(\rho z)$ curves using a simplex optimization. The dependence of the four parameters on electron energy, absorption edge energy and matrix atomic number and weight were then inferred from appropriate plots of the optimized parameters. Equations were then generated for the actual dependence. The values can be found in [12]. Robinson [13] later extended this same equation to lower energies while Buchner and Pitsch [6], on the basis of $\phi(\rho z)$ curves measured using a wedge specimen technique, proposed a similar equation.

Packwood and Brown [14] observed that by plotting $\phi(\rho z)$ versus $(\rho z)^2$, straight line plots were obtained (fig. 3), at least beyond the maxima in the $\phi(\rho z)$ versus $\phi(\rho z)$ curves. This implied that the $\phi(\rho z)$ were Gaussian in character but centered at the surface of the specimen, with an exponential term to account for the loss of intensity near the surface

$$\phi(\rho z) = \gamma_0 \, e^{-(\alpha \rho z)^2} \left[1 - \frac{(\gamma_0 - \phi(0)) \, e^{-\beta \rho z}}{\gamma_0} \right]. \tag{5}$$

Again, the four parameters which describe the modified Gaussian curve were fitted to the measured $\phi(\rho z)$ curves using the simplex optimization method. The dependence of these parameters on electron energy, absorption edge energy, and matrix was developed both by appropriate plots as well as theoretically by considering electron interactions with the specimen. The parameters were then optimized using error histograms to obtain best fit between measured and calculated k-ratios for a large body of data [15]. A real advantage of this equation is that it is possible to integrate it to obtain a rather simple form for the corrections for atomic number and absorption [16]. Bastin and Heijligers [17–19] used the same equation to optimize the parameters for low Z analysis and Tirira Saa et al. [20] have proposed other values for the β parameter. At this stage in development, the $\phi(\rho z)$ method would seem to be the best choice for routine analysis since it is based on as accurate as possible a description of the x-ray generation curves.

IV. Conclusions

The three basic corrections can be simply and accurately modelled in terms of $\phi(\rho z)$ equations. These equations are now known well enough that routine quantitative analysis is as accurate as the best of ZAF models, while for low atomic number (or low energy characteristic lines) higher accuracy can be achieved. $\phi(\rho z)$ curves are Gaussian, centered at the surface with the deviation from Gaussian behavior at the surface the result of nonrandom electron trajectories. Calculations using Gaussian curves are no more complex than the ZAF methods which they replace.

V. References

[1]   Castaing, R. (1960), Advances in Electronics and Electron Physics, Vol. XIII, Marton, L., ed., Academic Press, New York, 317.
[2]   Castaing, R. and Descamps, J. (1955), J. Phys. et Radium 16, 304.
[3]   Brown, J. D., Ph.D. (1966), Thesis, University of Maryland, College Park, MD, 167.
[4]   Castaing, R. and Henoc J. (1966), X-Ray Optics and Microanalysis, Castaing, R., Descamps, P. and Philibert, J., eds., Hermann, Paris, 120.
[5]   Brown, J. D. and Parobek, L. (1972), Proc. 6th Int. Conf. X-Ray Optics and Microanalysis, Shinoda, G., Kohra, K., and Ichinokawa, T., eds., U. of Tokyo Press, Tokyo, 163.
[6]   Buchner, A. R. and Pitsch, W. (1971), Z. Metallkunde 62, 393.
[7]   Buchner, A. R. and Pitsch, W. (1972), Z. Metallkunde 63, 368.
[8]   Rehbach, W. and Karduck, P. (1987), Proc. 11th Int. Conf. X-Ray Optics and Microanalysis, Brown, J. D. and Packwood, R. H., eds., UWO Graphics Serv., London, Ontario, Canada, 244.
[9]   Criss, J. W. (1968), NBS Spec. Publ. 298, 53.
[10] Wittry, D. B., (1957), Ph.D. Thesis, California Institute of Technology.
[11] Kyser, D. F. (1972), Proc. 6th Int. Conf. X-Ray Optics and Microanalysis, Shinoda, G., Kohra, K., and Ichinokawa, T., eds., 147.
[12] Parobek, L. and Brown, J. D. (1978), X-Ray Spectr. 7, 26.
[13] Brown, J. D. and Robinson, W. H. (1979), Microbeam Analysis, 238.
[14] Packwood, R. H. and Brown, J. D. (1981), X-Ray Spectr. 10, 138.
[15] Brown, J. D. and Packwood, R. H. (1982), X-Ray Spectr. 11, 187.
[16] Packwood, R. H. and Brown, J. D. (1980), Microbeam Analysis, 45.
[17] Bastin, G. F. and Heijligers, H. J. M. (1984 and 1985), Internal Reports, Univ. of Technology, Eindhoven, ISBN, 90-6819-002-4 and 90-6819-006-7.
[18] Bastin, G. F., Heijligers, H. J. M., and van Loo, F. J. J. (1984), Scanning 6, 58.
[19] Bastin, G. F., Heijligers, H. J. M. (1986), X-Ray Spectr. 15, 143.
[20] Tirira Saa, J. H., Del Giorgio, M. A., and Riveros, J. A. (1987), X-Ray Spectr. 16, 255.

A COMPREHENSIVE THEORY OF ELECTRON PROBE MICROANALYSIS

ROD PACKWOOD

Metals Technology Laboratories
CANMET/EMR, Ottawa, Canada

I. Introduction

An ideal theory for the electron probe microanalyzer would enable the analyst to obtain corrected concentrations that were precise and accurate for essentially any combination of elements and operating conditions. First consider the fundamental assumption of microprobe analysis, i.e., that the characteristic x-ray intensities generated by the electron beam in the specimen are proportional to the mass fraction of each emitting element present and that to determine these quantities entails correcting the observed data for x-ray absorption, backscatter losses and fluorescence effects. To do this requires that the depth distribution of x-ray production be known for the material in question. Therefore a logical starting point in the search would be to devise a theoretical model that would be able to predict the relative number and depth distribution of x-ray production, or $\phi(\rho z)$ as it is called, as a function of the relevant physical parameters: namely the accelerating potential, E_0, the critical excitation potential of the x rays of interest, E_c, the mean atomic number and atomic weight of the specimen, Z and A, and the orientation of the specimen's surface with respect to the electron beam. If this can be accomplished then immediately a number of other benefits accrue, not the least of which would be an equation for the electron range, valuable information when analyzing specimens that are thought to be nonuniform. That in turn brings up the next requirement: the theory should be able to predict the x-ray signal from various configurations of nonuniform specimens such as thin deposits or layered samples and thin film specimens such as are used in the TEM/STEM. Additional features such as the ability to predict backscattered and transmitted or absorbed electron fluxes from these various types of specimens would also be useful in particular circumstances. To predict the x-ray generation in specimens with two or even three dimensions smaller than the electron range would be the ultimate quest.

It is no exaggeration to say that the pursuit of the ideal microprobe theory can be considered to have begun almost simultaneously with the inception of the instrument and has continued unabated since then.

Castaing [1] in his doctoral thesis assumed that the electron beam was absorbed in an exponentially decreasing fashion, i.e., following Lenard's law. Subsequent measurements by Castaing and Descamps [2] revealed the presence of a transient increase in the x-ray generation just below the surface before something approaching the expected exponential decay set in. This rise was on account of the increased efficiency in x-ray production that resulted once the well-aligned incident beam began to become randomized as it penetrated the material of the specimen. The observed behavior was described by Wittry [3] as a Gaussian centered at the peak of the x-ray distribution and by Philibert [4] as a combination exponential decay/capacitor charging function. Other workers used other means to address the basic problem, the Monte Carlo method—Duncumb and Curgenven (see [5]); transport equations—Brown et al. [6]; or physical considerations—Borovskii and Rydnik [7] or Yakowitz and Newbury [8]. Realizing the importance of $\phi(\rho z)$ to all microprobe analyses Brown and Parobek [9] set about measuring $\phi(\rho z)$ curves for a significant range of values of E_0, E_c, and Z, and on the basis of their experimental results derived an empirical formula

Electron Probe Quantitation, Edited by K.F.J. Heinrich and
D. E. Newbury, Plenum Press, New York, 1991

that gave a relatively detailed and accurate description of the overall behavior of an electron beam striking a target. More recently Packwood and Brown [10,11] used a random-walk approach—again derived from experimental observations—as a basis for predicting variation, and as a parametric structure for analyzing the collection of $\phi(\rho z)$ measurements. These were subsequently refined against a collection of analytical data both by those workers and, with some modifications by Bastin et al. [12]. Brown and Packwood [13] and Packwood et al. [14] were able to extend this work into the area of thin deposits and layered samples, and to open the study of the electron backscatter and transmitted fluxes.

A very considerable effort along these general lines but employing a pair of parabolic arcs to model the shape of the $\phi(\rho z)$ curve, and using both physical determinations and Monte Carlo experiments has been carried out by Pouchou and Pichoir [15]. Their results are also applicable to layered samples.

A number of minimalist, nonphysical models [16] have also been tried, with quite reasonable results in so far as data correction is concerned, but which make no effort towards the other goals mentioned at the outset.

The purpose of this review is to outline the progress made towards the goal of a complete theory of the electron probe microanalyzer by those working on the modified-surface-centered Gaussian form for $\phi(\rho z)$ or the random-walk model.

II. Theory

The detailed analysis [11] of the random-walk model that leads to equations for the physical parameters is set out in the Appendix. For now it suffices to say that $\phi(\rho z)$ can be considered to be made up of a Gaussian distribution centered at the target surface that is in turn modified by an efficiency of generation term that gives increased x-ray production away from the immediate subsurface region. To a high degree of precision this situation can be described by the equation:

$$\phi(\rho z) = \gamma_0 \cdot \exp[-\alpha^2(\rho z)^2] - (\gamma_0 - \phi_0) \cdot \exp[-\alpha^2(\rho z)^2 - \beta \rho z] \ .$$

The parameters α, γ_0, β, and ϕ_0 can themselves be written as shown in table A1 (see Appendix).

The exact values that should be assigned to the various terms are still subject to debate; however, those suggested are more than adequate for the purposes of demonstrating the very considerable promise that the theory offers.

III. Microprobe Data Correction

The observed intensity from a standard semi-infinite sample is shown in the Appendix to be given by

$$I_{obs} = C_A(\sqrt{\pi}/2\alpha)\{\gamma_0 \cdot \exp(\chi/2\alpha)^2 \cdot \mathrm{erfc}(\chi/2\alpha) -$$

$$(\gamma_0 - \phi_0) \cdot \exp[(\chi + \beta)/2\alpha]^2 \cdot \mathrm{erfc}[(\chi + \beta)/2\alpha]\} \ .$$

Using this formula it is possible to project intensities for known compositions and compare them with experimentally determined values and so find how well the theory works [12]. Other authors in this volume show that the random-walk theory can give excellent agreement with error histograms below 2 percent spread.

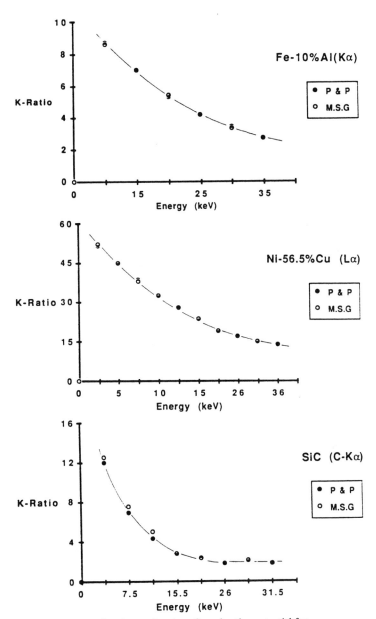

Figure 1. k-ratio as a function of accelerating potential for
a)   Al $K\alpha$ in Fe-10% Al.
b)   Cu $L\alpha$ in Ni-56.5% Cu.
c)   C $K\alpha$ in SiC.

Measurements by Pouchou and Pichoir

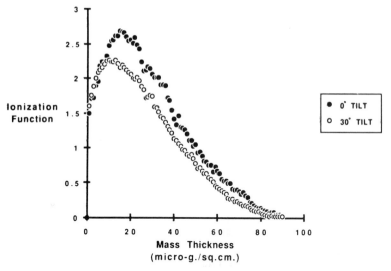

Figure 2. Monte Carlo predictions for Cu $K\alpha$ in copper at 0° and 30° tilt, reference layer at same orientation (20 kV).

These sorts of tests are applied to data gathered under relatively benign conditions and represent a balance of the old atomic number and absorption factors of the ZAF procedure. A different test comes when dealing with soft x rays or severe x-ray absorption; here more emphasis is placed on the ϕ_0 values and the near surface behavior of the formula. To show how well it does, three data sets from Ref. [15] have been analyzed by this method; Al $K\alpha$ in iron, Cu $L\alpha$ in nickel and C $K\alpha$ in SiC, shown in figures 1a–1c. Evidently there are no major difficulties.

IV. Tilted Samples

Although the work is still in its infancy, it may be that the tilted sample configuration is amenable to examination by this method. The random-walk model predicts that the maximum penetration by the electron flux is independent of the surface tilt angle with respect to the beam: in other words that α, related to the deepest penetration by $2/\alpha$ is essentially not a function of the tilt [17]. This prediction is upheld both by Monte Carlo calculations, see figure 2, and by the available experimental data. Once over the peak of the distribution the two curves differ in amplitude only, implying, as was found to be the case, equal α values for the curves. In this same study it was found that ϕ_0 was almost unaffected by the tilt angle Ψ, provided that the reference film was also tilted to the same angle. In figure 3 is a plot of $\phi_0(30°)$ and $\phi_0/\cos 30°$ as observed for Au $M\alpha$ at 20 keV for zero and 30° tilt angles. The increase in backscatter with tilt is evidently close to $\cos \Psi$ for these physical parameters, with the result that $\phi_0(\Psi)$ is closely equal to ϕ_0. $\gamma_0(\Psi)$ proved to vary with angle as:

$$\gamma_0(\Psi) = \gamma_0 \cdot \cos \Psi \cdot [1 - \eta(\Psi)]/(1 - \eta) \approx \gamma_0 \cos^2 \Psi$$

where the first cosine term is exact and due to the tilted configuration; the second term is only approximate and represents a measure of the ionization lost between the two orienta-

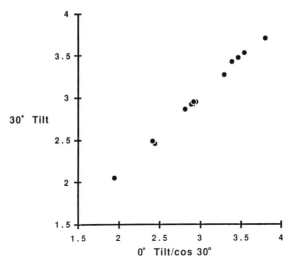

Figure 3. Comparison of surface x-ray intensities of Au $M\alpha$ at 20 keV for 0° and 30° tilt, reference layer at same orientation.

tions and happens in this case to be approximately equal to cos Ψ. The initial slopes of the depth distributions were also approximately equal to one another which means that $\beta(\Psi)$ for the tilted sample is given by:

$$\beta(\Psi) = \beta(\gamma_0 - \phi_0)/[\gamma_0(\Psi) - \phi_0(\Psi)] .$$

This follows from differentiation at $\rho z = 0$.

We have measured [17] $\phi(\rho z, \Psi)$ for a selection of matrix tracer layers, E_0 and E_c and for zero and 30° tilt angles. Figure 4 shows the results for Au $L\alpha$ in copper at 20 keV.

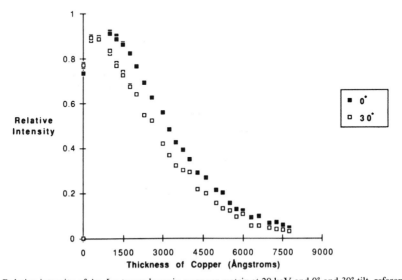

Figure 4. Relative intensity of Au $L\alpha$ tracer layer in a copper matrix at 20 keV and 0° and 30° tilt, reference layer at same orientation.

Table 1. MAC Probe Data: NBS Ag-Au SRM[a]

| | Ag% | | Au% | | |
Nominal	Observed	Predicted	Observed	Predicted	Nominal
80	67.2	66.8	22.9	18.0	20
60	41.2	43.1	36.7	36.9	40
40	24.3	24.8	57.7	57.0	60
20	11.4	11.1	78.4	77.6	80

[a] 20 keV, 35° TOA, 27.5° tilt, $L\alpha$ lines.

Using this scheme k-ratio data collected on NBS Ag-Au reference materials with a MAC probe are shown in table 1 together with the k-ratio predicted on the basis of the transformations spoken of above [18].

V. Electron Range

The Gaussian format has some minor drawbacks particularly concerning mathematical complexity; on the other hand it does have some distinct advantages not the least of which is its familiarity from statistics. The notion of standard deviation can be directly applied to the microprobe situation. In figure 5 is illustrated the concept of "extrapolated range," a tangent drawn to the curve at $1/e$ the height intersects the abscissa at $3/2\alpha$. A useful number that encompasses more than 90 percent of the area under the curve or in other words, more than 90 percent of the x-ray production. The latter neglects the loss of x rays in the subsurface region but is an excellent rule of thumb for estimating the x-ray distribution. The more exhaustive ultimate range is given by $2/\alpha$ although this would best be measured from the extrapolated range where the counting rate is still appreciable.

Table 2 shows measurements made by Reuter et al. [19], of electron range in a variety of targets; the exact fraction of the x-ray generation that these mass thicknesses represent is not constant, but varies from 93–99 percent, so the Gaussian $3/2\alpha$ range was deemed the most appropriate for comparison purposes. The agreement, whilst not perfect, is good enough for almost all practical applications.

VI. Surface Deposits

The so called "thin film" formula [20] is suitable for estimating area concentrations when the surface species are sufficiently slight that they can be thought of as contributing signal and nothing else, their presence not reducing the electron beam nor altering the surface ionization of the substrate material. Many real surfaces have contaminant layers, oxidized zones, or reactant layers such as might occur in sulphidation. There is every

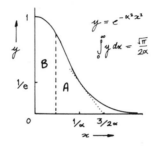

Figure 5. Extrapolated range for a Gaussian depth distribution.

Table 2. Electron range[a] versus Gaussian $3/2\alpha$ (μgm $\cdot$ cm^{-2})

		keV	Observed	Predicted
Al	$K\alpha$	3.0	31	27
		7.3	142	129
		10.8	286	249
		17.4	611	547
Ti	$K\alpha$	7.7	104	114
		11.3	236	254
Ni	$L\alpha$	4.5	72	66
		5.4	94	89
Zr	$L\alpha$	3.1	31	29
Nb	$L\alpha$	4.3	64	58
		8.6	218	201
Mo	$L\alpha$	4.6	72	65
Pd	$L\alpha$	15.8	475	558
Ta	$M\alpha$	3.1	44	46
Pt	$M\alpha$	4.3	84	79
		15.0	518	630

[a] Reuter et al. [19].

reason to suppose that in some cases the affected depth will have a diffusion profile or similar decreasing concentration with depth. And at the same time the Gaussian equation lends itself to being multiplied by other exponential terms such as other Gaussian factors or exponentials. So it is quite straightforward to write out the generated intensity for any given distribution of an additional element. If the concentration of interest is written:

$$C = A \cdot \exp(-\kappa\rho z) \qquad \text{or} \qquad C = B \cdot \exp[-\kappa^2(\rho z)^2]$$

then the observed intensity equation needs the substitutions:

$$C_A \rightarrow A \text{ and } \chi \rightarrow \chi + \kappa, \qquad \text{or} \qquad C_A \rightarrow B \text{ and } \alpha^2 \rightarrow \alpha^2 + \kappa^2 \ .$$

This method was used in Ref. [21], a study of tarnishing on Chalcopyrite adjacent to Argentite grains in cross sections through certain ore samples.

A freshly polished surface if put under strong light for a period of hours or left for a few days at ambient temperature, will start to exhibit yellow or gold colored areas that are found to be high in silver and sulphur and also to show an apparent concentration that increases as the probe voltage is decreased. The latter is a sure sign of a surface enrichment. Scanning Auger work had established that the affected zones were of the order of 50 to 150 Å deep. With this information it was possible to predict observed intensities for a selection of Gaussian Ag$_2$S concentration profiles. Table 3 illustrates this process.

Table 3. Observed, predicted and apparent concentrations: assuming 30% Ag in a 150 Å Gaussian surface layer specimen—Bankovsky deposit Siberia

keV	Observed	Predicted	Apparent bulk
	Orange colored chalcopyrite		
8	3.69	3.92	4.60
15	1.08	1.08	1.37
30	0.45	0.43	0.65
	Yellow colored chalcopyrite		
8	0.60	0.61	0.80
15	0.16	0.18	0.21
30	0.13	0.08	0.22

It should be remarked that the microprobe is able to detect quite thin layers on surfaces provided that there is some distinct composition difference and there is no reason why an ion gun could not be mounted on the specimen chamber and used for depth profiling just as in the SAM. The microprobe is capable of detecting <0.1 monolayer equivalent and whilst not solely surface-sensitive is both rapid and accurately quantitative when the true distribution is known or can be conjectured [13].

There is one small uncertainty, namely the choice of the correct absorption coefficient for the zone of varying concentration. When the zone extends an appreciable mass thickness in to the specimen, the absorption factor will be changing with depth as will the density. A simple weighting procedure using the exponential decrease will suffice for most applications. Alternatively numerical integration will solve the problem even if somewhat inelegantly.

VII. Layered Specimens

A general equation for the observed intensity from a layered specimen can be obtained by simply integrating the generated flux over the appropriate limits and correcting for absorption in the layer and in the overlayer if present [22,23]. It is shown in the appendix that I_{obs} is given by:

$$I_{obs} = (\sqrt{\pi}/2\alpha)[\gamma_0 \cdot \exp(\chi/2\alpha)^2 \cdot \{erf(\alpha\delta + \chi/2\alpha) - erf(\alpha\delta' + \chi/2\alpha)\}$$
$$- (\gamma_0 - \phi_0)\exp[(\chi+\beta)/2\alpha]^2 \cdot \{erf[\alpha\delta + (\chi+\beta)/2\alpha] - erf[\alpha\delta' + (\chi+\beta)/2\alpha]\}]$$
$$\cdot \exp[-\delta'(\chi'-\chi)]$$

where δ' and δ are the top and bottom mass thicknesses for the layer in question. Incidentally, the term "general" refers to the fact that this formula yields the observed intensities for both substrate and surface film as well as that for the conventional semi-infinite specimen, all by changing the depth parameters δ' and δ. This development assumes that the $\phi(\rho z)$ equation is not disrupted by the change in material parameters implicit in the presence of a layer. For layers with mean atomic numbers relatively near to one another in the periodic table this is a sound assumption. Figure 6 demonstrates the case for Cu $K\alpha$ from layers of copper on a nickel substrate at 15 and 20 keV [24]. Obviously there are no problems for this combination.

For layer-matrix pairs chosen further and further apart in their mean atomic number the "no change" approximation must eventually fail. Although α and γ_0 are not pronounced functions of Z, we can see from the formulae in table A1 that both ϕ_0 and β are strongly

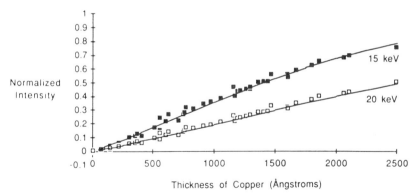

Normalized Intensity

Figure 6. Cu $K\alpha$ intensity vs. copper deposit thickness for copper on nickel substrate.

dependent on Z, indeed, strongly dependent upon the electron backscatter coefficient. The problem is to find some means of weighting the surface and substrate contributions to the backscattering that would permit the retention of the formula. The work by Hutchins [25] on η for gold on silicon proved to be most instructive. As shown in figure 7 the data increase smoothly from $\eta(Si)$ to $\eta(Au)$ in a fashion very reminiscent of the error function, i.e., the area under a Gaussian curve. It seemed only reasonable to weight the two contributions according to their respective areas under the $\exp(-\alpha^2(\rho z)^2)$ curve. This did not work, predicting a rate of change that was a factor of 2 too slow. Weighting by area under a 2α-Gaussian is shown in the same figure and this is a excellent match to the data. On reflection the basis for the agreement is obvious. To be backscattered, an electron must exit the specimen, therefore, on the average it can only travel half as far into the specimen as those electrons that will eventually be absorbed. In other words the backscattered flux will be found to originate in distribution characterized by a 2α- rather than a 1α-Gaussian. Hence, η for a layered specimen would be given by

$$\eta = \eta_{dep} \cdot \mathrm{erf}(2\alpha\rho z) + \eta_{sub} \cdot \mathrm{erfc}(2\alpha\rho z)$$
$$= \eta_{sub} + \mathrm{erf}(2\alpha\rho z) \cdot (\eta_{dep} - \eta_{sub}) \ .$$

Reasoning along this line is also supported by the work of Niedrig [26].

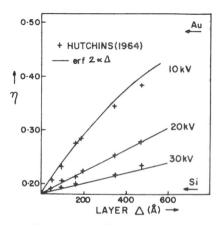

Figure 7. Electron backscatter coefficient vs. gold thickness for gold on a silicon substrate, after Hutchins.

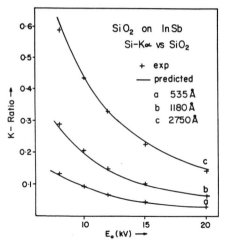

Figure 8. Observed and predicted Si $K\beta$ intensities from SiO_2 layers on InSb at various accelerating potentials, Remond et al.

For parameters that are not surface dependent such as α, the 1α-error function weighting is appropriate.

Reuter [27] in his study of thin deposits on substrates and their effect on ϕ_0, found that he could divide ϕ_0's behavior into two regions: i) a near-surface thin layer situation where the observed ϕ_0 was linearly related to the thickness of the deposit, and ii) a thick layer regime where the observed ϕ_0 approaches that of the deposit material in an exponential fashion. This is a striking experimental discovery of an error function type variation for the surface parameters from layered specimens.

This weighting scheme has been used in the prediction of Si $K\beta$ intensity observed from various thickness deposits of SiO_2 on InSb [28]; the plot shown in figure 8 is for data recorded at various accelerating potentials. The agreement between theory and experiment is good. A second example in figure 9 is for Ag $L\alpha$ from silver layers on silicon [24].

There is more direct evidence for the postulated backscatter behavior being true. Monte Carlo calculations made by Shinoda et al. [29] are shown schematically in figure 10

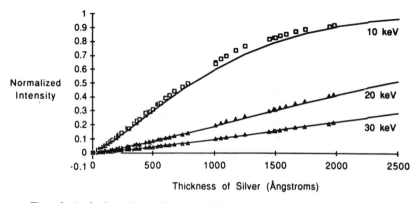

Figure 9. Ag $L\alpha$ intensity vs. silver deposit thickness for silver on a silicon substrate.

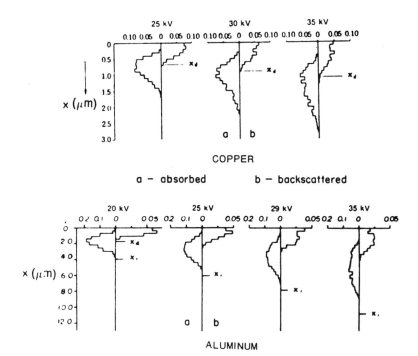

Figure 10. Monte Carlo predictions of the depth distributions of absorbed electrons and the maximum penetration of backscattered electrons for aluminum and copper at various accelerating potentials. Shinoda et al. 1967.

for aluminum and copper. These agree both in form and magnitude with the predictions made here.

VIII.   Electron Backscatter and Transmission

The question of electron fluxes is central to a fundamental understanding of x-ray generation in nonstandard specimen configurations. Looking again at the backscatter distribution, if it is in fact a 2α-Gaussian centered at the specimen surface then what does this

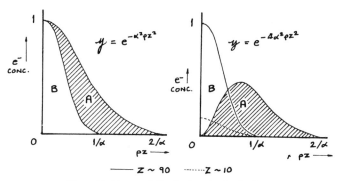

Figure 11. Proposed distributions in depth of
a)   backscattered electrons and
b)   absorbed electrons.

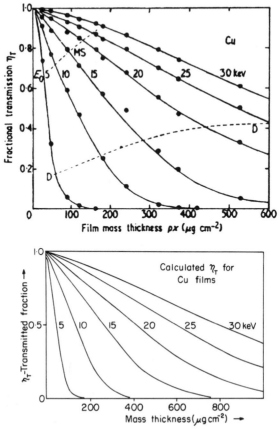

Figure 12. Fractional electron transmission coefficient for copper films at various accelerating potentials
  a) experimental data, after Cosslett and Thomas 1964.
  b) predicted behavior.

mean for the remainder of the electron flux, either absorbed or transmitted as the case may be? How do we combine this information with the observation that x-ray production is essentially a 1α-Gaussian and the original assumption that the overall electron distribution is also essentially Gaussian in form? At high atomic numbers the value of η is of the order of 0.5; the two electron fluxes are roughly equal. If it is assumed that the two fluxes are contained within the envelope of the 1α-Gaussian and the backscatter flux with depth given by a 2α-Gaussian, then the absorbed flux is given by a compound curve made up of a 1α- with a 2α-Gaussian subtracted from its surface region. These postulated distributions are shown schematically in figures 11a and 11b. Furthermore, we will assume that the only effect of changing the atomic number is to change the relative size of the backscattered portion, reducing it to zero at atomic number 1.

Examination of the curves calculated by Shinoda et al. shows that their Monte Carlo simulations support this sort of model distribution. In the extreme, we can interpret these 1α- and 2α-Gaussians to mean that in the case of thin specimens that the transmitted flux is simply the area under the two curves that extend beyond the thickness limit and in fact is given by some combination of error functions.

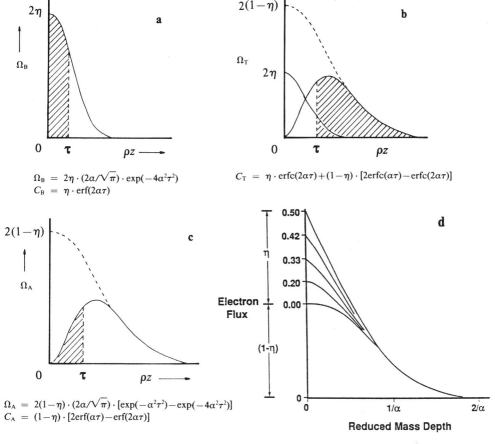

$$\Omega_B = 2\eta \cdot (2\alpha/\sqrt{\pi}) \cdot \exp(-4\alpha^2\tau^2)$$
$$C_B = \eta \cdot \mathrm{erf}(2\alpha\tau)$$

$$C_T = \eta \cdot \mathrm{erfc}(2\alpha\tau) + (1-\eta) \cdot [2\mathrm{erfc}(\alpha\tau) - \mathrm{erfc}(2\alpha\tau)]$$

$$\Omega_A = 2(1-\eta) \cdot (2\alpha/\sqrt{\pi}) \cdot [\exp(-\alpha^2\tau^2) - \exp(-4\alpha^2\tau^2)]$$
$$C_A = (1-\eta) \cdot [2\mathrm{erf}(\alpha\tau) - \mathrm{erf}(2\alpha\tau)]$$

Figure 13. Proposed distributions of electrons scattered by thick films
    a)   backscattered from the film
    b)   transmitted through the film and
    c)   absorbed by the film
    d)   transmitted electron flux versus reduced mass thickness as a function of η,
         the backscatter coefficient.

The electron scattering work of Cosslett and Thomas [30] dealt extensively with transmission through thin films. Their data for copper as a function of thickness and electron voltage are shown schematically in figure 12a; in figure 12b are shown transmission factors based on the ideas just mentioned. The general size and shape of the two sets of data are very similar; particularly encouraging is the manner in which the inflection at around 0.9 is reproduced. There is similar agreement with backscatter measurements. In that both backscattered and transmitted fluxes are well described then perforce the same must hold true for the absorbed flux in a thin film. Formulae for the proposed electron depth distributions and the related fluxes are shown in figures 13a–13c and table 4. The letter C has been used to denote the fluxes because, with Dr. Cosslett's permission, it is intended to call these the Cosslett functions. Figure 13d shows the electron flux as a function of reduced depth

Table 4. Electron depth distributions and Cosslett functions

Backscatter from film

$$\Omega_B = 2\eta \cdot (2\alpha/\sqrt{\pi}) \cdot \exp(-4\alpha^2\tau^2)$$
$$C_B = \eta \cdot \text{erf}(2\alpha\tau)$$

Absorption in film

$$\Omega_A = 2(1-\eta) \cdot (2\alpha/\sqrt{\pi}) \cdot [\exp(-\alpha^2\tau^2) - \exp(-4\alpha^2\tau^2)]$$
$$C_A = (1-\eta) \cdot [2\text{erf}(\alpha\tau) - \text{erf}(2\alpha\tau)]$$

Transmission through film

$$\Omega_T = \Omega_A + \Omega_B$$
$$C_T = \eta \cdot \text{erfc}(2\alpha\tau) + (1-\eta) \cdot [2\text{erfc}(\alpha\tau) - \text{erfc}(2\alpha\tau)]$$

τ is the mass thickness of the film. For x-ray production from conventional microprobe specimens, τ becomes ρz in the transmission equation. Notice the use of the complementary error function, $\text{erfc}(x)$, in this equation.

---

and backscatter fraction, a surrogate for the atomic number; this is the information needed in order to start calculating the x-ray generation as a function of depth. As noted the $(1-\eta)$ fraction is assumed to be the same for all Z values; the η fraction not only varies as shown but also in x-ray generating efficiency via the over voltage dependence which must be allowed for when trying to predict the $\phi(\rho z)$ curve.

IX. The Electron Distribution

The findings of the electron flux study reflect back upon the original suppositions concerning the distribution of x-ray production and its physical source. The Gaussian x-ray distribution cannot in fact be derived from a Gaussian electron distribution. As has been shown the electron depth distribution is, at least for high atomic numbers, approximately Gaussian; on the other hand, the x-ray generation is proportional to the electron flux at any particular depth and not to the final depth reached by the electrons. Therefore, the x-ray generation must in fact follow the compound error functions of the Cosslett equations in the previous section. At first sight this may seem to threaten the agreement between experiment and theory that has underpinned this whole work. However, a Gaussian and the particular combination of error functions in the transmission/absorption equation turn out to be remarkably similar, see figure 14. The chief difference is in the near-surface region just where they are masked in x-ray measurements by the rapid rise in generation efficiency due to randomization of the beam. Where visible the combined error function behaves very much like the Gaussian curve from which it is derived. Of course, in any real target there is a second contribution to the x-ray signal in this region from the backscattered electron flux. It happens that when η is large the energy carried away is also large and as a result only a relatively small amount of energy is available for x-ray production, when η is small even when a significant fraction of the electron energy is deposited in the target it does not in net give rise to a very large unaccounted for x-ray signal.

In essence the original single γ_0 turns into two terms; the first applies to the absorbed flux $(1-\eta)$ and will most likely be of the same form as γ_0: the electrons are assumed to be completely randomized both with respect to direction and energy. The second term concerns the backscattered flux; again the electrons will be thought of as random in direction but to have an energy distribution that is distinctly peaked around 0.75 to 0.85 E_0 as mea-

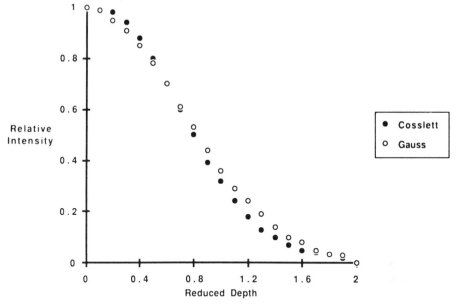

Figure 14. Comparison between a Gaussian and the proposed Cosslett function for the electrons absorbed by a semi-infinite specimen.

sured by Bishop [31]. The total x-ray generation by this part of the electron flux can be estimated from the Duncumb and Reed [32] tables of R-values. In all, the new $\phi(\rho z)$ formula will look something like this:

$$\phi(\rho z) = (1-\eta) \cdot [2\text{erfc}(\alpha\rho z) - \text{erfc}(2\alpha\rho z)] \cdot \{1 + (\gamma_0 - 1) \cdot \text{erf}[f_1(U,Z)\alpha\rho z]\}$$

$$+ \eta \cdot \text{erfc}(2\alpha\rho z) \cdot \{1 + [(\pi Q\hat{U})/QU_0] \cdot \text{erf}[f_2(U,Z)\alpha\rho z]\}$$

The amplitude factors now exhibit some atomic number dependence via the η term, thus justifying retroactively the use of a small Z dependence when fitting the parameter from analytical data bases. The functions f_1 and f_2 are thought to be relatively simple functions with magnitudes ranging from 0.5 to 10. The general shape of these curves is shown in figure 15.

X.   Summary and Discussion

The random-walk theory has proved itself to be a suitable means for examining the phenomena that take place when a nonrelativistic electron beam interacts with an homogenous target. With suitable parameterization, and an optimization based upon a wide range of reliable data, the theory is able to describe the depth distribution of x-ray production well enough to permit accurate microprobe analyses over a wide range of operating conditions. By this is meant an accelerating potential at least 10 to 30 keV, 10 to 92 in mean atomic number, 2 to 20 in overvoltage ratio and 0° to 30° in specimen tilt angle. In addition, it is able to predict the x-ray signal from specimens with surface layers that differ in mean atomic number by factors of 5. After some refinements the various electron fluxes from massive and film specimens can also be estimated provided that the scattering that occurs is

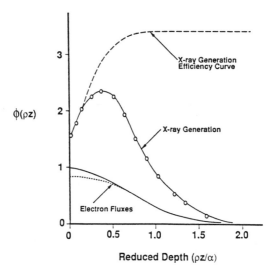

Figure 15. Schematic of the proposed error function $\phi(\rho z)$ for Cd $L\alpha$ in Al at 25 keV data, after Parobek and Brown.

a reasonable approximation to being randomized; very high voltages or single digit atomic number specimens will need individual attention until a more sophisticated version of the theory is developed. The improved understanding of the electron fluxes shown here will be one factor in this process.

One thing was plain from the first work of Packwood and Brown; the Gaussian form for $\phi(\rho z)$ was able to model a large amount of data collected under conditions typical of microprobe analysis at that time. However, that very success lead to the expansion of what was deemed to be normal and in extreme the model must fail and need modification to match the actual conditions employed or at least be reoptimized with more appropriate data. It comes as no surprise to find that Rehbach and Karduck [33] find difficulty with a surface-centered Gaussian distribution when experimenting in situations of relatively low

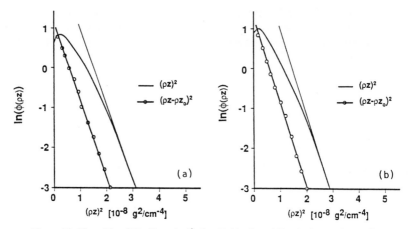

Figure 16. Plot of $\log_e[\phi(\rho z)]$ vs. $(\rho z)^2$ after Rehbach and Karduck, and $(\rho z - \rho z_0)^2$ for
a) C $K\alpha$ in carbon, and
b) C $K\alpha$ in aluminum.

scattering, i.e., $Z = 6$. In Ref. [11] aluminum at 30 keV showed similar behavior and yet could be encompassed within the parameters found for other combinations without going to any great lengths. However, there comes a time when it is simpler to move the midpoint of the Gaussian into the body of the target in a fashion first advocated by Wittry [3]. The extra term representing the displacement is very readily incorporated into the equations set out above; the only minor problem is exactly how this displacement should be modeled, undoubtedly a topic that will receive attention in the not too distant future. This process is illustrated in figure 16, with Monte Carlo data for $CK\alpha$ in Al at 7 keV. The simple subtraction of a suitable constant from the ρz values recovers the $\log_e$-square law, and also gives as might be expected, a plot that has the same slope as the first one at large depths.

A second criticism of the standard formulae by those same authors [34] concerns the γ_0 term at very high overvoltages, again a condition not covered in the original work. In this case the theory predicts γ_0 values that are may be 30 percent too low. Obviously something was overlooked in the first analysis. The presence of significant numbers of fast secondary electrons only happens in the microprobe for soft x rays, indeed the ionization cross sections used have obviously been collected under thin target conditions so as to avoid any such effects! It is a comparatively easy calculation to include fast secondary electrons in γ_0 using Gryzinski's semi-empirical formula [35].

XI. Acknowledgments

It is a pleasure to acknowledge the enthusiastic help of students Cathy Parker, Sandy Thomas, and Hasan Mahmud; the careful considerations of Guy Remond, BRGM, France; and Jim Brown, UWO, London, Ontario; and the, as always, dedicated work of Vera Moore, CANMET, together with numerous others who as critics and users ensured that the ideas presented here really work.

XII. Appendix

The depth distribution of x-ray production or $\phi(\rho z)$ was first measured by the tracer layer technique [2] which proceeds as follows: first determine the x-ray intensity from a thin layer of element A both on, and at various depths within, a specimen of element B; the results are then normalized against the intensity from the same layer of A isolated in space.

A typical $\phi(\rho z)$ curve is shown in figure A1, this is by Brown and Parobek for $Zn K\alpha$ in Cu at 25 keV. Castaing then plotted $\log_e\phi(\rho z)$ vs. ρz as is done in figure A2. The aim presumably to show a straight line dependence that would be the case if Lenard's law held in this circumstance. If it does hold at all it is only over very short intervals of ρz. In figure A2 the straight line has a slope σ calculated from the Philibert-Duncumb and

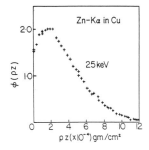

Figure A1. Depth distribution of x-ray production for $Zn K\alpha$ in copper at 25 keV, after Brown and Parobek.

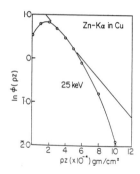

Figure A2. Castaing plot of $\log_e[\phi(\rho z)]$ vs. ρz for data in figure A1.

Shields-Heinrich [36] modification of the Lenard coefficient. The experimental points appear, to the naked eye, to be on a parabolic curve, with maybe a piece missing from the region just beneath the surface. Plot $\log_e\phi(\rho z)$ vs. $(\rho z)^2$ and a parabola will come out as a straight line. This is tried in figure A3. The extensive linear behavior seen here turns out to be typical of all of several hundred $\phi(\rho z)$ data sets when plotted in this way. The implications of the $\log_e$-squared law are simple: $\phi(\rho z)$ is a Gaussian centered at or near the specimen surface with a transient affecting the subsurface region. A suitable name would be; a modified surface-centered Gaussian or MSG theory. We can now venture the following deductions regarding the nature of events occurring in the electron bombarded volume.

i) A Gaussian implies that something random is happening, most likely to the electron paths.

ii) The relative lack of x-ray production in the surface region is caused by the beam requiring some finite distance or number of atomic interactions in order to become randomized.

These assumptions give an equation of the general form:

$$\phi(\rho z) = \gamma_0 \cdot \exp[-\alpha^2(\rho z)^2] \cdot \{1 - [(\gamma_0 - \phi_0)/\gamma_0]\exp(-\beta\rho z)\}$$

where γ_0 is the true height of the Gaussian, α, or rather $1/\alpha$, is a measure of the Gaussian's width, ϕ_0 is the surface ionization or normalized x-ray production in the surface and β is the exponent in the capacitor charging function used to model the efficiency transient.

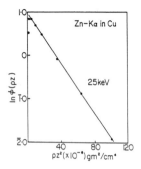

Figure A3. Plot of $\log_e[\phi(\rho z)]$ vs. $(\rho z)^2$ for data in figure A1.

Table A1. Formulae for Gaussian parameters

$$\alpha \approx 4.5 \times 10^5 \cdot [(Z-N)/Z] \cdot (Z/A)^{0.5} E_0^{-0.75} \times [(Z/A)\log_e\{1.166[(E_0+E_c)/2J]\}/(E_0^2-E_c^2)]^{0.5}$$

$$J \approx 11.5 \times 10^{-3} Z \quad \text{and} \quad N \approx 1.3$$

$$\gamma_0 \approx 10\pi \cdot [U_0/(U_0-1)] \cdot [1+(10/\log_e U_0) \cdot (U_0^{-0.1}-1)], \qquad\qquad n = 0.9$$
$$\approx 5\pi \cdot [U_0/(U_0-1)] \cdot [1+(5/\log_e U_0) \cdot (U_0^{-0.2}-1)], \qquad\qquad n = 0.8$$
$$\approx (\pi/2) \cdot [U_0/U_0-1)] \cdot \log_e U_0, \qquad\qquad\qquad\qquad\quad n = 1.0$$

for the ionization cross section given by $Q \cdot E_c^2 \propto (\log_e U_0)/U_0^n$, and $U_0 = E_0/E_c$

$$\beta \approx 10\eta\alpha$$
$$\approx 0.4\alpha Z^{0.6}$$

$$\phi_0 \approx 1+2.8\eta \cdot [1-(0.9/U_0)]$$
$$\approx 1+0.75\pi\eta \cdot \{1-\exp[(1-U_0)/2]\}$$

For compounds weight averaging is used for all appropriate variables: Z, Z/A, η, and $(Z/A)\log_e [1.166 \cdot (E_0+E_c)/2J)]$.

It is possible to gain some insight into the functional form taken by these physical parameters. Einstein [37] in the early part of the century worked out a formula for the probability distribution for a random walking particle in terms of the number of steps N, and step length, λ:

$$P(r) = [1/(2\pi N\lambda^2)] \cdot \exp[-r^2/(2N\lambda^2)]$$

a formula that holds even for quite small N.

Therefore, we can start by putting the step length equal to the scattering mean free path, λ, and assume that the number of steps will depend on the energy loss per step, ΔE, and the energy available for x-ray production (E_0-E_c). ΔE is of course just λ multiplied by the rate of energy loss per unit mass thickness given by the Bethe formula. X-ray production is a rare event and need not be taken into account in the energy balance. The tacit assumption being made is that the x rays are produced in the same depth distribution as the electrons, about which more is said in the main text. An approximate form for α is shown in table A1.

For γ_0, the height of the virtual Gaussian, we must think about how $\phi(\rho z)$ is measured and how x rays are generated. Only two terms really stand out. The ionization cross section, or $Q(U)$ is given in terms of the overvoltage ratio and describes x-ray generation as a function of electron energy. The second term is the average path length in the tracer layer, the longer the path the greater the chance of an x ray being generated. These may be called the Geometry and Ionization factors. For the isolated layer the electron beam is at E_0, and therefore the ionization will be $Q(U_0)$, and the beam perpendicular to the layer, i.e., a path length factor of unity, a geometrical factor of 1. On the other hand, inside the target the average energy has dropped below E_0 so some average value for $Q(U)$ is appropriate, i.e., $\overline{Q(U)}$, where the average is made over the whole available energy or overvoltage interval, U_0-1. Electron paths are no longer perpendicular to the layer and the effective path length goes from 1 to something of the order of 3.

The exact form of γ_0 depends on the $Q(U)$ function chosen. Various formulae are to be found in the literature [38], $Q(U)=(1/U^n) \cdot \log_e U$ with n ranging from 0.7 to 1, $n=0.9$ is satisfactory for most purposes. The corresponding γ_0 are shown in table A1.

Table A2. Observed intensities predicted for layered specimens

---

After Packwood and Milliken. For meaning of symbols δ' to δ see figure A4.

General equation for a buried layer, δ' to δ. (PM-1)

$$
\begin{aligned}
I_{obs} = {}& C_A(\sqrt{\pi}/2\alpha)(\gamma_0 \cdot \exp(\chi/2\alpha)^2 \cdot \{\mathrm{erf}[\alpha\delta + (\chi/2\alpha)] - \mathrm{erf}[\alpha\delta' + (\chi/2\alpha)]\} \\
& - (\gamma_0 - \phi_0) \cdot \exp[(\chi + \beta)/2\alpha]^2 \cdot \{\mathrm{erf}[\alpha\delta + (\chi + \beta)/2\alpha] - \mathrm{erf}[\alpha\delta' + (\chi + \beta)/2\alpha]\}) \\
& \cdot \exp[-\delta'(\chi' - \chi)]
\end{aligned}
$$

Equation for a surface layer, 0 to δ. (PM-2)

$$
\begin{aligned}
I_{obs} = {}& C_A(\sqrt{\pi}/2\alpha)(\gamma_0 \cdot \exp(\chi/2\alpha)^2 \cdot \{\mathrm{erf}[\alpha\delta + (\chi/2\alpha)] - \mathrm{erf}(\chi/2\alpha)\} \\
& - (\gamma_0 - \phi_0) \cdot \exp[(\chi + \beta)/2\alpha]^2 \cdot \{\mathrm{erf}[\alpha\delta + (\chi + \beta)/2\alpha] - \mathrm{erf}[(\chi + \beta)/2\alpha]\})
\end{aligned}
$$

Equation for a covered substrate, δ' to ∞. (PM-3)

$$
\begin{aligned}
I_{obs} = {}& C_A(\sqrt{\pi}/2\alpha)(\gamma_0 \cdot \exp(\chi/2\alpha)^2 \cdot \{1 - \mathrm{erf}[\alpha\delta' + (\chi/2\alpha)]\} \\
& - (\gamma_0 - \phi_0) \cdot \exp[(\chi + \beta)/2\alpha]^2 \cdot \{1 - \mathrm{erf}[\alpha\delta' + (\chi + \beta)/2\alpha]\}) \cdot \exp[-\delta'(\chi' - \chi)]
\end{aligned}
$$

Equation for a conventional, semi-infinite specimen, 0 to ∞. (PM-4)

$$
\begin{aligned}
I_{obs} = {}& C_A(\sqrt{\pi}/2\alpha)(\gamma_0 \cdot \exp(\chi/2\alpha)^2 \cdot [1 - \mathrm{erf}(\chi/2\alpha)] \\
& - (\gamma_0 - \phi_0) \cdot \exp[(\chi + \beta)/2\alpha]^2 \cdot \{1 - \mathrm{erf}[(\chi + \beta)/2\alpha]\})
\end{aligned}
$$

β and ϕ_0 are more difficult to calculate because they do not describe events in equilibrium, as a consequence we use empirical formulae for these terms as shown in table A1. It should be noted that both seem to depend on the backscatter coefficient η and in all likelihood the β factor is actually an error function involving 2α.

The observed x-ray intensity for a given target is simply

$$
I_{obs} = C_A \int_0^\infty \phi(\rho z) \cdot \exp(-\chi\rho z) \cdot d\rho z .
$$

Pressing the logic of the situation even further the limits of integration were altered from their usual zero and infinity settings and put equal to δ' and δ, two arbitrary depths inside the target. The result is a predicted intensity for a buried layer:

$$
I_{obs} = C_A \cdot \exp(-\delta'\chi') \int_{\delta'}^\delta \phi(\rho z) \cdot \exp[-\chi(\rho z - \delta')] \cdot d\rho z
$$

the separate exponential term allows for the absorption in the overlayer. (This was included in the version of the Packwood and Milliken equations quoted in Ref. [39], but subsequently removed, e.g., in the 1986 MAS Tour Talk. Waldo [40], whose derivation is quoted here obtains excellent results with the factor in place.) Performing the integration yields the general equation for predicting the intensity to be seen from a layered specimen. Changing the values of δ and δ' will in turn give the formula for the intensity from i) a buried layer, ii) a surface deposit, iii) a covered substrate, and iv) a conventional, semi-infinite target, see table A2 and figure A4.

XIII. Error Functions

For some small computer applications it may be necessary to use a calculated form for $\mathrm{erf}(x)$. For most microprobe needs the approximation found in Hastings [41] will suffice:

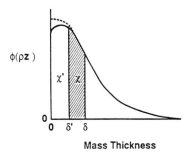

$\phi(\rho \mathbf{z})$

Mass Thickness

$$I_{obs} = C_A(\sqrt{\pi}/2\alpha)(\gamma_0 \cdot \exp(\chi/2\alpha)^2 \cdot \{erf[\alpha\delta+(\chi/2\alpha)]-erf[\alpha\delta'+(\chi/2\alpha)]\}$$
$$-(\gamma_0-\phi_0) \cdot \exp[(\chi+\beta)/2\alpha]^2 \cdot \{erf[\alpha\delta+(\chi+\beta)/2\alpha]-erf[\alpha\delta'+(\chi+\beta)/2\alpha]\})$$
$$\cdot \exp[-\delta'(\chi'-\chi)]$$

Figure A4. Schematic of $\phi(\rho z)$ for a buried layer, after Packwood and Milliken 1985.

$$erf(x) = 1-(a_1t+a_2t^2+a_3t^3)\exp(-x^2)+\epsilon(x)$$

$$t = 1/(1+px) \text{ and } \epsilon(x) \leqslant 2.5\times10^{-5}$$

$$p = 0.47047, \ a_1 = 0.3480242, \ a_2 = -0.0958798, \ a_3 = 0.7478556 \ .$$

In the particular equations used herein the $\exp(-x^2)$ is quite often canceled by an $\exp(x^2)$ factor with the result that only the term in braces has to evaluated. Thus avoiding possible difficulties with over- and underflow of numbers in the exponential terms.

XIV. References

[1]   Castaing, R. (1951), Thesis, Univ. of Paris, Paris, France.
[2]   Castaing, R. and Descamps, J. (1955), J. Phys. Radium **16**, 304.
[3]   Wittry, D. B. (1958), J. Appl. Phys. **29**, 1543.
[4]   Philibert, J. (1963), Proc. 3rd Int. Cong. X-Ray Optics and Microanalysis, Pattee, Cosslett, Engstrom, eds., Academic Press, New York, 379.
[5]   Duncumb, P. (1971), Electron Microscopy and Analysis, Nixon, ed., Conf. Series Inst. Phys., London 1971, 132.
[6]   Brown, D. B., Wittry, D. B., and Kyser, D. F. (1969), J. Appl. Phys. **40**, 1627.
[7]   Borovskii, I. B. and Rydnik, V. I. (1967), Quantitative Electron Probe Microanalysis, NBS Spec. Publ. 298, Heinrich, ed., 35.
[8]   Yakowitz, H. and Newbury, D. E. (1976), SEM (1976) I: 151.
[9]   Brown, J. D. and Parobek, L. (1976), X-Ray Spectrometry **5**, 36.
[10] Packwood, R. H. and Brown, J. D. (1980), Microbeam Analysis—1980, Wittry, ed., 45.
[11] Packwood, R. H. and Brown, J. D. (1981), X-Ray Spectrometry **10**, 138.
[12] Bastin, G. F., Heijligers, H. J. M., and van Loo, F. J. J. (1984), Scanning **6**, 58.
[13] Brown, J. D. and Packwood, R. H. (1986), Appl. Surf. Sci. **26**, 294.
[14] Packwood, R. H., Remond, G., and Holloway, P. H. (1988), Surf. Interf. Anal. **11**, 127.
[15] Pouchou, J. L. and Pichoir, F. (1984), Rech. Aerosp., 167 and 349.
[16] Bishop, H. E. (1974), J. Phys. D, Appl. Phys. **7**, 2009.
[17] Packwood, R., Parker, C., and Moore, V. E. (1988), Microbeam Analysis—1988, Newbury, ed., 258.
[18] Packwood, R. and Pringle, G., to be published.
[19] Reuter, W., Kuptsis, J. D., Lurio, A., and Kyser, D. F. (1978), J. Phys. D: Appl. Phys. **11**, 2633.
[20] Duncumb, P. and Melford, D. A. (1966), Optiques des Rayons X et Microanalyse, Castaing, Descamps, Philibert, eds., Hermann, Paris, 153.

[21] Remond, G., Giraud, R., Holloway, P. H., and Packwood, R. H., SEM-1984, I, 151.

[22] Packwood, R. H. and Milliken, K. S. (1985), CANMET Report No. PMRL/85-25(TR), May, 1985.

[23] Packwood, R., Microbeam Analysis—1986, Romig, Chambers, eds., 268.

[24] Packwood, R., Moore, V. E., and Thomas, S. E., Microbeam Analysis—1989.

[25] Hutchins, G. A. (1964), The Electron Microprobe, McKinley, Heinrich, Wittry, eds., Washington, 1964, J. Wiley, New York, 390.

[26] Niedrig, H. (1978), SEM (1978) I, 841.

[27] Reuter, W. (1971), Proc. 6th Int. Cong. X-Ray Optics and Microanalysis, Shinoda, Kohra, Ichinokawa, eds., Osaka 1971, Univ. Tokyo, 121.

[28] Packwood, R. H., Remond, G., and Holloway, P. H. (1988), Surf. Interf. Anal. 11, 127.

[29] Shinoda, G., Murata, K., and Shimizu, R. (1967), Quantitative Electron Probe Microanalysis, NBS Spec. Publ. 298, Heinrich, ed., 155.

[30] Cosslett, V. E. and Thomas, R. N., Brit. J. Appl. Phys. 15, 1964, 883 and 16, 1965, 779.

[31] Bishop, H. E. (1966), Optiques des Rayons X et Microanalyse, Castaing, Descamps, Philibert, eds., Hermann, Paris, 153.

[32] Duncumb, P. and Reed, S. J. B. (1967), Quantitative Electron Probe Microanalysis, NBS Spec. Publ. 298, Heinrich, ed., 133.

[33] Rehbach, W. and Karduck, P. (1986), Proc. 11th Int. Cong. X-Ray Optics & Microanalysis, Brown, Packwood, eds., London-Ontario, 244.

[34] Rehbach, W. and Karduck, P., Microbeam Analysis—1988, Newbury, ed., 285.

[35] Gryzinski, M. (1965), Phys. Rev. 138A, 301, 335, and 358.

[36] Heinrich, K. F. J. (1968), Adv. X-Ray Anal. 11, 40.

[37] Einstein, A. (1963), The Feynmann Lectures on Physics, R. B. Feynmann, R. B. Leighton, M. Sands, Addison Wesley, Reading, MA, 68.

[38] Abe, H., Murata, K., Cvikevich, S., and Kuptsis, J. D. (1985), Microbeam Analysis—1985, Armstrong, ed., 85.

[39] Packwood, R., Remond, G., and Brown, J. D. (1986), Proc. 11th Int. Cong. X-Ray Optics and Microanalysis, Brown, Packwood, eds., London, Ontario, 274.

[40] Waldo, R. (1988), Microbeam Analysis—1988, Newbury, ed., 310.

[41] Hastings, C., Jr. (1955), Approximations for Digital Computers, Princeton Univ. Press, Princeton, NJ.

A FLEXIBLE AND COMPLETE MONTE CARLO PROCEDURE
FOR THE STUDY OF THE CHOICE OF PARAMETERS

J. HENOC* AND F. MAURICE**

*CAMECA, 103, boulevard Saint-Denis
Courbevoie Cedex, France

**Centre de Energie Nucleaire de Saclay
91191 Gif sur Yvette Cedex, France

Most early Monte Carlo programs which intended simulation of electron trajectories with depth distribution profiles put emphasis on the elastic scattering responsible for large angle deflection. The problem is the following: find an effective d(ρs) elementary path associated with every d(ρz) layer imbedded in the target. There is only one statistical variable besides those of the traditional ZAF data reduction procedure: the scattering angle.

Taking advantage of the properties of the convolution of Goudsmit and Sanderson, the first attempts at simulating electron trajectories had been achieved using a multiple scattering theory. The elementary path length may be an order of magnitude greater than the mean free path; consequently the number of steps in the calculation is reduced as well as the processor time with respect to the single scattering procedure. Preparation of the set of angles of equal scattering probability represents the major part of the code; on the other hand, the sampling of the scattering angle is straightforward. This model is known as the multiple scattering model.

The scattering cross section generally employed is the screened Rutherford cross section. Its associated distribution function is easily reversed. Because of this the mean free path has been chosen as the elementary path length in conjunction with the single scattering cross section. This model is known as the single scattering model. Many Monte Carlo programs are based on the above two models—the multiple scattering and the single scattering theories. As far as x-ray production is concerned, good agreement exists between calculated and experimental results.

The field of application was later extended to physical phenomena related to scanning electron microscopy and spectroscopy; this leads to a revision of basic assumptions. The Mott elastic scattering cross section has been substituted for the Rutherford as more realistic in heavy targets. Utilization of the continuous slowing down approximation may be questionable when accurate electron energy distribution is needed; therefore the so called energy "straggling" was investigated. At this point, the concept of secondary electrons is not used in the basic treatment. Later, the inelastic interaction was introduced to account for them. Up to now we have also supposed that atoms are independent but most of the time we study solids, so collective excitations may play a role as well. Many of these improvements had been added progressively in the programs until the appropriate time came to treat independently every possible interaction.

The different techniques will be illustrated by the description of three programs employing respectively the multiple scattering, the single scattering, and individual interactions.

Electron Probe Quantitation, Edited by K.F.J. Heinrich and
D. E. Newbury, Plenum Press, New York, 1991

I. Single and Multiple Scattering—Background

As stated before, elastic scattering is the main phenomenon governing the progressive scattering of electrons inside the target. According to quantum mechanics [1], the differential elastic scattering cross section may be written

$$\sigma(\theta) = (Ze^2/2E)^2[1/(1+2\alpha-\cos\theta)^2] \tag{1}$$

with the following notations:

Z	the atomic number of the target
e	the electron charge
E	the kinetic electron energy
θ	the scattering angle
α	the screening parameter $\alpha = (\hbar^2/8mEa_0^2)Z^{2/3}$
$\hbar = h/2\pi$	the Planck constant
a_0	the Bohr radius of the hydrogen atom
m	the electron mass

This expression is reduced to the classical Rutherford elastic cross section when the screening of the nucleus by peripheral electrons is neglected. The total elastic scattering cross section is obtained by integrating this function over the solid angle:

$$S(E) = (Ze^2/2E)^2\pi/[\alpha(1+\alpha)]. \tag{2}$$

The ratio of the total cross section to the differential cross section gives the scattering factor $f(\theta)$ representing the probability density

$$f(\theta) = \alpha(1+\alpha)/[\pi(1+2\alpha-\cos\theta)^2]. \tag{3}$$

The associated distribution function is

$$F(\theta) = \int_0^\theta f(\theta)d\Omega = \frac{(1+\alpha)(1-\cos\theta)}{1+2\alpha-\cos\theta} \tag{4}$$

where Ω is the solid angle.

A. SINGLE SCATTERING MODEL

The average number of scattering events in the unit length is, by definition

$$\Lambda/\Lambda(E) = NS(E)$$

where N is the number of atoms in the unit volume. The average mean free path Λ is its inverse

$$\Lambda(E) = 1/NS(E). \tag{5}$$

According to the flow chart in figure 1, starting with the ith step of trajectory number j, the electron has the residual energy E. The mean free path is calculated for this energy and the free path is sampled by

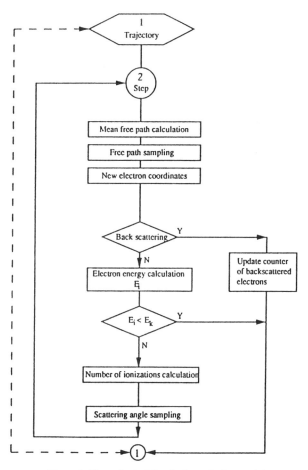

Figure 1. Flow chart of the single scattering model.

$$\lambda = -\Lambda \ln R_1 \qquad (6)$$

where R_1 is a pseudo random number with uniform distribution in the interval (0,1). The electron moves along this path without scattering; the final position is located and tested: if the electron is backscattered, a new trajectory is started. Otherwise, the residual electron energy is calculated, employing a continuous energy loss formula, typically the Bethe deceleration law

$$dE/d\rho s = -2\pi e^4 N_A [(Z/A)/E] \ln(1.166E/J), \qquad (7)$$

where

ρs is the mass thickness
e is the electron charge
N_A is the Avogadro number
Z is the atomic number of the target

A is the atom mass
E is the kinetic electron energy
J is the mean ionization potential.

If the energy E_i is lower than the excitation energy E_k the trajectory is ended; otherwise the number of ionizations is calculated, usually by the Bethe ionization function

$$Q(E)=(\pi e^4/EE_k)z_{nk}b\ \ln(E/E_k) \tag{8}$$

where

z_{nk} is the number of electrons in the shell
b is a constant.

Then the scattering angle is sampled from formula 4 as

$$\cos\theta=1-2\alpha R_2/(1+\alpha-R_2) \tag{9}$$

where R_2 is a pseudo random number. The azimuthal angle is uniformly distributed. We then proceed to the next step.

These sequences of calculations had been first utilized by Reimer [2] in the late sixties for investigating the beam broadening in a scanning electron microscope. They were then used by Murata et al. [3] and Henoc and Maurice [4] for simulating x-ray depth distribution in a semi-infinite or thin targets, as well as Newbury et al. [5] with particular emphasis on nonstandard specimen configurations. Finally, Kyser and Murata [6] have applied the technique to the calculation of x rays emitted by thin layers deposited on a substrate. The Reimer model differs from others both by employing the classical Rutherford cross section and by adding an inelastic scattering cross section instead of introducing the corrective factor $(Z+1)/Z$ to account for the difference in Z dependence observed between elastic and inelastic scattering. Moreover, Murata et al. prefer the Worthington and Tomlin [7], rather than the Bethe ionization cross section

$$Q(E)=0.7(\pi e^4/EE_k)z_{nk}\ \ln\{(4E/E_k)/[1.65+2.35\exp(1-E/E_k)]\}. \tag{10}$$

The program MCSING given in Appendix 1 has, been written by the authors as an illustration of the model for simulating x-ray depth distribution. The algorithm is simple but consumes more processor time than the multiple scattering model.

B.   MULTIPLE SCATTERING MODEL

Starting from the convolution technique developed by Goudsmit and Sanderson [8], Bishop [9] proposed a multiple scattering model. This approach replaces the effect of a number of single scattering events suffered by an electron traveling each free path in the target by a single event selected at random within the elementary path or step. The step length may be an order of magnitude greater than the mean free path and represents a constant fraction of the electron range. The ratio of the processor time used, after a reasonable number of trajectories, between the single and the multiple scattering models can be greater than 1000. The derivation of the convolution formula is rather tedious and it has been detailed in Bishop's thesis [9]. It is also reported in [10]. The final result is the distribution function $F(\theta)$ of the elastic scattering angle:

$$F(\theta)=1/2+(1/2)\exp(pa_1-p)-\sum_{n=1}^{\infty}(1/2)[\exp(pa_{n-1}-p)-\exp(pa_{n+1}-p)]P_n(\cos\theta) \tag{11}$$

where

$p = NS(E)\mathrm{d}s$ is the mean number of collisions suffered by the electron of energy E while traveling the step length $\mathrm{d}s$

a_n is the coefficient of expansion of the single scattering cross section in Legendre polynomials

P_n is the nth order Legendre polynomial.

Figure 2 is the flow chart of the algorithm we have written for simulating electron trajectories by the multiple scattering theory. This algorithm is named E10 in the source listing given in Appendix 2. After calculating the range by integrating the reverse of Bethe's deceleration formula and selecting a convenient step length, half- and full-step electron

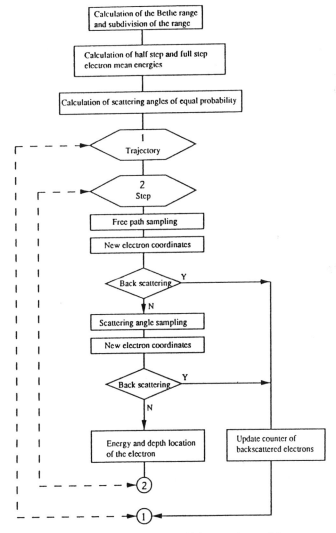

Figure 2. Flow chart of the multiple scattering model.

energies are determined by the Runge Kutta method. Then the distribution function of the scattering angle is calculated by a recursive process for a discrete set of angles and for the energy corresponding to every step. A step of angles of equal probability is built by interpolation and stored for each above-mentioned energy value. The simulation of the trajectories then proceeds step by step. Each step consists of two linear segments: the first segment is the free path selected at random, as a fraction of the step length; the length of the second segment is the difference between the step length and the free path, the angle between the two segments is selected at random among the set of angles of equal probability. Meanwhile, the penetration depth is tested after the electron has moved to the end of the first and second segment. If the depth is negative the electron is backscattered and the corresponding counter is incremented, otherwise the matrix describing the position and the energy of the individual electron is updated. We then proceed to the next step.

Shimizu et al. [11] have employed the transport equation of Lewis to describe the angular distribution of electrons. They claimed that this approach fits the Goudsmit and Sanderson deconvolution for small step lengths. On the other hand they have chosen a variable step length which decreases with the electron energy.

II. Further Improvements

Despite the simplifications implicit in the two basic assumptions, namely that the elastic scattering is the main phenomenon governing the progressive diffusion of electrons and that the energy deposit is continuous, the resulting treatment was quite successful in reproducing and even predicting the shape and the magnitude of signals induced by electron penetration into amorphous targets. The x-ray depth distribution is reasonably well represented but the backscattering coefficient is generally overestimated. The agreement between experimental and simulated measurements on electron transmission by thin foils is, at best, qualitative. To correct these inaccuracies, some refinements have been introduced: the Dirac equation has been substituted for the Schroedinger equation, leading to the Mott elastic scattering cross section. Furthermore, electron energy losses are not continuous; the x-ray production being an example. Therefore a stepwise description of electron energy losses is preferable in principle.

A.   MOTT ELASTIC SCATTERING CROSS SECTION

Reimer et al. [12] have replaced the screened Rutherford elastic cross section by the Mott form to calculate backscattering of 10–100 keV electrons. The comparison between the two functions in the range 0–10 keV appears in the papers of Kotera et al. [13]; the Rutherford equation underestimates the cross section at large angles and overestimates it at medium angles. For gold, Rutherford gives a greater average scattering angle than Mott in the same range of electron energies. The utilization of the Mott cross section seems to lead to more accurate results [14,15].

B.   ENERGY LOSSES

Experimental evidence shows that electrons of a given incident energy have a statistical energy distribution after crossing a thin layer of known thickness. Therefore, the introduction of a statistical energy loss is more realistic than the continuous energy loss. The treatment of statistical energy losses was developed by Landau. Henoc and Maurice [16] introduced this concept as an alternative in the calculation of the $\phi(\rho z)$ distribution. The agreement between experiments and calculation is slightly improved for great thickness.

Shimizu et al. [17] have defined the inverse mean free path for inelastic collisions by taking for the number of these interactions the ratio of dE/ds given by the Bethe's equation to any mean energy loss ΔE per inelastic event. The total inverse mean free path is the summation of the elastic and inelastic inverse mean free paths. The simulation is performed, as explained in section III, by choosing an exponential form for the probability density of energy loss. The technique employed compares with the earlier treatment of Reimer [2]; they are both attempts to introduce several competing interactions.

C. SECONDARY ELECTRON EMISSION

The introduction of the concept of inelastic processes is necessary to study the behavior of secondary electron emission, which applies, for example, to electron spectroscopy [18]. The generation of secondary electrons is caused either by single excitation of deep levels or by collective excitation of conduction electrons. In the next section we will describe a comprehensive treatment of interactions between the primary or secondary electrons with the solid.

III. Simulation of Individual Excitations

Various attempts have been made to describe realistic models which simulate individual events. An extensive investigation originates from the Japanese team [19]. Rather than presenting an exhaustive review, we will fully describe the program provided by Terrissol [20] which was written for this purpose. The description of the algorithm presented below will demonstrate its flexibility: the probability density for each possible event may be easily updated when a better knowledge of the production of that particular event is reached.

A. GENERAL DESCRIPTION

Electron trajectories in the target may be seen as a chain of linear segments (free path) between successive interactions with the target atoms. After each interaction, both the energy and the direction of the primary electron change; a secondary electron may also be created. The problem consists of finding the length of the segment, the type of interaction, and the parameters of this interaction using the individual cross section. We believe that the proposed model will be particularly useful for testing the effects of parameters.

1. Free Path

The probability $p(x)$ for observing one scattering event while the primary electron, of energy E, is crossing the medium along the path limited by x and $x+dx$ is

$$p(x)=S(E){\cdot}N{\cdot}\exp[-S(E){\cdot}N{\cdot}x] \tag{12}$$

where

$S(E)$ total interaction cross section;

N number of atoms per unit volume;

$p(x)$ being the density of probability, $\displaystyle\int_0^\infty p(x)dx =1$.

Alternatively we can say that $p(x)$ represents the probability of observing a free path x of the particle.

2. Mean free path

The mean free path $\Lambda(E)$ is defined by

$$\Lambda(E) = \int_0^\infty xp(x)dx = 1/[S(E)\cdot N] \tag{13}$$

If we assume that the target is pure and that one possible interaction i has a total cross section $S_i(E)$

$$\Lambda(E) = 1/\left[N \sum_i S_i(E) \right] \tag{14}$$

the summation is made over all possible interactions.

3. Free path sampling

The free path λ is obtained from the mean free path, $\Lambda(E)$, after selecting a uniformly distributed R random number $(0 < R < 1)$, according to the formula

$$R = \int_0^\lambda p(x)dx = 1 - \exp[-S(E)\cdot N\cdot\lambda].$$

Let $R' = 1 - R$, then

$$R' = \exp[-S(E)\cdot N\cdot\lambda]$$

$$\ln(R') = -S(E)\cdot N\cdot\lambda$$

$$\lambda = -\Lambda(E)\ln(R'). \tag{15}$$

4. Interaction sampling

Let $\{1, ..., j, ..., n\}$ be the set of possible interactions for an electron of energy E and $S_j(E)$, the total section corresponding to the interaction with subscript j. The interaction i which will occur satisfies the relation

$$\left[\sum_{j=1}^{i-1} S_j(E)/\sum_{j=1}^{n} S_j(E) \right] < R_1 < \left[\sum_{j=i+1}^{n} S_j(E)/\sum_{j=1}^{n} S_j(E) \right] \tag{16}$$

where R_1 is a pseudo random number.

5. Sampling the values for the statistical variable of the previously selected interaction

An interaction may depend on several statistical variables. Let u be one of them and $s(E,u)$ be the corresponding differential cross section. The selected value u must satisfy

$$R_2 = \int_{u_1}^{u} s(E,u)du / \int_{u_1}^{u_2} s(E,u)du \tag{17}$$

with R_2 a random number of uniform distribution and (u_1, u_2) the definition interval of interest of u. The variable u stands for the defined variables $(dE, x, y, z, \Theta, \phi)$, where dE is the energy loss, x, y, z are the Cartesian coordinates, Θ, ϕ are spherical coordinates. The result is often expressed as:

$$u = F^{-1}(R_2) \text{ with } F(u) = \int_{u_1}^{u} s(E, u)du \bigg/ \int_{u_1}^{u_2} s(E, u)du. \tag{18}$$

The main difficulty of the calculation lies in finding the inverse of F. Many sampling methods such as the composition method or the variable interval size stratified method, avoid this complication. A review of these techniques and some references may be found in the thesis of Terrissol [20].

6. Electron trajectory

After sampling the free path, the kind of interaction, the energy loss, and the scattering angle, the new electron coordinates are calculated by the well-known spherical trigonometry formulae. A schematic diagram of the description of one trajectory is represented on figure 3. The program includes the treatment of secondary electrons.

B. CROSS SECTIONS

1. Elastic scattering cross section

Mott cross sections are computed separately by means of a program written by Raynal [21] for nuclear physics computations. The differential cross section is calculated for a discrete set of scattering angles and electron energies. The total scattering cross section is calculated for the same energies. The computer output is directed to a file used as input by the Monte Carlo program. Figure 4 shows the variation of the differential cross section for aluminum at some typical electron energies. In figure 5 we have plotted the total cross section obtained by us for aluminum at electron energies in the range 1–30 keV. On the same graph we have represented the values given by Reimer [14]. A polynomial fit of $\ln(S)$ versus $\ln(E)$ is convenient to calculate the mean free path. The scattering angle is obtained by stratified sampling with variable interval size, which is particularly efficient for a density known on a discrete set of points.

2. Ionization of deep levels

Total ionization cross section.   The total ionization cross section employed by Terrissol is that of Gryzinski [22]:

$$S = \sigma_0 z_{nk}/(E^2_k U)[(U-1)/(U+1)]^{3/2}\{1+(2/3)[1-1/(2U)]\ln[2.7+(U-1)^{1/2}]\} \tag{19}$$

where

E_k is the energy of the ionized level
z_{nk} the number of electrons of the k shell
U the overvoltage
$\sigma_0 = 6.56 \cdot 10^{-14} \text{ eV}^2 \text{ cm}^2$.

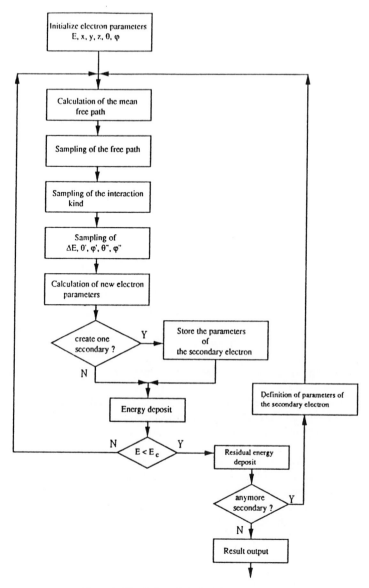

Figure 3. Simulation of the individual excitations.

Differential energy loss cross section. If the amplitude of the energy loss is sufficient, a secondary electron may be created. This possibility will be treated later under "The production of secondary electrons." The differential cross section for energy loss chosen by Terrissol is the Moller expression reported by Berger [23]:

$$s(E,\epsilon)=(C/E)\{(1/\epsilon^2)+1/(1-\epsilon)^2+[\tau/(\tau+1)]^2-(2\tau+1)/[\epsilon(1-\epsilon)(\tau+1)^2]\} \tag{20}$$

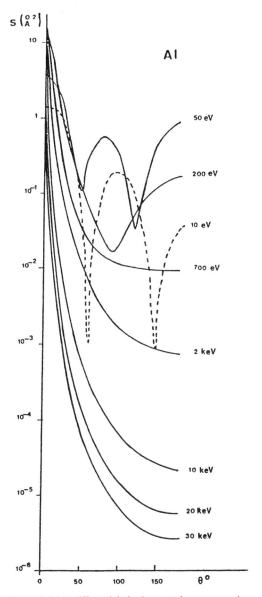

Figure 4. Mott differential elastic scattering cross section.

where

$C = 2\pi e^4/mv^2$

v electron speed

E electron kinetic energy

ϵ ratio of energy transferred to the kinetic electron energy at that point

τ kinetic electron energy in m_0c^2 unit.

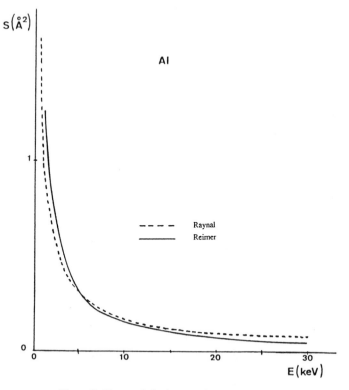

Figure 5. Mott total elastic scattering cross section.

The probability density is obtained by dividing this expression by the total cross section for energy loss. The limits of integration are commonly taken as ϵ_c the reduced excitation energy ($\epsilon_c = E_k/E$), and 1/2 according to the principle of the undiscernibility of the electrons. The following form, used in the program, is more suitable for applying the composition method sampling:

$$s(E,\epsilon) \propto \frac{1-2\epsilon_c}{2}\left(\frac{\tau}{\tau+1}\right)^2 \cdot \frac{2}{1-2\epsilon_c} + \frac{1-2\epsilon_c}{1-\epsilon_c} \cdot \frac{1-\epsilon_c}{1-2\epsilon_c} \cdot \frac{1}{(1-\epsilon)^2}$$

$$+ \frac{1-2\epsilon_c}{\epsilon_c} \cdot \frac{1-\left(1+\frac{2\tau+1}{(\tau+1)^2}\right)\epsilon_c}{1-\epsilon_c} \cdot \frac{1}{\epsilon^2} \cdot \frac{\epsilon_c}{1-2\epsilon_c} \cdot \left\{\frac{1-\left(1+\frac{2\tau+1}{(\tau+1)^2}\right)\epsilon}{1-\epsilon} \middle/ \frac{1-\left(1+\frac{2\tau+1}{(\tau+1)^2}\right)\epsilon_c}{1-\epsilon_c}\right\} \quad (21)$$

Gryzinski [22] has also proposed an expression for the energy loss cross section. For the practical purpose of finding a set of equal probability of energy losses, it is convenient to write this expression as

$$S(E,\epsilon) \propto \frac{1}{\epsilon^3}\left\{\frac{\epsilon}{\epsilon_c}(1-\epsilon_c) + \frac{4}{3}\ln\left[2.7+\left(\frac{1-\epsilon}{\epsilon_c}\right)^{1/2}\right]\right\}(1-\epsilon)^{\epsilon_c/(\epsilon+\epsilon_c)}. \quad (22)$$

using the same notation as above.

Angular deflection.   The angular deflection Θ' of the primary electron from its original direction is given by the classical formula

$$\cos^2\Theta' = [(E - T)(E + 2)]/[2E + (E - T)E] \tag{23}$$

where

E is the energy of the primary electron before the interaction;
T is the energy transfer during the interaction.

3.   Electron hole pair

Mean free path.   According to Ritchie [24], if the probability $\tau(\epsilon,\epsilon')$ for an electron of energy ϵE_F to create an electron hole pair by losing the energy $\epsilon'E_F$ in the unit path length is multiplied by $(\epsilon+1)$, the product $(\epsilon+1)\tau(\epsilon,\epsilon')$ is a universal function for $\epsilon'<\epsilon$ and for a given value of r_s. The radius r_s is defined by

$$n^{-1} = (4/3)\pi\,(a_0 r_s)^3$$

where n is the electron density of the Fermi gas for the metal and a_0 is the Bohr radius. All data are quoted in E_F unit (Fermi energy). The Ritchie function is plotted in figure 6. The inverse mean free path is then

$$\Lambda^{-1} = \int_0^{\epsilon} a_0\,(\epsilon+1)\tau(\epsilon,\epsilon')d\epsilon'/[a_0(\epsilon+1)]. \tag{24}$$

When $\epsilon'>5$ the ϵ'^{-2} law can be substituted for the Ritchie function.

Energy loss.   The energy loss is sampled from the probability density:

$$p(\epsilon') = \tau(\epsilon,\epsilon') \tag{25}$$

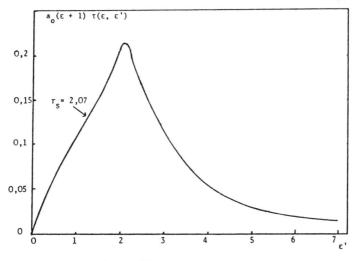

Figure 6. Ritchie Universal Function.

The variable interval size stratified sampling is employed when $0 < \epsilon' < 5$. When $\epsilon' > 5$ the sampling is done directly [eq (18)].

Angular deflection. For a sampled ϵ', the probability density $T(\epsilon',z)$ of the transfer wave vector is (Ritchie [24])

$$T(\epsilon',z) = k(\alpha^2 - 4z^2)^{1/2} f_2(\epsilon',z) / [(z^2 + \chi^2 f_1(\epsilon',z)) + \chi^4 f_2^2(\epsilon',z)] \qquad (26)$$

with

$\alpha = (E/E_F)^{1/2}$
$z = k/(2k_F)$ k: transfer wave vector
                k_F: Fermi wave vector

$[(1+\epsilon')^{1/2} - 1]/2 < z < [(1+\epsilon')^{1/2} + 1]/2$

$\chi^2 = e^2/(\pi h v_F)$ v_F: Fermi velocity

$f_1(\epsilon',z) = 0.5 + \{(1-u^2)\ln|(u+1)/(u-1)| + (1-v^2)\ln(v+1)/(v-1)|\}/8z$

$f_2(\epsilon',z) = \pi(1-u^2)/8z$

$u = z - \epsilon'/4z$
$v = z + \epsilon'/4z$

$f_1(\epsilon',z)$ is the real part of the dielectric function of the metal,

$f_2(\epsilon',z)$ the imaginary part of the same function.

When z is sampled, the deflection angle Θ is given by

$$\cos\Theta = (2\alpha^2 - 4z^2 - \epsilon')/[2\alpha(\alpha^2 - \epsilon')^{1/2}]. \qquad (27)$$

The production of secondary electrons. For the production of one secondary electron, the primary electron must transfer to the electron gas at least the amount of energy E_G, where

$$0 < E_G < E_F.$$

Kittel [25] gives the state density $d(E_G)$ of free electrons with energy E_G

$$d(E_G) = K \cdot E_G^{1/2}.$$

The energy E_G of the electron gas is directly sampled from the selection of the random number R as

$$E_G = E_F R^{2/3}. \qquad (28)$$

The energy transfer to the secondary electron is then

$$\epsilon'' E_F = \epsilon' E_F - E_G.$$

The same situation arises in ionizations of deep levels; if the energy loss dE exceeds the ionization energy E_k, the energy E_S of the secondary electron is

$$E_S = dE - E_k.$$

The angle Θ'' between the unit vector of the secondary electron and the unit vector of the primary electron before interaction is given by:

$$\cos^2\Theta'' = (2dE + E\,dE)/(2E + E\,dE) \tag{29}$$

where dE is the energy loss and E is the primary electron energy.

4. Volume plasmon

Mean free path. Ashley and Ritchie [26] give for the mean free path

$$\Lambda^{-1} = (e^2\omega/\hbar v^2)\ln[v/(1.13v_F)] \tag{30}$$

e electron charge
ω resonance pulse of the electron gas
$\hbar$ Planck constant divided by 2π
v incident electron velocity
v_F Fermi velocity of the electron gas.

Energy loss. Sevier [27] writes the dE energy loss induced by a volume plasmon creation as follows

$$dE = \hbar\omega h + 0!(\theta^2) \tag{31}$$

where θ is the scattering angle. If the scattering angle is small, then the second term is neglected in the program. It is assumed that the energy loss is a constant depending only on the target metal.

Angular deflection. The differential cross section $s(E,\theta)$ may be written

$s(E,\theta) = K/(\theta^2 + \theta^2_E)$ for $\theta < \theta_c$ (K is a constant)

$\theta_E = \hbar\omega/(2E)$

$\theta_c = \hbar\omega/(2(EE_F)^{1/2})$ with the same notation as above.

The distribution function F is then

$$F(\theta)=\left[\int_0^\theta d\theta/(\theta^2+\theta^2{}_E)\right]\Big/\left[\int_0^{\theta_c} d\theta/(\theta^2+\theta^2{}_E)\right]$$

$$=\text{arctg }(\theta/\theta_E)/\text{arctg }(\theta_c/\theta_E)$$

with the selected R random number, the sampled scattering angle is

$$\theta=\theta_E\text{tg}[R\cdot\text{arctg}(\theta_c/\theta_E)].$$

5. Surface plasmon

At the boundary vacuum/solid the probability $p(E)$ of exciting a surface plasmon is (Ritchie et al. [28])

$$p(E)=\pi(R/E)^{1/2}/(1+\epsilon_0)$$

ϵ_0 dielectric constant of the metal
R Rydberg energy
E kinetic electron energy.

If the selected random number R and the probability of surface plasmon excitation obey the relation

$$R\leqslant p(E)$$

the surface plasmon is created with a corresponding energy loss for the primary electron. Angular scattering may be neglected: there is, at most, only one surface plasmon creation in an electron trajectory.

C. Depth Distribution of K Ionizations

The original program, which was designed to simulate energy deposit, has been modified by the authors to simulate the depth distribution of K-shell ionizations.* The scattering medium was assumed to be infinite; we have introduced the vacuum/target boundary and the backscattering concept. When the primary electron energy falls below the K-shell excitation, the description of the trajectory may be ended.

A first run of the program was performed to simulate depth distribution of K ionizations in an aluminum target at 20 keV. A special version of the program gives the number of K ionizations produced in a thin layer of aluminum isolated in vacuum. The order of magnitude of K ionizations produced in a layer of 0.01 mg/cm is 0.001. This implies that a great number of trajectories must be described to get meaningful results. A run of 1000

---

* The modified program is available free of charge from the authors, but previous authorization must be requested from Dr. M. Terrissol, Centre de physique atomique, Université Paul Sabatier, 118 route de Narbonne, 31062 Toulouse Cedex France.

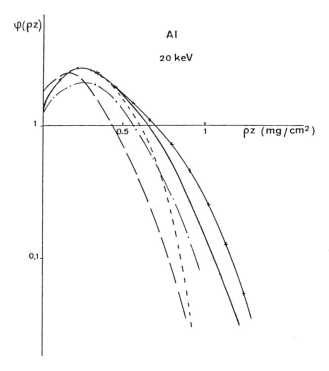

Figure 7. Depth distribution of Al K ionization at 20 keV.

×—×   experiment (Castaing, Henoc) [30]

– – –   ⌈multiple scattering model (Henoc, Maurice) [29]
          ⌊single scattering model (Henoc, Maurice) [4]

—·—   Reimer, Krefting [14]

— —   Terrissol with Moller energy loss [20]

_____   modified Terrissol with Gryzinski energy loss

trajectories on a MICROVAX computer lasts about one hour. Figure 7 shows the results, they are comparable with results obtained by other techniques. The multiple scattering model (Henoc and Maurice [29]) and the single scattering model [4] (programs MCMUL and MCSING, respectively) give exactly the same results. The curve published by Reimer [14] shows a faster decrease with increasing depth; the Terrisol treatment gives a much faster decrease. The elastic cross section is not involved: both use the Mott formula. For the sake of homogeneity, we have substituted the Gryzinski energy loss cross section for the Moller cross section. The agreement with experimental results is not improved. At least, we have assumed that the energy loss is constant, equal to the binding energy. The mean ionization depth is then greater than that observed. This suggests that the distribution function must increase faster than that expected from the formulation of Moller and Gryzinski.

The results presented above put emphasis on the influence of the choice of parameters. The way these parameters are utilized in computer programs is not always quite clear because publications give little information on this subject. Round Robin tests would be helpful, as well as more cross-section comparisons.

The authors wish to express their gratitude to J. Raynal for supplying the adaptation to electrons of his nuclear physics computation program, and to M. Terrissol for supplying his program.

IV. References

[1]   Microanalysis and Scanning Electron Microscopy (1978), 121, Les editions de physique, Orsay, France.
[2]   Reimer, L., Optik **27**, 86 (1986).
[3]   Murata, K., Matsukawa, T., and Shimizu, R., Japan J. of Appl. Phys. **10**, 67B (1971).
[4]   Henoc, J. and Maurice, F., J. Micros. Spect. Elect. **5**, 347 (1980).
[5]   Newbury, D. E. and Myklebust, R., Ultramicroscopy **3**, 391 (1979).
[6]   Kyser, D. and Murata, K., IBM J. Res. Devel. **18**, 352 (1976).
[7]   Worthington, C. R. and Tomlin, S. G., Proc. Phys. Soc. **69**, 401 (1956).
[8]   Goudsmit, S. and Sanderson J. L., Phys. Rev. **57**, 24 (1940).
[9]   Bishop, H., Thesis, Cambridge University (1965).
[10]   Henoc, J. and Maurice, F., NBS Spec. Publ. 460, 61 (1975).
[11]   Shimizu, R., Murata, K., and Shinoda, G., ICXOM Orsay, 127, Hermann, Paris (1966).
[12]   Reimer, L., Badde, H. G., and Seidel, H., Angew Z. Physik **31**, 145 (1971).
[13]   Kotera, M., Murata, K., and Nagami, K., J. Appl. Phys. **52**, 997 (1981).
[14]   Reimer, L. and Krefting, E. R., NBS Spec. Publ. 460, 45 (1976).
[15]   Murata, K., Cvikevich, S., and Kupsis, J. D., ICXOM Toulouse. J. de Physique, colloque C2, 45, 13 (1984).
[16]   Henoc, J. and Maurice, F., J. Phys. D: Appl. Phys. **8**, 1542 (1975).
[17]   Shimizu, R., Kataoka, Y., Matsukawa, T., Ikuta, T., Murata, K., and Hashimoto, H., J. Phys. D: Appl. Phys. **8**, 820 (1975).
[18]   Shimizu, R., Aratama, M., Ichimura, S., Yawazaki, Y., and Ikuta, T., Appl. Phys. Letters **31**, No. 10, 692 (1977).
[19]   Shimizu, R., Kataoka, Y., Ikuta, T., Hoshikawa, T., and Hashimoto, H., J. Phys. D: Appl. Phys. **9**, 101 (1976).
[20]   Terrissol, M., Thesis, Université Paul Sabatier, Toulouse (1978).
[21]   Raynal, J., Methods in Computational Physics **6**, 1 (1966). Adler, R., Ferbach, S. and Rotenberg, M., eds., Academic Press, New York.
[22]   Gryzinski, M., Phys. Rev. **A138**, 305 (1965).
[23]   Berger, M. J., Methods in Computational Physics, 1, 135 (1966). Adler, R., Fernbach, S., Rotenberg, M., eds., Academic Press, New York.
[24]   Ritchie, R. H., Phys. Rev. **114**, 3, 644 (1959).
[25]   Kittel, C., Introduction to Solid State Physics (1968), John Wiley, New York.
[26]   Ashley, J. C. and Ritchie, R. H., Phys. Stat. Sol. **40**, 623 (1970A); **38**, 425 (1970B).
[27]   Sevier, D., Low Energy Electron Spectroscopy (1972), Wiley Interscience, New York.
[28]   Ritchie, R. H., Garber, F. W., Nakai, M. Y., and Birkhoff, R. D., Advance in Radiation Biology III (1969), Academic Press, New York.
[29]   Henoc, J. and Maurice, F., Rapport CEA-R-4615 (1975).
[30]   Castaing, R. and Henoc, J., ICXOM Orsay 1965, 120, Hermann, Paris (1966).

Appendix 1

mcsing.for   **Fri Nov 17 08:00:53 1989**   **1**

```
C
C Monte Carlo simulation of electron trajectories - Single scattering
C                         screened Rutherford
C Special edit for simulating phi(roz) for pure target
C
C    Input    EO    :  acceleration voltage ( kV )
C             NUM   :  number of trajectories
C             NPAS  :  number of subdivisions of the range ( up to 45 )
C                      ( according to parameter MP )
C             NZ    :  atomic number of the target
C             A     :  atom weight of the target
C             EK    :  excitation voltage of the target atom ( kV )
C             BLA   :  comments ( up to 80 characters )
C
C    Logical units for I/O ( DATA statement )
C             LU5   :  terminal input ( 5 )
C             LU6   :  terminal output ( 5 )
C             LU7   :  output file for results ( 7 )
C
C    Constants of interest
C             NS    :  Switch employed in the multipurpose routine RUNGE
C                      in this program the value 2 is assigned in a DATA
C             ME    :  Also in RUNGE          "   1      "          "
C
      PROGRAM MCSING
      PARAMETER (MP=45)
      PARAMETER (MEP=1)
      PARAMETER (MEPP=2)
      PARAMETER (NRD=3)
      CHARACTER*80 BLA
      DIMENSION EN(MEPP),FP(MEP),GA(4),PHI(MP),RD(NRD),ROZ(MP)
      COMMON/MCF/AB,AJ
      DATA NS,PI/2,3.14159/
      DATA ME/1/
      DATA AN,E4S4/6.02E23,5.2E-21/
C
C lun assignement
C
      DATA LU5,LU6,LU7/5,5,7/
C
      MD=MP
      MED=MEP
      MEDD=MEPP
C
C Input
C
      WRITE(LU6,110)
110   FORMAT(' EO,NUM,NPAS ?')
      READ(LU5,100) EO,NUM,NPAS
100   FORMAT(F,2I)
      WRITE(LU6,120)
120   FORMAT(' NZ,A,EK ? ')
      READ(LU5,2000) NZ,A,EK
2000  FORMAT(I,2F)
      WRITE(LU6,510)
510   FORMAT(/' Comments ( up to 80 characters ? ')
      READ(LU5,520) BLA
520   FORMAT(A80)
      WRITE(LU7,520) BLA
```

```
C
C Calculation of voltage independant factors
C
      Z=NZ
      AJ=(9.76*Z+58.5/Z**.19)/1000.
      AJ=SQRT(EXP(1.)/2.)/AJ
      AB=A/Z/7.85E4
C
C Range calculation
C
      CALL RANRED(AJ,AB,EO,EK,RANGE)
C
C Heading of the output file
C
      WRITE(LU7,130) NZ,EO,NPAS,NUM,RANGE
130   FORMAT(10X,'FULL SINGLE SCATTERING FOR :'/
     &5X,'NZ=',I2,2X,'EO=',F6.2,2X,'NPAS=',I3,2X,'NUM=',I5,2X
     &,'RANGE=',E15.8)
      RPAS=NPAS
      DS=RANGE/RPAS
C
C Calculation of energy independant factors ( screen , total cross section )
C
C WENTZEL
      ALPHME=3.4E-3*Z**(2./3.)
C THOMAS-FERMI
C**   ALPHME=3.4E-3/.8853/.8853*Z**(2./3.)
      CSIG=AN*E4S4/A*Z*(Z+1.)*PI
C
C Trajectory sampling
C
      CALL SAMPTJ(A,ALPHME,CSIG,DS,EK,EN,EO,FP,GA,IBAC
     &,MD,ME,MED,MEDD,NPAS,NS,NUM,NZ,PHI,PI,RD,NRD)
      ANUM=NUM
C
C Backscattering coefficient
C
      ETA=IBAC
      ETA=ETA/ANUM
C
C Calculation of phi(roz)
C
      CALL FIROZ(EK,EO,MD,NPAS,NPASS,NUM,PHI,ROZ,DS)
C
C Output results to file
C
      WRITE(LU7,210) ETA
210   FORMAT(/10X,'ETA=',F6.3)
      WRITE(LU7,200) (ROZ(I),PHI(I),I=1,NPASS)
200   FORMAT(/(5X,'ROZ=',E15.8,2X,'ALOPHI=',E15.8))
      STOP
      END
C
C range calculation
C
      SUBROUTINE RANRED(AJ,AB,EO,EK,RANGE)
      RANGE=AB/AJ/AJ*(EI(2.*ALOG(AJ*EO))-EI(2.*ALOG(AJ*EK)))
      RETURN
      END
```

```
C
C Exponential integral function
C
      FUNCTION EI(X)
      EI=ALOG(X)+X
      TN=X
      N=1
300   AN=N
      IF(AN.GT.2.*X.AND.TN.LT.1.E-06) GO TO 400
      TN=TN*X*AN/(AN+1.)**2
      EI=EI+TN
      N=N+1
      GO TO 300
400   RETURN
      END
C
C Function employed in the calculation of half step & full step energy
C
      FUNCTION FXY(X)
      COMMON/MCF/AB,AJ
      FXY=-1./AB/X*ALOG(AJ*X)
      RETURN
      END
C
C Calculation of half step & full step energy
C         Runge kutta method
C
      SUBROUTINE RUNGE(XO,YO,EN,FP,GA,FXY,NS,ME,MD,MDD)
      DIMENSION FP(MD),EN(MDD),GA(4)
      XD=XO
      YD=YO
      DO 1 I=1,ME
      H=FP(I)/2.
      AA=H/2.
      BB=H/6.
      DO 3 K=1,2
      DO 2 IP=1,4
      GO TO (51,52,53,54),IP
51    GO TO (11,12,13),NS
11    GA(1)=FXY(XD)
      XC=XD+AA
      GO TO 2
12    GA(1)=FXY(YD)
      YC=YD+AA*GA(1)
      GO TO 2
13    GA(1)=FXY(XD,YD)
      XC=XD+AA
      YC=YD+AA*GA(1)
      GO TO 2
52    GO TO (21,22,23),NS
21    GA(2)=FXY(XC)
      GO TO 2
22    GA(2)=FXY(YC)
      YC=YD+AA*GA(2)
      GO TO 2
23    GA(2)=FXY(XC,YC)
      YC=YD+AA*GA(2)
      GO TO 2
53    GO TO (31,32,33),NS
31    GA(3)=FXY(XC)
      XC=XD+H
      GO TO 2
```

```
32      GA(3)=FXY(YC)
        YC=YD+H*GA(3)
        GO TO 2
33      GA(3)=FXY(XC,YC)
        XC=XD+H
        YC=YD+H*GA(3)
        GO TO 2
54      GO TO (41,42,43),NS
41      GA(4)=FXY(XC)
        XD=XD+H
        GO TO 2
42      GA(4)=FXY(YC)
        XD=XD+H
        GO TO 2
43      GA(4)=FXY(XC,YC)
        XD=XD+H
2       CONTINUE
        YD=YD+BB*(GA(1)+2.*(GA(2)+GA(3))+GA(4))
        L=I+(K/2)*ME
        EN(L)=YD
3       CONTINUE
1       CONTINUE
        RETURN
        END
C
C Calculation of the mean free path
C
        FUNCTION FRIPAV(NZ,ALPHME,CSIG,ED)
C
C       Screen factor
C WENTZEL or THOMAS-FERMI
        ALPH=ALPHME/ED
C MOLIERE
C**     Z=NZ
C**     ALPH=ALPHME/ED*(1.13+3.76*3.69*3.69*Z*Z/ED/1.E3)
        FRIPAV=CSIG/ED/ED/ALPH/(1.+ALPH)
        FRIPAV=1./FRIPAV
        RETURN
        END
C
C Calculation of phi(roz)
C
        SUBROUTINE FIROZ(EK,EO,MD,NPAS,NPASS,NUM,PHI,ROZ,DS)
        DIMENSION PHI(MD),ROZ(MD)
        ANUM=NUM
C
C Unsupported film
C
        COISO=1./EO/EK*ALOG(EO/EK)*ANUM
C
C Surface layer
C
        ROZ(1)=DS/4.
        PHI(1)=ALOG10(PHI(1)*2./COISO)
```

```
C
C inbedded layer
C
      DO 1 I=2,NPAS
      IF(PHI(I).LE.0.) GO TO 2
      ROZ(I)=I-1
      ROZ(I)=ROZ(I)*DS
      PHI(I)=ALOG10(PHI(I)/COISO)
1     CONTINUE
2     NPASS=I
      RETURN
      END
C
C Trajectory sampling
C
      SUBROUTINE SAMPTJ(A,ALPHME,CSIG,DS,EK,EN,EO,FP,GA,IBAC
     &,MD,ME,MED,MEDD,NPAS,NS,NUM,NZ,PHI,PI,RD,NRD)
      DIMENSION EN(MEDD),FP(MED),GA(4),PHI(MD),RD(NRD)
C
C Seed of the pseudo random number selector
C
      DATA ISTART/490623069/
C
      EXTERNAL FXY
C
C Statement function calculating the ionization cross section
C        ( constant ignored )
C
      Q(E)=1./E/EK*ALOG(E/EK)
C
C Statement function calculating sine from cosine in argument
C
      SI(X)=SQRT(1.-X*X)
C
C IBAC is the backscattering coefficient
C PHI is phi(roz)
C H is the incidence angle
C
      IBAC=0
      DO 1 I=1,NPAS
      PHI(I)=0.
1     CONTINUE
      H=0.
C
C Loop on trajectories
C
      DO 2 K=1,NUM
C
C      Initialize the current variables
C
      PATH=0.
      ASI=0.
      ACO=1.
      AZC=1.
      Z1=0.
      AA=COS(H)
      BB=SIN(H)
      E1=EO
```

```
C
C Start one step
C
C          calculate mean free path
C
5     FPAV=FRIPAV(NZ,ALPHME,CSIG,E1)
C
C          Sample 3 random numbers
C
      CALL URAN(ISTART,NRD,RD)
C
C          Sample free path
C
      FPA=-FPAV*ALOG(RD(1))
C
C          free path in range fraction unit
C
      FPAR=FPA/DS
C
C          Electron new coordinates
C
      AA1=AA*ACO-BB*ASI*AZC
      BB1=SI(AA1)
      Z2=Z1+FPAR*AA1
      IF(Z2.LE.0) GO TO 10
C
C          Electron location
C
      IZ=Z1+FPAR*AA1/2.+1.5
C
C Half step & full step energy calculation
C
      FP(1)=FPA
      CALL RUNGE(PATH,E1,EN,FP,GA,FXY,NS,ME,MED,MEDD)
      IF(EN(1).LE.EK) GO TO 2
C
C          phi(roz) calculation
C
      PHI(IZ)=PHI(IZ)+Q(EN(1))*FPAR
C
      IF(EN(2).LE.EK) GO TO 2
      PATH=PATH+FPA
      E1=EN(2)
C
C     Screen factor
C WENTZEL or THOMAS-FERMI
      ALPH=ALPHME/E1
C MOLIERE
C**      Z=NZ
C**      ALPH=ALPHME/E1*(1.13+3.76*3.69*3.69*Z*Z/E1/1.E3)
C
C Sample scattering angle & azimuth. angle then update current variables
C
      ACO=1.-2.*ALPH*RD(2)/(1.+ALPH-RD(2))
      ASI=SI(ACO)
      AZC=COS(2.*PI*RD(3))
      BB=BB1
      AA=AA1
      Z1=Z2
C
C End of step
C
      GO TO 5
```

```
C
C Increment counter of backscattered electrons
C
10     IBAC=IBAC+1
2      CONTINUE
       RETURN
       END
C
C Select three random numbers - RAN is a system library function
C         ISTART  : seed
C         NRD     : number of sampled random numbers
C         RD      : random numbers in the range ( 0.,1. ), limits excluded
C
       SUBROUTINE URAN(ISTART,NRD,RD)
       DIMENSION RD(NRD)
       DO 1 I=1,NRD
2      RD(I)=RAN(ISTART)
       IF((RD(I).GT.0.).AND.(RD(I).LT.1.)) GO TO 1
       GO TO 2
1      CONTINUE
       RETURN
       END
```

Appendix 2

mcmul.for   **Fri Nov 17 08:01:18 1989**   1

```
      PROGRAM MCMUL
C
C From   *** RAPPORT CEA R4615 ***   special edit for vax
C
      PARAMETER (ME=2)
      PARAMETER (MP=45)
      PARAMETER (MPP=91)
      CHARACTER*1 QU
      DIMENSION NZ(ME),C(ME),A(ME),VK(ME)
      DIMENSION EN(MPP),IXF(MP,MP),IXB(MP,MP),IBAC(MP),IBB(MP)
     &,ROZ(MP),PHI(MP)
      COMMON NZ,A,VK,C,EO,RANGE,E14,EN,IXF,IXB,IBAC,IBB,ROZ,PHI,ISTART
      COMMON/LUI/LU5,LU6,LU7
      DATA LU5,LU6,LU7/5,5,7/
      DATA ISTART/490623069/
C
C Input
C
1     WRITE(LU6,100)
100   FORMAT(' Acceleration voltage (kV) ? ')
      READ(LU5,200,ERR=1) EO
200   FORMAT(F12.0)
      WRITE(LU6,101)
101   FORMAT(' Input constants of elements present in the target'/
     &'      the simulation will be done for the first input element')
C
C Initialization for pure target
C
      JO=1
      JJA=1
      JJM=1
      C(1)=1.
C
C simulated element
C
2     WRITE(LU6,102)
102   FORMAT(' Atomic number ? ')
      READ(LU5,201,ERR=2) NZ(1)
201   FORMAT(I)
3     WRITE(LU6,103)
103   FORMAT(' Atom weight ? ')
      READ(LU5,200,ERR=3) A(1)
4     WRITE(LU6,104)
104   FORMAT(' Excitation voltage (kV) ? ')
      READ(LU5,200,ERR=4) VK(1)
5     WRITE(LU6,105)
105   FORMAT(' Is it a pure target ( Y or N ) Def[Y] ? ')
      READ(LU5,202,ERR=5) QU
202   FORMAT(A1)
      IF((QU.NE.'N').AND.(QU.NE.'n')) GO TO 20
C
C Complex target
C
7     WRITE(LU6,106)
106   FORMAT(' Weight fraction of the calculated element ? ')
      READ(LU5,200,ERR=7) C(1)
8     WRITE(LU6,107) ME
107   FORMAT(' Number of companions ( < ',I2,' ? ')
      READ(LU5,201,ERR=8) JJD
      IF((JJD.LE.0).OR.(JJD.GT.ME-1)) GO TO 8
      JJM=JJD+1
```

```
C
C Companions
C
      DO 9 J=1,JJD
      WRITE(LU6,108) J
108   FORMAT('        Companion no.',I2)
10    WRITE(LU6,102)
      READ(LU5,201,ERR=10) NZ(J+1)
11    WRITE(LU6,103)
      READ(LU5,200,ERR=11) A(J+1)
12    WRITE(LU6,104)
      READ(LU5,200,ERR=12) VK(J+1)
13    WRITE(LU6,109)
109   FORMAT(' Weight fraction ? ')
      READ(LU5,200,ERR=13) C(J+1)
9     CONTINUE
C
C Parameters of the simulation
C
C    Xray range is employed  -> nswr=2
C
20    NSWR=2
C
C NUM   is the number of simulated trajectories
C NPAS  is the number of steps dividing the range
C
21    WRITE(LU6,110) MP+1
110   FORMAT(' Number of steps ( < ',I3,' ? ')
      READ(LU5,201,ERR=21) NPAS
      IF((NPAS.LE.0).OR.(NPAS.GT.MP)) GO TO 21
22    WRITE(LU6,111)
111   FORMAT(' Number of trajectories ? ')
      READ(LU5,201,ERR=22) NUM
      ANUM=NUM
C
C Display input
C
      WRITE(LU6,112)
112   FORMAT('       Target identification'/
     &' no.    Z        A       EK          C')
      WRITE(LU6,113)
113   FORMAT(/)
      WRITE(LU6,203) (J,NZ(J),A(J),VK(J),C(J),J=1,JJM)
203   FORMAT(2X,I2,3X,I3,4X,F6.2,4X,F7.4,4X,F6.4)
      WRITE(LU6,204) EO,NPAS,NUM
204   FORMAT('       Simulation parameters'/
     &' Acceleration voltage   :',F6.3/
     &' Number of steps        :',I3/
     &' Number of trajectories :',I6)
C
C Proceed to mc calculation
C
      CALL RESU(JO,JJM,JJA,ANUM,NPAS,NSWR)
      STOP
      END
C
C E10
C                     Routine***E10***
C
C          Argument list
C
C input : JO       index of the first element of the target
C        : JJA      index of the calculated element
C        : JM       index of the last element of the target
C        : ANUM     number of trajectories (real)
```

```
C          : NPAS     number of intervals dividing the range
C          : NSWR     if 1 : full range ; if 2 : xrays range
C output : RETRO1     backscattering coefficient
C          : RETRO2   fraction of electrons entering again the target
C                          ( forw -> back -> forw )
C
C
C            common
C
C scalar : E14       electron energy at the surface layer
C          : ISTART   seed of the pseudo random number selector
C array  : NZ        atomic numbers
C          : A        atom weights
C          : VK       excitation potentials
C          : C        weight fractions
C          : EO       incident electron energies
C          : EN       half step and full step electron mean energies
C          : IXF      table of transmited electrons
C          : IXB       "        backscattered electrons
C          : IBAC     energy distribution of backscattered electrons
C          : IBB      the same as before ( 2th backscat. )
C          : ROZ      depth ( in mass thickness unit )
C          : PHI      phi-ro-z
C
C            parameters
C
C            ME       maximum number of elements in the target
C            MP       number of steps
C            MPP      dimension of the energy table
C            MAG      number of angles of equal scattering probability
C
C            function
C
C            EI       exponential integral
C            URAN     pseudo random number selector
C
        SUBROUTINE E10(JO,JM,JJA,ANUM,NPAS,NSWR,RETRO1,RETRO2)
        PARAMETER (ME=2)
        PARAMETER (MP=45)
        PARAMETER (MPP=91)
        PARAMETER (MAG=128)
        PARAMETER (NRD=3)
        DIMENSION NZ(ME),A(ME),VK(ME),C(ME),EN(MPP),AB(ME),
       &AC(ME),GA(4),RD(NRD)
        DIMENSION IXF(MP,MP),IXB(MP,MP),IBAC(MP),IBB(MP),ACO(MAG,MP)
       &,ASI(MAG,MP),P(MP),B(MAG),AP(MAG),RES(MP),ROZ(MP),PHI(MP)
        COMMON NZ,A,VK,C,EO,RANGE,E14,EN,IXF,IXB,IBAC,IBB,ROZ,PHI,ISTART
        COMMON/LUI/LU5,LU6,LU7
        DATA NMAX1,NAN,PI,P180/128,128,3.14159265,.174532925E-1/
C
C ECRAN is a constant used in the calculation of the screening parameter
C
        DATA ECRAN/1./
C
C Statement function calculating sine from cosine in argument
C
        SI(X)=SQRT(1.-X**2)
100     FORMAT('   WARNING - Slow convergence for step no.',I3)
C
C       NUM : number of sampled trajectories
C       ED  : the popular constant of the Bethe's formula (1.166)
C       BB  : real conversion of the atomic number of the emitting element
```

```
C       AA   : mean ionization potentiel of the emitting element
C
        NUM=ANUM
        ED=SQRT(EXP(1.)/2.)
        BB=NZ(JJA)
        AA=(9.76*BB+58.5/BB**.19)/1000.
C
C Determination of a minimum energy limit for calculating the Bethe,s range
C       first find the greater mean ionization potential
C
        DO 15 JJ=JO,JM
        BB=NZ(JJ)
        AB(JJ)=A(JJ)/(7.85E4*BB*C(JJ))
        AC(JJ)=(9.76*BB+58.5/BB**.19)/1000.
        IF(AC(JJ).GT.AA) AA=AC(JJ)
        AC(JJ)=ED/AC(JJ)
15      CONTINUE
        GO TO (16,17),NSWR
C
C Lower limit for full range calculation
C
16      ZA=1.03*AA
        GO TO 18
C
C Lower limit for xrays range calculation
C
17      ZA=VK(JJA)
C
C Range calculation
C
18      SM=.0
        ED=.0
        DO 19 JJ=JO,JM
        ED=ED+ALOG(AC(JJ))/AB(JJ)
        SM=SM+1./AB(JJ)
19      CONTINUE
        ED=ED/SM
        V=EXP(ED)
        RANGE=1./SM/V**2*(EI(2.*(ED+ALOG(EO)))-
       &EI(2.*(ED+ALOG(ZA))))
C
C Calculation of half step and full step electron mean energies
C
        H=NPAS
        H=RANGE/H/4.
        AA=H/2.
        BB=H/6.
        EA=EO
        NPAS2=2*NPAS
        DO 20 I=1,NPAS2
        DO 11 J=1,2
        EB=EA
        DO 25 IP=1,4
        GA(IP)=.0
        DO 26 JJ=JO,JM
        GA(IP)=GA(IP)+ALOG(AC(JJ)*EB)/AB(JJ)/EB
26      CONTINUE
        GO TO (27,27,28,25),IP
27      EB=EA-AA*GA(IP)
        GO TO 25
28      EB=EA-H*GA(IP)
25      CONTINUE
```

```
          EA=EA-BB*(GA(1)+2.*(GA(2)+GA(3))+GA(4))
          IF(I.NE.1) GO TO 11
          IF(J.EQ.1) E14=EA
11        CONTINUE
          EN(I)=EA
20        CONTINUE
          IF(EN(NPAS2).LT.VK(JJA)) EN(NPAS2)=VK(JJA)
C
C Energy sorting
C         EN(I)    I  1 -> NPAS          : half step mean energy
C         EN(NPAS+1)                     : incident electron energy
C         EN(I)    I  NPAS+2 -> 2*NPAS+2 : full step mean energy
C
          DO 12 I=2,NPAS
          EN(NPAS2+1)=EN(I)
          DO 12 J=I,NPAS2
          EN(J)=EN(J+1)
12        CONTINUE
          EN(NPAS2+1)=EN(NPAS+1)
          EN(NPAS+1)=EO
C
C Preliminary for calculating the inverse mean free path
C              ( energy independant)
C
C  AC  : partial inverse mean free path
C  AB  : screen factor
C
          DO 7 JJ=JO,JM
          AA=NZ(JJ)
          AC(JJ)=C(JJ)*6.02*4.79E5*AA**(1./3.)*RANGE*(AA+1.)/NPAS
         &/A(JJ)
          AB(JJ)=(3.4E-3*AA**(2./3.))/ECRAN/ECRAN
7         CONTINUE
C
C              Calculation of the set of angles
C              of equal scattering probability
C              _____
C
C  some initialization
C      I is the energy index
C      J is the index of the set of angles
C
          NMAX3=3
          DO 8 I=1,NPAS
          NMAX=NMAX1
          P(I)=.0
          DO 9 J=1,NMAX1
          ACO(J,I)=.0
          ASI(J,I)=.0
          AP(J)=.0
9         CONTINUE
C
C calculation of the inverse mean free path and of the coefficients
C   of the Legendre expension ( energy dependant )
C         P   inverse mean free path
C         B   expension coefficients
C
          DO 42 JJ=JO,JM
          BB=AC(JJ)/EN(I)
          AA=AB(JJ)/EN(I)
          P(I)=P(I)+BB
          B(1)=1.
          B(2)=1.+2.*AA+2.*AA*(1.+AA)*ALOG(AA/(1.+AA))
```

```
      DO 41 J=3,NMAX1
      C1=2*J-3
      C2=J-1
      C3=J-2
      B(J)=(C1*(1.+2.*AA)*B(J-1)-C2*B(J-2))/C3
41    CONTINUE
C
C calculation of coefficients of the Legendre expension
C            of the distribution function
C
      DO 42 J=1,NMAX1
      AP(J)=AP(J)+B(J)*BB
42    CONTINUE
      AP(1)=0.5*EXP(AP(1)-P(I))
      AP(2)=0.5*EXP(AP(2)-P(I))
      RES(I)=.5+AP(2)
      DO 43 J=3,NMAX1
      AP(J)=EXP(AP(J)-P(I))*.5
      AP(J-2)=AP(J-2)-AP(J)
      IF(AP(J-2).LE.1.E-5) GO TO 44
43    CONTINUE
      WRITE(LU6,100) I
      GO TO 52
44    NMAX=J
52    NMAX2=NMAX-2
      IF(NMAX3.LT.NMAX2) NMAX3=NMAX2
C
C coefficients of the expension are stored into ACO(J,I)
C            I : energy index
C            J : coefficient index
C
      DO 75 J=1,NMAX2
      ACO(J,I)=AP(J)
75    CONTINUE
8     CONTINUE
C
C calculation of the distribution function for a given set
C                                        of known angles
C
      IH=1
      AMAX=NAN
      AMAX=1.-.5/AMAX
      DO 78 J=2,90
      TETA=J-1
      TETA=TETA*PI/90.
      Y=COS(TETA)
      B(1)=Y
      B(2)=(3.*Y**2-1.)/2.
      DO 77 N=3,NMAX3
      C1=2*N-1
      C2=N-1
      C3=N
      B(N)=(C1*Y*B(N-1)-C2*B(N-2))/C3
77    CONTINUE
      I1=IH
      DO 78 I=I1,NPAS
      AA=RES(I)-ACO(1,I)*Y
      DO 79 N=2,NMAX3
      AA=AA-ACO(N,I)*B(N)
      IF(ACO(N,I).LE.1.E-5) GO TO 80
79    CONTINUE
C
C the value of distribution function for a given angle ( index J )
C    and for a given energy ( index I ) is stored into ASI(J,I)
C
```

```
80      ASI(J,I)=AA
        IF(AA.LT.AMAX) GO TO 78
        IF(IH.EQ.NPAS) GO TO 64
        IH=IH+1
        NMAX3=N
78      CONTINUE
C
C calculation of a set of angles of equal scattering probability
C
64      B(1)=.0
        B(91)=1.
        DO 65 I=1,NPAS
C
C B is the temporary storage of distribution function values
C       for a given energy
C
        DO 66 J=2,90
        B(J)=ASI(J,I)
66      CONTINUE
C
C calculation by interpolation of the set of angles
C                   of equal probability
C
        IMIN=3
        DO 54 J=1,NAN
        AJ=J
        RET=NAN
        RET=(AJ-.5)/RET
        DO 55 K=IMIN,90
        IF(B(K).GT.RET) GO TO 56
55      CONTINUE
56      IMIN=K
        K=K-1
        IF(B(K).EQ.B(K-1)) GO TO 29
        IF(B(K).EQ.B(K+1)) GO TO 29
        AP(J)=K-1
        AA=B(K+1)-B(K-1)
        H=RET-B(K)
        BB=2.*(B(K+1)+B(K-1)-2.*B(K))*H
        C1=AA**2+BB
        C2=C1+BB
        C3=H/AA*2.
        AP(J)=(AP(J)+C3*C1/C2)*2.
        GO TO 54
29      AP(J)=K-1
54      CONTINUE
C
C cosine(theta) is stored into ACO & sine(theta) into ASI
C
        DO 30 J=1,NAN
        AP(J)=AP(J)*P180
        ACO(J,I)=COS(AP(J))
        ASI(J,I)=SIN(AP(J))
30      CONTINUE
65      CONTINUE
C
C calculation of a set of azimuthal angles
C         ( uniform distribution )
C
        C2=NAN
        DO 57 J=1,NAN
        C1=J
        B(J)=COS((C1-.5)/C2*2.*PI)
57      CONTINUE
```

```
C
C          trajectory sampling
C          ----------
C
C table initialization
C
      DO 58 I=1,NPAS
      IBAC(I)=0
      IBB(I)=0
      DO 58 J=1,NPAS
      IXF(J,I)=0
      IXB(J,I)=0
58    CONTINUE
C
C H is the incidence angle
C
      H=.0
      DO 59 IT=1,NUM
C
C start one trajectory
C      NSW1  1 -> electron moving in the forward direction
C      NSW1  2 -> backscattered electron
C
C      NSW2  1 -> the counter of backscattered has already been incremented
C      NSW2  1 -> the counter of backscattered must bee incremented
C
      NSW1=1
      NSW2=1
      Y=.0
      AA=COS(H)
      DO 60 I=1,NPAS
C
C  "free path" sampling
C
      CALL URAN(ISTART,NRD,RD)
      BB=RD(1)
      Y=Y+AA*BB
      IF(Y.GT.0.) GO TO 47
C move backward
      IF(NSW1.EQ.2) GO TO 62
      NSW1=2
      NSW2=2
      GO TO 62
C move forward
47    IF(NSW1.EQ.2) GO TO 61
C
C deflection sampling
C
62    J=1.+RD(2)*FLOAT(NAN)
      C1=ACO(J,I)
      C2=ASI(J,I)
      J=1.+RD(3)*FLOAT(NAN)
      C3=B(J)
      AA=AA*C1+SI(AA)*C2*C3
      Y=Y+(1.-BB)*AA
      IF(Y.LE.0.) GO TO (49,48),NSW1
      GO TO (63,61),NSW1
C
C update the table of electron position inside the target
C
63    J=Y+1.5
      IXF(J,I)=IXF(J,I)+1
      GO TO 60
```

```
C
C bacscattered electron
C
49    NSW1=2
      NSW2=2
C
C update the table of electron position outside the target
C           ( simulation of the Derian experiment )
C
48    J=Y-1.5
      J=-J
      IXB(J,I)=IXB(J,I)+1
      GO TO (60,67),NSW2
67    IBAC(I)=IBAC(I)+1
      NSW2=1
60    CONTINUE
      GO TO 59
C
C again in the target after being backscattered
C
61    IF(NSW2.EQ.2) IBAC(I)=IBAC(I)+1
      IBB(I)=IBB(I)+1
59    CONTINUE
C
C calculation of the backscattering coefficient
C
      IF(NSWR.EQ.2) GO TO 23
      RETRO1=.0
      RETRO2=.0
      DO 33 I=1,NPAS
      C1=IBAC(I)
      RETRO1=RETRO1+C1
      C1=IBB(I)
      RETRO2=RETRO2+C1
33    CONTINUE
      RETRO1=RETRO1/ANUM
      RETRO2=RETRO2/ANUM
23    RETURN
      END
C
C Calculation of Xray depth distribution in the target
C     Output :
C
C               ALOPHI  log(phi(roz))
C               RXF     total xray generation inside the target
C               RXFO    total xray that would be generated inside
C                       the target in the absence of backscattering
C               NPASS   index of the last layer giving a significant
C                       contribution
C
      SUBROUTINE XRAF(JJA,ALOPHI,NPAS,ANUM,NPASS,RXF,RXFO)
      PARAMETER ME=2,MP=45,MPP=91 ,MAG=128
      DIMENSION NZ(ME),A(ME),VK(ME),C(ME),EN(MPP),IXF(MP,MP),
     &IXB(MP,MP),IBAC(MP),IBB(MP),ROZ(MP),PHI(MP),ALOPHI(MP)
      COMMON NZ,A,VK,C,EO,RANGE,E14,EN,IXF,IXB,IBAC,IBB,ROZ,PHI,ISTART
      COISO=1./EN(NPAS+1)/VK(JJA)*ALOG(EN(NPAS+1)/VK(JJA))
      NSW=1
      NPASS=NPAS
      PHI(1)=1./E14/VK(JJA)*ALOG(E14/VK(JJA))
      RPAS=NPAS
      ROZ(1)=RANGE/RPAS/4.
      RXFO=.0
      DO 9 I=1,NPAS
      K=NPAS+1+I
      IF(EN(K)/VK(JJA).LE.1.) GO TO 30
```

```
          AA=IXF(1,I)
          PHI(1)=PHI(1)+AA/ANUM/EN(K)/VK(JJA)*ALOG(EN(K)/VK(JJA))
9         CONTINUE
          I=NPAS
30        NPAS1=I
          NPAS2=I-1
          ALOPHI(1)=ALOG10(PHI(1)/COISO)
          RXF=PHI(1)/2.
          DO 10 J=2,NPAS1
          PHI(J)=.0
          J1=J-1
          RXFO=RXFO+1./EN(J1)/VK(JJA)*ALOG(EN(J1)/VK(JJA))
          GO TO (4,10),NSW
4         ROZ(J)=J1
          ROZ(J)=ROZ(J)*RANGE/RPAS
          DO 11 I=J1,NPAS2
          K=NPAS+1+I
          AA=IXF(J,I)
          PHI(J)=PHI(J)+AA/ANUM/EN(K)/VK(JJA)*ALOG(EN(K)/VK(JJA))
11        CONTINUE
          IF(PHI(J).EQ.0.) GO TO 24
          ALOPHI(J)=ALOG10(PHI(J)/COISO)
          RXF=RXF+PHI(J)
          GO TO 10
24        NPASS=J-1
          NSW=2
10        CONTINUE
          IF(EN(NPAS1)/VK(JJA).GT.1.) RXFO=RXFO+1./EN(NPAS1)/VK(JJA)
         &*ALOG(EN(NPAS1)/VK(JJA))
          AA=RANGE/COISO/RPAS
          RXF=RXF*AA
          RXFO=RXFO*AA
          RETURN
          END
C
C Calculation of Xray depth distribution in the Derian target
C     Output :
C                ALOPHI  log(phi(roz))
C                RXB     total xray generation inside the  Derian target
C                RXBO    total xray that would be generated inside
C                        the Derian target in the absence of electrons
C                        scattered again
C                RXBBO   contribution of electrons going back to the
C                        target
C                NPASS   index of the last layer giving a significant
C                        contribution
C
          SUBROUTINE XRAB(JJA,ALOPHI,NPAS,ANUM,NPASS,RXB,RXBO,RXBBO)
          PARAMETER ME=2,MP=45,MPP=91 ,MAG=128
          DIMENSION NZ(ME),A(ME),VK(ME),C(ME),EN(MPP),IXF(MP,MP),
         &IXB(MP,MP),IBAC(MP),IBB(MP),ROZ(MP),PHI(MP),ALOPHI(MP)
          COMMON NZ,A,VK,C,EO,RANGE,E14,EN,IXF,IXB,IBAC,IBB,ROZ,PHI,ISTART
          COISO=1./EN(NPAS+1)/VK(JJA)*ALOG(EN(NPAS+1)/VK(JJA))
          NPASS=NPAS
          NSW=1
          PHI(1)=.0
          RPAS=NPAS
          ROZ(1)=RANGE/RPAS/4.
          RXBO=.0
          RXBBO=.0
          DO 9 I=1,NPAS
          K=NPAS+1+I
          IF(EN(K)/VK(JJA).LE.1.) GO TO 30
          AA=IXB(1,I)
          PHI(1)=PHI(1)+AA/ANUM/EN(K)/VK(JJA)*ALOG(EN(K)/VK(JJA))
```

```
9       CONTINUE
        I=NPAS
30      NPAS1=I
        NPAS2=I-1
        ALOPHI(1)=ALOG10(PHI(1)/COISO)
        RXB=PHI(1)/2.
        DO 10 J=2,NPAS1
        PHI(J)=.0
        J1=J-1
        B=IBAC(J1)
        BB=IBB(J1)
        ROZ(J)=J1
        ROZ(J)=ROZ(J)*RANGE/RPAS
        DO 11 I=J1,NPAS2
        GO TO (4,6),NSW
4       K=NPAS+1+I
        AA=IXB(J,I)
        PHI(J)=PHI(J)+AA/ANUM/EN(K)/VK(JJA)*ALOG(EN(K)/VK(JJA))
6       RXBO=RXBO+B/EN(I)/VK(JJA)/ANUM*ALOG(EN(I)/VK(JJA))
        RXBBO=RXBBO+BB/ANUM/EN(I)/VK(JJA)*ALOG(EN(I)/VK(JJA))
11      CONTINUE
        IF(NSW.EQ.2) GO TO 10
        IF(PHI(J).EQ.0.) GO TO 24
        ALOPHI(J)=ALOG10(PHI(J)/COISO)
        RXB=RXB+PHI(J)
        GO TO 10
24      NPASS=J-1
        NSW=2
10      CONTINUE
        IF(EN(NPAS1)/VK(JJA).LE.1.) GO TO 7
        B=IBAC(NPAS1)
        BB=IBB(NPAS1)
        RXBO=RXBO+B/ANUM/EN(NPAS1)/VK(JJA)*ALOG(EN(NPAS1)/VK(JJA))
        RXBBO=RXBBO+BB/ANUM/EN(NPAS1)/VK(JJA)*ALOG(EN(NPAS1)/VK(JJA))
7       AA=RANGE/COISO/RPAS
        RXB=RXB*AA
        RXBO=RXBO*AA
        RXBBO=RXBBO*AA
        RETURN
        END
C
C Control the simulation and output some balance
C
        SUBROUTINE RESU(JO,JM,JJA,ANUM,NPAS,NSWR)
        PARAMETER ME=2,MP=45,MPP=91,MAG=128
        CHARACTER*2 SYMB(ME),ATSY(100)
        DIMENSION NZ(ME),A(ME),VK(ME),C(ME),EN(MPP),IXF(MP,MP),
       &IXB(MP,MP),IBAC(MP),IBB(MP),ROZ(MP),PHI(MP),ALOPHI(MP)
        COMMON NZ,A,VK,C,EO,RANGE,E14,EN,IXF,IXB,IBAC,IBB,ROZ,PHI,ISTART
        COMMON/LUI/LU5,LU6,LU7
        DATA ATSY/'H','He','Li','Be','B','C','N','O','F','Ne','Na','Mg'
       &,'Al','Si','P','S','Cl','Ar','K','Ca','Sc','Ti','V','Cr','Mn'
       &,'Fe','Co','Ni','Cu','Zn','Ga','Ge','As','Se','Br','Kr'
       &,'Rb','Sr','Y','Zr','Nb','Mo','Tc','Ru','Rh','Pd','Ag','Cd','In'
       &,'Sn','Sb','Te','I','Xe','Cs','Ba','La','Ce','Pr','Nd','Pm','Sm'
       &,'Eu','Gd','Tb','Dy','Ho','Er','Tm','Yb','Lu','Hf','Ta','W','Re'
       &,'Os','Ir','Pt','Au','Hg','Tl','Pb','Bi','Po','At','Rn','Fr','Ra'
       &,'Ac','Th','Pa','U','Np','Pu','Am','Cm','Bk','Cf','Es','Fm'/
```

```
C
      WRITE(LU7,200) JM
200   FORMAT(1X,I2,'      <- Number of elements')
      WRITE(LU7,201)
201   FORMAT(' Atom symbol(s) and weight fraction(s)')
      DO 1 J=JO,JM
      I=NZ(J)
      SYMB(J)=ATSY(I)
1     CONTINUE
      WRITE(LU7,202) (SYMB(J),C(J),J=JO,JM)
202   FORMAT(10(3X,A2,':',F6.4))
      WRITE(LU7,203) NPAS
203   FORMAT(1X,I3,'      <- Number of steps')
      NUM=ANUM
      WRITE(LU7,204) NUM
204   FORMAT(1X,I6,' <- Number of trajectories')
      WRITE(LU7,205) EO
205   FORMAT(1X,F5.2,'  <- Acceleration voltage')
C
C Trajectory sampling
C
      CALL E10(JO,JM,JJA,ANUM,NPAS,NSWR,RETRO1,RETRO2)
C
      RANG3=1000.*RANGE
      WRITE(LU7,206) RANG3
206   FORMAT(1X,E15.8,' <- Range ( mg/cm2 )')
      WRITE(LU7,207)
207   FORMAT(' Half step and full step energies')
      DO 2 I=1,NPAS
      J=I+NPAS+1
      WRITE(LU7,208) I,EN(I),EN(J)
2     CONTINUE
208   FORMAT(I3,3X,2(4X,F6.3))
      NPASS=2*NPAS+1
C
C Output electron distribution
C
      WRITE(LU7,209)
209   FORMAT('   Electron distribution')
      CALL PRINT(NPAS)
C
      WRITE(LU7,210) (IBB(I),I=1,NPAS)
210   FORMAT(' BB',20(3X,I3))
      IF(NSWR.EQ.2) GO TO 23
      WRITE(LU7,211) RETRO1,RETRO2
211   FORMAT(1X,F7.4,4X,F7.4,'  <- Retro1 , Retro2')
C
C Xray from Derian target
C
23    CALL XRAB(JJA,ALOPHI,NPAS,ANUM,NPASS,RXB,RXBO,RXBBO)
C
      WRITE(LU7,212)
212   FORMAT('   Derian target distribution')
      DO 3 J=1,NPASS
      WRITE(LU7,213) J,ROZ(J),ALOPHI(J)
3     CONTINUE
213   FORMAT(1X,I3,3X,2(6X,E15.8))
      WRITE(LU7,214) RXB,RXBO,RXBBO
214   FORMAT(1X,3(E15.8,4X),'<- RXB,RXBO,RXBBO')
```

```
C
C Xray from the regular target
C
      CALL XRAF(JJA,ALOPHI,NPAS,ANUM,NPASS,RXF,RXFO)
C
      WRITE(LU7,215)
215   FORMAT('   Phi(roz) distribution')
      DO 4 J=1,NPASS
      WRITE(LU7,213) J,ROZ(J),ALOPHI(J)
4     CONTINUE
C
C Total amount
C
      WRITE(LU7,216) RXF,RXFO
216   FORMAT(1X,2(E15.8,4X),' <- RXF,RXFO')
      RXFO=RXF+RXB+RXBBO
      RXBO=RXB+RXBBO
      WRITE(LU7,217) RXFO,RXBO
217   FORMAT(1X,2(E15.8,4X),' <- RXF+RXB+RXBBO , RXB+RXBBO')
      RETURN
      END
C PRINT
      SUBROUTINE PRINT(NPAS)
      PARAMETER ME=2,MP=45,MPP=91
      DIMENSION NZ(ME),C(ME),A(ME),VK(ME),EN(MPP),IXF(MP,MP)
     &,IXB(MP,MP),IBAC(MP),IBB(MP),ROZ(MP),PHI(MP)
      COMMON NZ,A,VK,C,EO,RANGE,E14,EN,IXF,IXB,IBAC,IBB,ROZ,PHI,ISTART
      COMMON/LUI/LU5,LU6,LU7
103   FORMAT(1X,3X,20(3X,I3))
101   FORMAT(1X,I3,20(1X,I5))
111   FORMAT(' BS',20(1X,I5))
      IS=20
      NTA=NPAS/IS
      IF(NTA.NE.0) GO TO 10
      IMIN=1
      GO TO 50
10    DO 15 IT=1,NTA
      IMIN=(IT-1)*IS
      IMAX=IS+IMIN
      IMIN=IMIN+1
      WRITE(LU7,103) (I,I=IMIN,IMAX)
      DO 31 J=1,NPAS
      K=NPAS+1-J
      WRITE(LU7,101) K,(IXB(K,I),I=IMIN,IMAX)
31    CONTINUE
      WRITE(LU7,111) (IBAC(I),I=IMIN,IMAX)
      DO 33 J=1,NPAS
      WRITE(LU7,101) J,(IXF(J,I),I=IMIN,IMAX)
33    CONTINUE
15    CONTINUE
      IF(NPAS-IS*(NPAS/IS).EQ.0) GO TO 20
      IMIN=IMAX+1
50    WRITE(LU7,103) (I,I=IMIN,NPAS)
      DO 34 J=1,NPAS
      K=NPAS+1-J
      WRITE(LU7,101) K,(IXB(K,I),I=IMIN,NPAS)
34    CONTINUE
      WRITE(LU7,111) (IBAC(I),I=IMIN,NPAS)
      DO 32 J=1,NPAS
      WRITE(LU7,101) J,(IXF(J,I),I=IMIN,NPAS)
32    CONTINUE
20    RETURN
      END
```

```
C
C Exponetial integral
C
      FUNCTION EI(X)
      EI=ALOG(X)+X
      TN=X
      N=1
300   AN=N
      IF(AN.GT.2.*X.AND.TN.LT.1.E-06) GO TO 400
      TN=TN*X*AN/(AN+1.)**2
      EI=EI+TN
      N=N+1
      GO TO 300
400   RETURN
      END
C
C Select three random numbers - RAN is a system library function
C         ISTART  : seed
C         NRD     : number of sampled random numbers
C         RD      : random numbers in the range ( 0.,1. ), limits excluded
C
      SUBROUTINE URAN(ISTART,NRD,RD)
      DIMENSION RD(NRD)
      DO 1 I=1,NRD
2     RD(I)=RAN(ISTART)
      IF((RD(I).GT.0.).AND.(RD(I).LT.1.)) GO TO 1
      GO TO 2
1     CONTINUE
      RETURN
      END
```

QUANTITATIVE ELECTRON PROBE MICROANALYSIS OF

ULTRA-LIGHT ELEMENTS (BORON-OXYGEN)

G. F. BASTIN AND H. J. M. HEIJLIGERS

Laboratory for Physical Chemistry-Center for Technical Ceramics
University of Technology
P.O. Box 513, NL-5600 MB Eindhoven
The Netherlands

I. Introduction

Due to the rapid development and expansion of materials science into new areas such as high-temperature materials, wear-resistant coatings and modern ceramics the interest in quantitative electron probe microanalysis of ultra-light elements, e.g., boron, carbon, nitrogen and oxygen, elements which are usually present in considerable quantities in such materials, is also rapidly growing. In comparison with the conventional analysis of medium-to-high Z-elements ($Z > 11$) the requirements for the successful analysis of ultra-light elements are much more stringent for all the steps involved in the complete procedure, which begins with the specimen preparation, followed by the actual intensity measurement and ending with the matrix correction.

As far as the specimen preparation and the intensity measurements are concerned there are a number of specific problems which have been discussed at length on previous occasions [1–5], the major problem being the very low count rates and frequently low peak-to-background ratios, especially for an element such as nitrogen. Besides, there is a theoretical possibility that x-ray emission for ultra-light elements exhibits systematic differences from one compound to another. Evidence to this effect has recently been found for, e.g., B $K\alpha$ in Ni-compounds [4,5]. On top of this there may be additional practical problems such as a lack of electrical conductivity which can strongly affect the intensity measurements, as is gradually becoming apparent now.

Regarding the matrix correction necessary in order to convert the measured intensity ratios (k-ratios) into concentration units, it is very important to realize that the necessary corrections for ultra-light elements can be an order of magnitude larger than in conventional analysis. It goes without saying that as a consequence the demands imposed upon a specific matrix correction program are correspondingly much more severe than usual. Besides, it is vital that certain physical quantities such as mass absorption coefficients (macs) be available with an accuracy of ~ 1 percent relative. Unfortunately it is highly unlikely that such values are actually available with the required precision, considering the practical difficulties involved in the measurement (or calculation) of these quantities. In light of these uncertainties it is almost impossible to make definite statements on the performance of correction programs, especially on their correction for absorption. This would only be possible if large databases were available for each of the light elements. Once such data are available they can be used in an iterative process to test diverse programs in conjunction with various sets of published macs. A next iteration step could provide improved equations in a specific program and/or an improvement in the consistency of the macs used. Ultimately, a good program can, in our opinion, be used nowadays to test the consistency of the mac values.

Electron Probe Quantitation, Edited by K.F.J. Heinrich and
D. E. Newbury, Plenum Press, New York, 1991

However this may be, the first important step is the collection of good light-element data for a wide range of compounds and accelerating voltages. Our laboratory has been very active in this field since the beginning of 1983. The work on carbon, boron and nitrogen has now been successfully completed [1–5] and the work on oxygen is in full progress.

Fortunately, considerable improvements in microanalysis have been realized in the last decade both in the hardware as well as in software, which will make life much easier for the microanalyst attempting to do quantitative EPMA of ultra-light elements.

A very important step forward is the introduction of the new synthetic multi-layer crystals which can provide 2–15 times higher peak intensities and, on top of that, usually suppress higher-order reflections very strongly, thus reducing the problems of background determination considerably. Equally important is the strong progress in matrix correction programs that has been realized in the past 10 years. Especially the introduction of the so-called $\phi(\rho z)$ programs [6–8] must be mentioned here. Because these programs are based on attempts to describe as accurately as possible the number of ionizations or x-ray photons (ϕ) as a function of the mass depth ρz in the specimen they must be considered the most direct, genuine, and straightforward approach to bulk matrix correction. They also provide the best possible starting base for thin-film analysis and in-depth profiling procedures. Some of the items mentioned so far will now be discussed in more detail.

II. New Synthetic Multilayers

As mentioned before, the introduction of commercially available (e.g., Ovonics Corp, Troy, Michigan, U.S.A.)[1] synthetic multilayer crystals has brought considerable improvements in the analysis of ultra-light elements. On the one hand these improvements consist of a significant increase in the peak count rates for light elements; on the other hand, the sometimes strong suppression of higher-order reflections, which are usually a big nuisance in the background determination, is in our opinion at least as important. At present our microprobe (JEOL 733, 4 W.D-spectrometers + EDX system)[1] is equipped with two of such new crystals, next to our conventional Pb-stearate crystal.

The first one (LDE) consists of 200 pairs of alternating W (a few Å) and Si ($\sim$60 Å) layers, $2d$-spacing 59.8 Å, and has been designed specially for the analysis of nitrogen, oxygen, and fluorine. The second one (OVH) is a Mo/B_4C multilayer, $2d = 149.8$ Å, optimized for boron and beryllium.

In the past 2 years these crystals have been extensively tested and their performance compared to that of the conventional stearate (STE) crystal. In fact, the quantitative work on oxygen, which is almost completed now, is almost exclusively being done with the LDE crystal. The work on nitrogen has been done with both STE and LDE crystals simultaneously, which makes a straightforward comparison under identical circumstances easy.

Figure 1 shows a typical example of the performance of the LDE crystal compared to that of the STE crystal. The increase in the peak count rate is obvious; perhaps even more important is the fact that the remnants of higher-order Zr-reflections on the STE crystal are completely suppressed on the LDE crystal, thus yielding a smooth background which is extremely important in view of the low peak-to-background ratios.

An impressive example of the sensitivity of the LDE crystal for $O\,K\alpha$ is given in figure 2 where an oxygen peak could be recorded on pure gold. As a matter of fact an oxygen peak can be found with the LDE crystal on virtually all elemental standards. This shows,

---

[1] Mention of specific commercial products does not imply their endorsement.

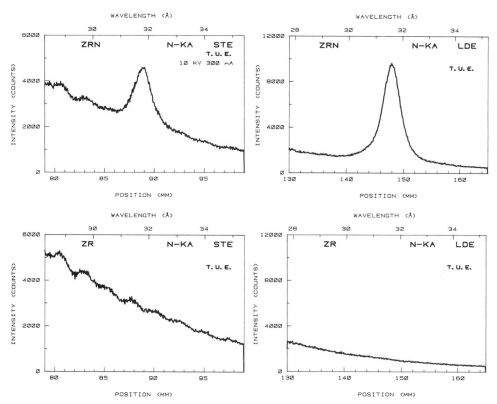

Figure 1. Nitrogen spectra recorded from ZrN with conventional STE crystal (left) and new LDE crystal (right). Lower half gives backgrounds from elemental zirconium. Exp. conditions: 10 kV, 300 nA. Note the difference in the vertical scale for LDE vs. STE by a factor of 2.

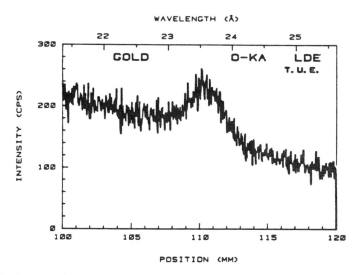

Figure 2. Oxygen peak recorded on pure gold with the LDE crystal. Exp. conditions: 10 kV, 50 nA.

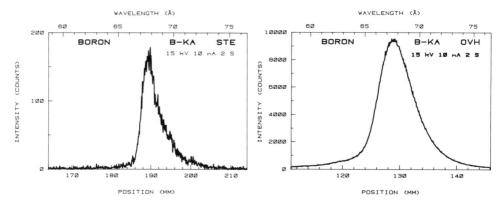

Figure 3. Performance of the OVH multilayer (Mo/B₄C) as compared to that of the conventional stearate crystal for B $K\alpha$ x rays, emitted from elemental boron. Note the difference in the vertical scale for OVH vs. STE of a factor of 50. Exp. conditions: 15 kV, 10 nA.

by the way, how surface-sensitive a microprobe can be, because the solid solubility of oxygen in gold is negligible.

An example of the performance of the OVH crystal is given in figure 3. The huge increase in peak count rate on the OVH crystal gives in this case the possibility to measure B $K\alpha$ x rays under the same conditions (15 kV, 10 nA) which are normally used for much heavier element radiations like Cu $K\alpha$, thereby greatly improving the compatibility of conditions optimal for light-element radiation and those for heavier-element radiation. Normally such cases require conflicting conditions, e.g., in terms of beam current, in order to avoid excessive dead-time corrections.

Another interesting feature of the OVH crystal is that it allows an extension of the possibilities towards Be $K\alpha$, an element which so far could only be measured (with very low count rates) with a lead-cerotate crystal. A more than 20-fold increase in the count rate has been realized here.

A typical advantage of the new multilayers, which has so far hardly been discussed in literature, is that they are much less sensitive to peak shape alterations in the light-element x-ray emission peaks than the conventional Pb-stearate crystal. This must be attributed to the somewhat poorer resolution of these crystals (see e.g., fig. 3). These peak shape alterations, which are in fact one of the biggest problems in the quantitative analysis of ultra-light elements, will now be discussed first.

III. Peak Shape Alterations

In electron-probe microanalysis it has always been common practice to measure the intensities of peaks at the spectrometer settings with the maximum count rates. It is usually not realized that, in principle, integral intensities are required. Peak measurements are only correct under the tacit assumption that the peak intensity is proportional to the integral emitted intensity. Fortunately, this assumption is correct for the K- and L-radiations of medium-to-high atomic number elements. However, for ultra-light element emissions, this assumption is no longer justified as figure 4 shows, and peak measurements, even when peak shifts are taken into account, can and will lead to large errors [1–4].

In order to elucidate this effect the net C $K\alpha$ peak intensities emitted by glassy carbon and TiC have been scaled to the same value, which would yield a peak k-ratio of 1. One

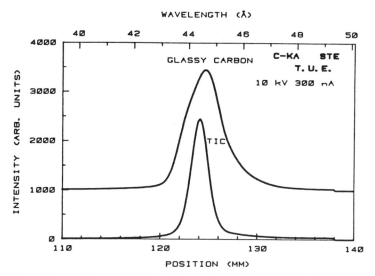

Figure 4. C $K\alpha$ emission profiles recorded from glassy carbon and TiC at 10 kV and 300 nA (stearate crystal). Spectra have been scaled to yield the same net peak count rates. Note the peak shift and the extreme peak shape alteration.

glance at figure 4, however, shows that this approach cannot be correct because it does not do justice to the fact that the peak emitted by glassy carbon is approximately twice as broad as that from TiC. So, actually, the true k-ratio for TiC, relative to glassy carbon, should rather be 0.5 and not 1.0. In this particular case an error of 50 percent is made when the effects of peak shape alterations are neglected; an error which can never be repaired afterwards by any correction program in the world.

One step back in the periodic system takes us to the element boron which is almost the lightest element which can be measured in the electron probe microanalyzer. In this case the situation is even more complex. Apart from the peak shifts and peak shape alterations which are, of course, to be expected, two new complications turn up:

1. The peak position in a specific boride can vary strongly from one crystal in the specimen to another; the peak position is apparently strongly dependent on the crystallographic orientation of the specimen.

2. The peak shape varies synchronously with the peak shifts.

To give an idea about the magnitude of these effects: Typical peak shifts can be of the order of 1 mm (0.357 Å); the associated variation in peak shape is best demonstrated with the aid of figure 5.

The broadest peak is always found at the longest wavelength and the narrowest one at the shortest wavelength. The particular effect can be observed in one single crystal of the specimen, merely by rotating it in its own plane under the electron beam. The boron peak will shift back and forth during this rotation and at the same time it will change its shape continuously. This peculiar phenomenon is not unlike the effects observed when viewing a specimen under a polarization microscope with crossed Nicol prisms while rotating the specimen table. In fact, the origin of both phenomena is the same: the peak shape alterations in the B $K\alpha$ peak are caused by the presence of polarized components [9] in the

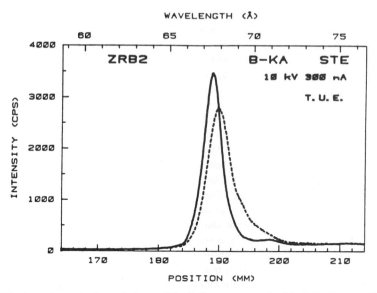

Figure 5. The two extremes in the peak shapes of the B $K\alpha$ spectrum emitted by ZrB$_2$. Exp. conditions: 10 kV, 300 nA, stearate crystal.

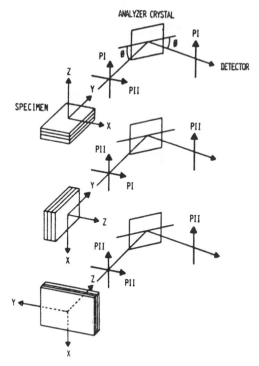

Figure 6. Schematic drawing showing the interaction of the analyzer crystal with the polarized components in the emitted B $K\alpha$ radiation. The electron beam is considered to hit the specimen in the origin of the coordinate system X, Y, Z. (After Wiech [9].)

emitted radiation and the filtering action that can be exercised on specific components by the Pb-stearate crystal under certain conditions.

This effect can, in principle, be expected in all compounds with a crystal symmetry lower than cubic and higher than triclinic. Polarization takes place (see fig. 6) in two mutually perpendicular planes (PI and PII) which are aligned along the principal crystallographic directions of the crystal lattice. The filtering action of the analyzer crystal is optimal when the angle of incidence of the x rays on the crystal equals 45°. In our spectrometer this angle happens to be 42.5° for B $K\alpha$, which is very close to the optimum angle.

The most pronounced shape alteration effects have so far been found in hexagonal compounds such as TiB_2 and ZrB_2. As expected, the relatively few cubic borides did not show an orientation dependence of the peak shape and peak position.

Approximately 50 percent of the 28 investigated borides (including elemental boron) did not show this effect although, from a theoretical point of view, it could be present. This does not mean, however, that such borides can be measured on the peak. As figure 7 shows there are still large differences in peak shapes which have to be taken into account. A striking example in this context is the spectrum emitted by hexagonal BN: The two satellites on both sides of the main peak actually belong to the integral B $K\alpha$ spectrum and as such their intensity has to be included in the measurements. Afterwards we had the opportunity also to record the B $K\alpha$ spectrum emitted by cubic BN. The left-hand satellite was almost gone; the integral intensity, however, was exactly the same.

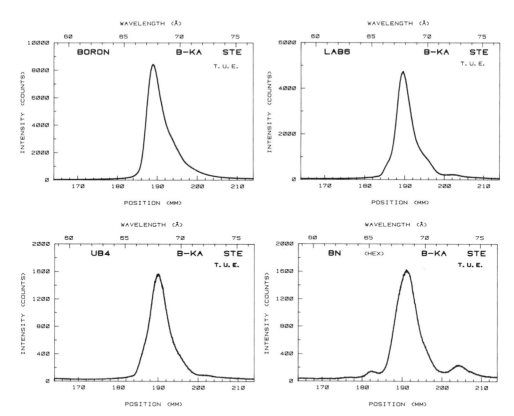

Figure 7. B $K\alpha$ peak shapes in B, LaB_6, hex. BN, and UB_4. Exp. conditions: 10 kV, 300 nA, stearate crystal.

Large errors in the intensity measurements can be made when peak shape alterations are ignored. In the case of boron these errors can even be made when the same compound is used as a standard; merely by the influence of the crystallographic orientation the variations in peak intensity can be in the order of several tens of percent (see fig. 5).

The effects discussed here are worst for the lightest element, boron, and slightly less for carbon, where at least the crystallographic effects and polarization phenomena are absent. For nitrogen and oxygen the effects of shape alterations are rapidly decreasing; in the latter case the effect is hardly noticeable, at least for the LDE crystal.

The conclusion which can be drawn from this section is: the microanalyst trying to deal with ultra-light elements has to be constantly aware of peak shape alteration effects and the lighter the element, the more caution should be exercised.

In most cases it is imperative to perform the intensity measurements in integral fashion which is a very frustrating prospect for on-line analysis with a wavelength-dispersive spectrometer.

On a number of occasions [1–5] we have shown that a considerable reduction in time and effort can be obtained by the use of the area-peak factor concept, which will be discussed in the next section.

IV. The Area-Peak Factor Concept and Its Use

The area-peak factor (APF) has been defined as the ratio between the correct integral (or area) k-ratio and the peak k-ratio. Clearly, this factor is only valid for a given compound with respect to a given standard and for a given spectrometer with its own typical resolution.

Once an APF has been determined, future measurements on the compound in question can simply be carried out on the peak; subsequent multiplication of the peak k-ratio with the APF will then yield the correct integral k-ratio.

To a large extent the APF can be regarded as a relative width-to-height ratio (relative to the standard) or as a kind of weight factor which has to be assigned to a certain peak intensity.

APF values measured on one particular microprobe cannot blindly be transferred to another one because the APF is strongly related to the resolution of the analyzer crystal and spectrometer.

As a matter of fact the only quantity that should be independent of the particular instrument or crystal used is the integral k-ratio for a given accelerating voltage and take-off angle.

This is nicely demonstrated in table 1 where a number of area and peak k-ratios are presented for C $K\alpha$ in near-stoichiometric TiC and ZrC, relative to Fe_3C as a standard for both the Pb-stearate and the LDE crystal.

Table 1. Area k-ratios (AKR), peak k-ratios (PKR) and area-peak factors (APF) for C $K\alpha$ in near-stoichiometric TiC and ZrC at 10 kV for stearate and LDE crystal. Fe_3C standard

| Crystal | TiC | | | ZrC | | |
	PKR	AKR	APF	PKR	AKR	APF
STE	5.749	4.279	0.744	1.367	0.961	0.703
LDE	4.958	4.297	0.867	1.123	0.978	0.870

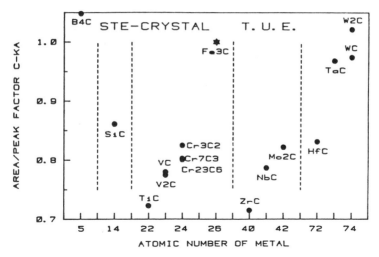

Figure 8. Area-peak factors for C $K\alpha$ radiation emitted from binary carbides, relative to Fe_3C, as a function of atomic number of the metal partner. The value for Fe_3C is equal to one, by definition (marked by the asterisk). Stearate crystal. Dashed vertical lines mark the transition from one period to another in the periodic system.

Two things should be noticed in table 1: in the first place the large deviations, especially for the STE crystal, of the APF from unity. The much lower values for the STE crystal as compared to the LDE crystal indicate that the resolution of the former is substantially better. In the second place it is interesting to point out that the integral k-ratios are the same, within experimental uncertainty, thus showing that whereas the peak k-ratios are more or less the accidental results of the use of a particular crystal, the integral k-ratios are the more fundamental quantities.

Reports from colleagues with instruments comparable in terms of crystal and resolution indicate that the same APF values apply within experimental error. This should, however, be checked in each case.

The two carbides in table 1 represent the cases with the largest shape alterations, expressed in the lowest APF values, which so far have been found for carbon.

It must be realized that Fe_3C has been used as a standard here which has a peak width of approximately 70 percent of that of the carbon peak emitted by glassy carbon. Measurements relative to this elemental form of carbon would result in APF values of 0.7 times those relative to Fe_3C. This illustrates again how large the errors are that can be made when shape alterations are neglected.

One of the results of our work on carbon is that the APF for C $K\alpha$ in binary carbides, when plotted vs. the atomic number of the metal partner, shows a pronounced sawtooth-like variation throughout the periodic system (fig. 8). The minimum values are held by notoriously strong carbide formers, such as titanium, zirconium, and hafnium, while weaker carbide formers such as iron have values much closer to unity.

Using these APF values a data file of 117 carbon measurements in 13 binary carbides between 4 and 30 kV could be collected and this file has been used for a comparison of the performances of various correction programs, together with various sets of macs.

For boron the situation was much more complex. The 50 percent of the borides with shape alterations independent of the crystallographic orientation exhibited largely the same

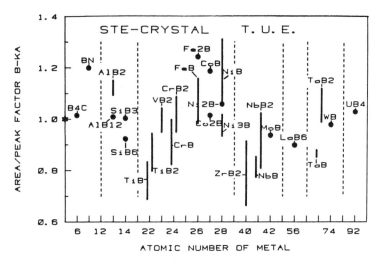

Figure 9. Area-peak factors, relative to elemental boron (asterisk), as a function of the atomic number of the metal partner for 27 binary borides. Solid circles indicate that no dependence on crystallographic orientation exists. Bars indicate the variation in APF with peak position (crystall. orientation) for the orientation-dependent cases. Stearate crystal. Dashed vertical lines mark the transition from one period to another in the periodic system.

pattern as found in the case of carbon: a pronounced sawtooth-like variation with atomic number of the metal partner (fig. 9). These "fixed" APFs have been indicated by solid circles. The vertical bars indicate the cases with an orientation-dependent APF. The length of the bars indicates the magnitude of the variation of APF observed upon rotation of the specimen. A typical example of this group is ZrB_2, the two extreme peak shapes of which have been shown in figure 5. The associated variation in area and peak k-ratio as a function of peak position is represented in figure 10. Once such variations are known the intensity measurements can be done on the peak. After a peak search for each individual grain in the specimen the peak position is used in a graph such as figure 10 to find the appropriate APF and thus the correct area k-ratio.

In this way a total number of 180 boron measurements, relative to elemental boron, in 27 borides between 4 and 15 kV (in some cases up to 30 kV) have been collected.

Our experiences with nitrogen indicate that the individual APF values, relative to Cr_2N as a standard, are much closer to unity than in the previous cases. The maximum deviations are approximately 8 percent and in many cases peak shape alterations are only a few percent and can sometimes even be neglected altogether.

In this case a data file containing 144 measurements in 18 binary nitrides between 4 and 30 kV has been collected using both STE and LDE crystals simultaneously.

The work on oxygen is almost completed at present. Our experiences so far indicate that peak shape alterations are almost negligible, at least for the LDE crystal. Maximum deviations found are of the order of 2 percent which is almost within experimental error. The effects for the stearate crystal are probably twice as high; however, due to problems in the background determination on the short wavelength side of the oxygen peak as a result of the proximity of the short wavelength limit of the spectrometer, it is impossible to make more accurate statements. At the moment a collection of 344 measurements (on LDE), relative to Fe_2O_3 as a standard, for accelerating voltages between 4 and 40 kV has been realized. These measurements have been carried out on more than 30 oxides, carefully selected for sufficient electrical conductivity, as was the standard used. Nonconducting oxides, which present a number of specific problems difficult to deal with, are at the moment the subject of a separate investigation.

More details of the work on nitrogen and oxygen will be published in the near future.

V.   What Can Be Achieved

In the introduction we mentioned that good quantitative results can only be expected from the combination of good measurements and a good matrix correction procedure, in conjunction with the use of a consistent set of macs.

Tests of matrix correction procedures are usually carried out by comparing the calculated k-ratio (k') for the given composition of a compound to the measured k-ratio (k). The ratio k'/k is usually displayed in a histogram showing the number of analyses vs. k'/k.

The narrowness of the histogram, expressed in the relative root-mean-square deviation (r.m.s. in %) and the final average k'/k value are used as a measure of success of a particular program.

Numerous tests with many correction programs available today in combination with various sets of published macs for light elements have shown that with the given sets it is impossible for any program to obtain r.m.s. values below 10 percent and this applies to all the databases for the elements boron through oxygen.

This observation is very surprising in view of the considerable progress in matrix correction procedures that has been realized in the last decade.

It is therefore more likely that the bad results are rather an indication of the average relative accuracy of the published macs than of the quality of the average correction program. We have collected quite some evidence to support this assumption.

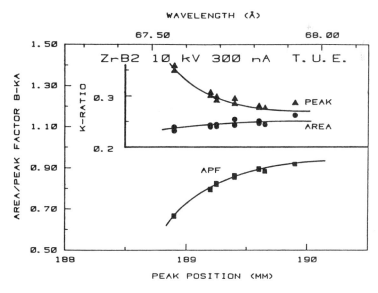

Figure 10.   Variation of peak k-ratio (triangles), area (integral) k-ratio (dots) and area-peak factor (squares) with detected peak position in ZrB_2 (hexagonal structure) as a typical representative of the non-cubic borides. Stearate crystal; 10 kV 300 nA. Reference: elemental boron.

In the extensive collection of light-element data we have realized so far it is easy to find inexplicable discrepancies between the results obtained for similar compounds of neighboring elements in the periodic system.

A good example in this respect is the series of carbides ZrC, NbC, Mo_2C. With the macs for carbon suggested by Ruste [10] satisfactory results are obtained from NbC while much too low k-ratios are calculated for ZrC and much too high ones for Mo_2C. In the absence of systematic differences in x-ray emission (a possibility hardly discussed so far) this must be taken as a strong indication that the mac for C $K\alpha$ x rays is much too high in Zr and much too low in Mo. Such discrepancies between alloys of neighboring elements can never be explained by improper functioning of any correction program because the latter are invariably based on smooth functions of atomic number Z and atomic weight A, either explicitly or implicitly.

We feel that it is the particular strength of our databases that they contain many measurements on compounds in which series of consecutive elements in the periodic system are involved, sometimes forming rather complete periods in the periodic system. This makes a consistent evaluation of the results much easier because once it has been established that the combination of program and macs works satisfactorily for a number of compounds the results can be extrapolated a few atomic numbers further. If then suddenly a discrepancy turns up (in the absence of absorption edges or other physical restrictions) it is very likely that something is wrong with the macs rather than with the program. In this light the bad figure for the r.m.s. of 10 percent and worse is, in our opinion, more indicative of the poor quality of the macs than that of a correction program. This is all the more so when similar discrepancies are obtained with a number of completely different programs, using the same macs as is frequently the case.

In practice a too low value of the mac leads to too high calculated k-ratios and the discrepancies are observed to increase gradually with accelerating voltage. The opposite is, of course, true for too high mac values. When a series of good measurements is available over a wide range of accelerating voltages it is possible to adjust the value of the mac in steps and to evaluate in each step the average k'/k ratio and the r.m.s. value obtained. In the vast majority of cases the best r.m.s. figure is obtained in combination with an average k'/k ratio closest to unity. In this way consistent sets of macs can be produced for each of the light elements boron, carbon, nitrogen, and oxygen.

Another method of producing consistent macs is to measure the relative intensity of the light-element radiation emitted as a function of accelerating voltage [7]. The use of a highly consistent program (and an accurate set of measurements) will produce highly consistent mass absorption coefficients.

In some cases we have found evidence that the mac in a particular element has been obtained in literature by inadvertent extrapolation of the used equations across an absorption edge. A nice example of this is the case of O $K\alpha$ in SnO_2. Our measurements for O $K\alpha$ in RuO_2 between 4 and 40 kV produced excellent agreement with calculations based on the Henke et al. [11] mac of 19700; average k'/k ratio 1.000, r.m.s. $=4.32\%$.

The use of the value of 23100 proposed by Henke et al. for O $K\alpha$ in Sn produced very large deviations for SnO_2 between measurements and calculations: average k'/k ratio 0.675, r.m.s. $=7.011$ percent. The discrepancies were found to increase rapidly with accelerating voltage. A series of tests showed that with a value of 15050 both an optimum value for the r.m.s. was obtained (3.25%) as well as a k'/k ratio closest to unity (0.997). Obviously, the equations used in the calculations of the mac have been extrapolated across the M_5 edge.

In general it was found that the macs for light-element radiations in elements such as zirconium, niobium, and molybdenum are usually unsatisfactory. The reason for this is

Table 2.   Results that can be achieved for light-element data

Element	No. of meas. ts	Accel. voltage (kV)	k'/k	r.m.s. (%)
B	180	4–15 (30)	0.977	6–7[a]
C	117	4–30	0.983	4
N	144	4–30	1.007	4
O[b]	344	4–40	0.998	2.4

[a] Significantly better results are obtained when the three nickel-borides, which exhibit approximately 15% too low B $K\alpha$ emission, are eliminated.
[b] Only sufficiently conductive oxides have been included.

simple: the vicinity of the M_5 edge in these elements produces a strongly nonlinear variation of mac with wavelength, which makes inter- and extrapolation very dangerous.

In many cases the new values we propose are intermediate between the older values of Henke and Ebisu [12] and the more recent ones of Henke et al. [11]; quite frequently the older values seem to be better.

Of course, as soon as an absorption edge is passed the evaluation process has to start on a fresh basis. Only after a few new cases of satisfactory results is it possible to start making extrapolations. This is yet another reason why so many measurements are necessary on many different systems.

In a number of cases it is even necessary to carry out this process when the composition of the specimen is not known with absolute certainty, which is sometimes the case with carbides and nitrides with wide homogeneity regions. In such circumstances the composition is another variable which has to be taken along in the iterative evaluation process. The most likely composition is usually the one giving the best r.m.s. value in combination with a k'/k ratio closest to unity. At the same time the mac value should be consistent with that of neighboring elements. A great help is in these cases the analysis of the x-ray lines of the metal partners over the widest possible range of accelerating voltages. This is the reason why in addition to the light elements themselves we usually also measure the metal lines. With a number of uncertain parameters at the same time it is virtually imperative to use all available data. Extensive use is also being made of the latest information on the phase diagrams of the systems in question. The ultimate goal in this process is to achieve the highest possible degree of consistency for all parameters involved.

Finally we will give a brief survey (table 2) of what can be achieved nowadays with a modern correction program (PROZA, for details see Appendix) in combination with consistent sets of macs.   It should be noted that the boron data file contains data for the three Ni borides Ni_3B, Ni_2B, and NiB which exhibit a systematically too low ($\sim 15\%$) emission of B $K\alpha$ x rays. The mac appears to be satisfactory (r.m.s. 2–3.7%); however, the averages in k'/k vary between 0.84 and 0.87.

The conclusion that can be drawn from table 2 is obvious: with sufficient care in the measurements and a good correction program in combination with consistent macs it is possible to obtain surprisingly good quantitative results for light elements, especially when one takes into consideration that in many cases very extreme conditions have been used, like an accelerating voltage of 40 kV for oxygen.

In milder conditions, e.g., in the range of 8–20 kV, a range which we prefer for a number of reasons, the results would be much better still. At the same time, however, the actual procedures involved in the collection of light-element data will always be more time-consuming than those for medium-to-high Z-elements, due to the pronounced peak shape alteration effects.

Perhaps a future introduction of new multilayer crystals, which have been shown to be less sensitive to these effects, can produce a further improvement here.

VI. References

[1] Bastin, G. F. and Heijligers, H. J. M. (1986), X-Ray Spectr. **15**, 135–150.
[2] Bastin, G. F. and Heijligers, H. J. M. (March 1984), Quantitative Electron Probe Microanalysis of Carbon in Binary Carbides, Internal Report, Eindhoven University of Technology.
[3] Bastin, G. F. and Heijligers, H. J. M. (1984), Microbeam Analysis, Romig, A. D., Jr. and Goldstein, J. I., eds., 291.
[4] Bastin, G. F. and Heijligers, H. J. M. (March 1986), Quantitative Electron Probe Microanalysis of Boron in Binary Borides, Internal Report, Eindhoven University of Technology.
[5] Bastin, G. F. and Heijligers, H. J. M. (1986), J. Microsc. Spectrosc. Electron. **11**, 215–228.
[6] Packwood, R. H. and Brown, J. D. (1981), X-Ray Spectr. **10**, 138–146.
[7] Pouchou, J. L. and Pichoir, F. (1984), Réch. Aérospat. **3**, 13–38.
[8] Bastin, G. F., van Loo, F. J. J., and Heijligers, H. J. M. (1984), X-Ray Spectr. **13**, 91–97.
[9] Wiech, G. (1981), X-Ray Emission Spectroscopy, Nato Adv. Study Inst., Day, P., ed., Emission and Scattering Techniques, Ser. C, 103–151.
[10] Ruste, J. (1979), J. Microsc. Spectrosc. Electron. **4**, 123.
[11] Henke, B. L., et al. (1982), Atomic Data and Nuclear Data Tables **27**, 1.
[12] Henke, B. L. and Ebisu, E. S. (1974), Adv. in X-Ray Analysis **17**, 150.
[13] Bastin, G. F., Heijligers, H. J. M., and van Loo, F. J. J. (1984), Scanning **6**, 58–68.
[14] Bastin, G. F., Heijligers, H. J. M., and van Loo, F. J. J. (1986), Scanning **8**, 45–67.
[15] Pouchou, J. L. and Pichoir, F. (August 1986), Proc. 11th ICXOM Conference, Brown, J. D. and Packwood, R. H., eds., Graphic Services, UWO, London, Canada, 249–253. (It should be noted that there is a nasty typing error on page 251: In the equation for the primary intensity I the ratio $\left(\dfrac{U_0}{V_0}\right)^{P_k}$ behind the Σ-term should read: $\left(\dfrac{V_0}{U_0}\right)^{P_k}$.

Appendix. Short Lay-Out of PROZA (Phi-Rho-Z and A correction program)

Introduction

In this new program a drastic change has been made compared to previous versions [4,8,13,14] in which the $\phi(\rho z)$ parameterizations were based on independent equations for the Gaussian parameters α, β, γ, and ϕ_0 in the Packwood-Brown model [6].

The equation for the parameter β is now no longer calculated using an independent equation but through a procedure based on the atomic number correction of Pouchou and Pichoir [15].

This atomic number correction provides the value of the integral of $\phi(\rho z)$ (i.e., the generated intensity in the specimen) which will be called F. Using this F-value and new expressions for α and γ the value of β is mathematically adapted in such a way as to ensure that the integral equals F and that the peak of the $\phi(\rho z)$ curve has the "correct" position as well as the "correct" height.

This means that the parameters α, β, γ, and ϕ_0 are now forced to cooperate in a consistent way in order to provide a specified value for the total generated intensity in the specimen.

Procedure

Step #1. Calculation of Primary Intensity (P.I.) and Integral of $\phi(\rho z)$ (F) according to Pouchou and Pichoir [15]

$$P.I. = R \cdot 1/S$$

$(R =$ backscattering factor, $1/S =$ stopping power$)$

F is proportional to: $P.I. /Q(E_0)$

$[Q(E_0)$ is ionization cross-section$]$.

The equations for η (backscatter coefficient), $\bar{W}$ (averaged reduced energy of backscattered electrons), R (backscattering factor), $Q(E_0)$ and J (ionization potential) are those used by Pouchou and Pichoir [15].

Step #2. Parameterization of Gaussian $\phi(\rho z)$ Curves

The object of this step is to find the α, β, γ, and ϕ_0 parameters which will provide the correct integral of $\phi(\rho z)$ $(\equiv F)$.

The equation for ϕ_0 is that used by Pouchou and Pichoir [15].

The equation for α is:

$$\alpha = \frac{2.1614 \cdot 10^5 \cdot Z^{1.163}}{(U_0-1)^{0.5} \cdot E_0^{1.25} \cdot A} \cdot \left[\frac{\ln(1.166\ E_0/J)}{E_c} \right]^{0.5}$$

in which Z, A, and J are atomic number, atomic weight, and ionization potential of the matrix element and E_0, E_c, and U_0 are accelerating voltage, critical excitation voltage, and overvoltage ratio for the x-ray line in question. For a compound target a matrix of $\alpha_{i,j}$ values (α for element i-radiation in interaction with each element j of the matrix) is calculated and the α_i-value in the compound target is composed as follows:

$$(1/\alpha_i)\ \text{comp.} = \sum_j c_j \cdot \frac{Z_j}{A_j} \cdot 1/\alpha_{i,j} / \sum_j c_j \cdot \frac{Z_j}{A_j}.$$

The equation for γ is:

for $U_0 \leqslant 6$:

$$\gamma = 3.98352 \cdot U_0^{-0.0516861} \cdot (1.276233 - U_0^{-1.25558 \cdot Z^{-0.1424549}})$$

for $U_0 > 6$:

$$\gamma = 2.814333 \cdot U_0^{0.262702 \cdot Z^{-0.1614454}}.$$

In order to accommodate the change in ionization cross section with atomic number for ultra-light element radiations, proposed by Pouchou and Pichoir, it is necessary in these cases to multiply γ further by the term:

$$E_c/(-4.1878 \cdot 10^{-2} + 1.05975 \cdot E_c).$$

This is only necessary if $E_c < 0.7$ keV.

For a compound target the weight-fraction averaged atomic number is substituted for Z.

The calculation of β proceeds in the following way: we have shown before [13] that the total intensity generated in a specimen ($\equiv F$) can be expressed by:

$$F = \frac{[\gamma - (\gamma - \phi_0) \cdot R(B/2\alpha)] \cdot \sqrt{\pi}}{2\alpha}.$$

in which $R(\beta/2\alpha)$ is the 5th degree polynomial used in the approximation of the erfc $(\beta/2\alpha)$ function. The latter equation is the formal solution in closed form of the Gaussian integral of $\phi(\rho z)$ between zero and infinity. After rearranging it follows that:

$$R(\beta/2\alpha) = [\gamma - 2\alpha \cdot F/\sqrt{\pi}]/[\gamma - \phi_0].$$

Contrary to our previous versions this time F is known first and the problem is now to find the value of β using the known values of α, γ, and ϕ_0 through the latter equation. This means that the function $R(\beta/2\alpha)$ has to be used backward: i.e., the function value is known and the argument $(\beta/2\alpha)$ has to be determined.

The simplest way to solve this problem was to cut the function into nine different regions and to fit these regions with much simpler geometric functions. If for a moment we substitute x for $R(\beta/2\alpha)$ we obtain as the best fits:

$0.9 \leqslant x < 1$	$\beta/2\alpha = 0.9628832 - 0.9642440 \cdot x$
$0.8 < x < 0.9$	$" = 1.122405 - 1.141942 \cdot x$
$0.7 < x \leqslant 0.8$	$" = 13.43810 \cdot \exp(-5.180503 \cdot x)$
$0.57 < x \leqslant 0.7$	$" = 5.909606 \cdot \exp(-4.015891 \cdot x)$
$0.306 < x \leqslant 0.57$	$" = 4.852357 \cdot \exp(-3.680818 \cdot x)$
$0.102 < x \leqslant 0.306$	$" = (1 - 0.5379956 \cdot x)/(1.685638 \cdot x)$
$0.056 < x \leqslant 0.102$	$" = (1 - 1.043744 \cdot x)/(1.604820 \cdot x)$
$0.03165 < x \leqslant 0.056$	$" = (1 - 2.749786 \cdot x)/(1.447465 \cdot x)$
$0 < x \leqslant 0.03165$	$" = (1 - 4.894396 \cdot x)/(1.341313 \cdot x)$

As a result of the fitting procedure the value of $R(\beta/2\alpha)$ thus obtained will never be exactly the same as the one calculated before, especially near the transition points of one function to another. In an extra loop in the program the approximated value of $R(\beta/2\alpha)$ can be compared to the formal one and β can be adjusted in order to meet a specified relative precision.

Once α, β, γ, and ϕ_0 are known the usual procedure [13] can again be followed.

A special precaution had to be taken at extremely low overvoltages. In such cases it is virtually impossible to ensure a correct parameterization of $\phi(\rho z)$ curves due to the extreme delicacy involved in the balance of parameters which are still expected to produce the specified F-value. Thus, it can happen occasionally that $R(\beta/2\alpha)$ values are calculated which are negative or larger than 1.

It is obvious though that $R(\beta/2\alpha)$ can only have values between 0 and 1, which means that β is between infinity and zero. When $R(\beta/2\alpha)$ is outside these limits then the normal parameterization route cannot be used and an auxiliary procedure has to be followed.

In these (rare) cases the calculated value of α is dropped and for a start it is assumed that the $\phi(\rho z)$ curve starts halfway the values of ϕ_0 and γ. Using the known value for F a new (and usually higher) value for α is calculated through:

$$\alpha = \frac{[\phi_0 + \gamma] \cdot \sqrt{\pi}}{4\,F}.$$

The value of $R\,(\beta/2\alpha)$ is thereby set at exactly 0.5. Although the $\phi(\rho z)$ curves in such cases may not be fully realistic, the answers returned by the program are still very good because the atomic number correction is still consistent and absorption effects under these conditions are usually negligible.

The advantage of the new program is that it can now be used down to the lowest possible overvoltages (if one insists on working under these difficult conditions).

Its performance on a data file of 877 measurements (see Ref. 14, file supplemented with metals analyses in borides [4]) of medium-to-high Z-elements is:

Av. k'/k ratio: 0.9955,
r.m.s. (%)   : 2.44.

The results for light elements are represented in table 2.

NONCONDUCTIVE SPECIMENS IN THE ELECTRON PROBE MICROANALYZER—A HITHERTO POORLY DISCUSSED PROBLEM

G. F. BASTIN AND H. J. M. HEIJLIGERS

Laboratory for Physical Chemistry
Centre for Technical Ceramics
University of Technology
P.O. Box 513, NL-5600 MB Eindhoven
The Netherlands

I. Introduction

Electron probe microanalysis and scanning electron microscopy of nonconducting specimens usually present the operator with problems. Many of these problems are related to the image quality: nonconducting specimens usually yield highly unstable images, often characterized by extremely bright areas at locations with the worst electrical conductivity and sudden shifts in the image with respect to the specimen from time to time. The bright areas in the image are the result of an excessive production of secondary electrons which in turn leads to a very poor resolution of surface details by the production of a halo.

It is usually assumed that this excessive production of secondary electrons leads to a positive electrical charge in the surface layers of the specimen. This positive charge, however, must be compensated by a negative space charge in the vacuum at the specimen surface/vacuum interface. The solution to this problem is well known: the application of a conductive surface layer of carbon or gold prevents in a simple way the surface charging effects and high-quality SEM pictures can be obtained, provided that a low-ohmic contact from specimen to earth is present.

The microanalyst is, unfortunately, not only interested in high-quality SEM pictures; his primary concern is usually the quality of the x-ray signals emitted by the specimen and these inevitably are also strongly affected by the conductivity of the specimen and by charging effects. The usual solution is also in this case the application of a conductive coating, for which in the majority of cases carbon is selected because it can easily be applied and is one of the lightest materials, exhibiting at the same time a relatively good electrical conductivity.

Many mineralogists have been and still are using this technique in order to analyze their usually nonconducting minerals. It must be mentioned, though, that they hardly ever measure a light element like oxygen, which is usually calculated "by difference" or from stoichiometry. However, for light elements such as nitrogen and oxygen a carbon layer is a very bad choice because the characteristic x rays of these elements are excessively absorbed in carbon. Therefore, copper has been used [1–3] in layers of 100–400 Å thick on some occasions, on the grounds that it can be applied in very thin layers and still has good conductivity. Besides, copper does not produce interfering x-ray lines in the spectral region of interest of these elements.

With one exception [1] no one has ever discussed the influence of a surface coating on the quality of the measurements. As far as we know Weisweiler [1] has been the only one to discuss this in detail, in connection with his extensive work on oxygen analysis in a large number of nonconducting oxides. In his discussion, which is mainly based upon the influence of the thickness of the coating in relation to the x-ray generation depth in the sample, he comes to the conclusion that the application of a conductive copper coating can, in some extreme cases of heavy absorption, lead to differences of 10–15 percent in the intensity ratios (k-ratios) as compared to uncoated samples. Strange enough, the author did not touch, in our opinion, upon the real problem in such cases, which is underneath the conductive layer: the nonconductive specimen itself. It is highly unlikely that the mere application of a conductive layer, which is an extrinsic measure, can change something in the electrical conductivity of the specimen itself, which is an intrinsic quality.

One should not forget that the situation in electron probe microanalysis is completely different from that of surface analysis techniques such as ESCA, Auger and SIMS, where one is only interested in a few atomic layers of the specimen surface and where charging effects can easily be relieved by using an aperture or grid surrounding the point of impact of the beam on the specimen. In EPMA, on the other hand, the x-ray signal comes from electrons penetrating into the specimen microns deep and it cannot a priori be expected that a conducting layer, tens of thousands of Ångstroms away from where it all has to happen, can take care of all unwanted effects.

The central question is whether the process of electron deceleration in a nonconductive specimen proceeds along the lines predicted by the various models used so far in matrix correction programs. Especially vital is the question whether the $\phi(\rho z)$ curves expressing the number of produced x-ray photons ϕ as a function of mass depth (ρz), which have been shown to work highly satisfactorily for conducting materials, can still be used for nonconducting specimens.

In fact, a completely new parameter, not taken into account by any model used so far, is introduced in the latter materials: the electrical field built up inside the specimen. Calculations about this field [4] induced by a charged beam inside an insulator show that field strengths of $\simeq 10^4$ V/cm and more are possible. In fact, the same author suggested that a better solution to this problem would be to take a thin specimen and to coat it underneath.

Summarizing, it seems to us that there are three vital questions:

1. How large is the effect of surface charging and how can it be relieved?
2. What is the influence of a coating on the reliability of quantitative intensity measurements once surface charging has been prevented?
3. Can we assume that everything is all right from then on and that no distortion of $\phi(\rho z)$ curves takes place under these circumstances?

In the following sections we will describe a number of experiments which may cast some light on these questions.

II.   Experimental

A.   SURFACE CHARGING AND ITS EFFECTS ON THE X-RAY SIGNALS

The first experiment described here concerns the simple investigation of a nonconducting and noncoated yttrium-aluminum garnet (YAG, $Y_3Al_5O_{12}$) specimen at 10 kV and a beam current of 50 nA. Figure 1a shows the EDX spectrum recorded from this specimen. At first sight it appears quite normal; the Al and Y peaks are there, just as expected. The problems turn up only after reducing the vertical scale by a factor of 30 (fig. 1b). Then it becomes visible that the short-wavelength cut-off is at $\simeq 5.9$ keV instead of the expected nominal value of 10 keV. This means that an appreciable voltage drop occurs at the specimen surface/vacuum interface and that, in fact, the useful accelerating voltage for the electrons is only 5.9 keV at the moment of entering the specimen. This voltage drop in itself

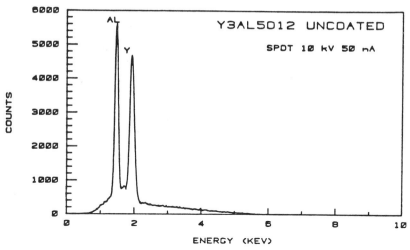

Figure 1a. EDX spectrum recorded from an uncoated $Y_3Al_5O_{12}$ specimen at 10 kV and 50 nA in the spot mode.

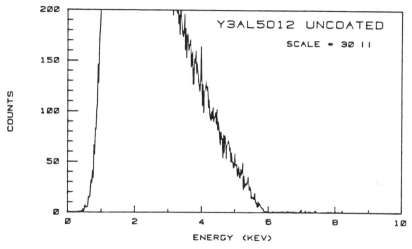

Figure 1b. As a; however, vertical scale reduced by a factor of 30 in order to show the value of the short-wavelength cut-off.

can already be a serious problem in the qualitative identification of elements present in the specimen because as soon as the voltage drops below the excitation threshold for a particular x-ray line this line will no longer be detected. It is quite possible that if, e.g., copper was supposed to be present the Cu $K\alpha$ line would not be found at 10 keV; however, the Cu $L\alpha$ line would still be present, thus leading to considerable interpretation problems. If an EDX-system is available it is always wise in suspect cases to check the short-wavelength cut-off, which only takes a few seconds.

In order to alleviate surface charging effects and the associated voltage drop a conductive coating of carbon (or sometimes copper) is commonly used. However, according to

Table 1. Actual voltage (kV), absorbed current (nA) and detectability of Cu $K\alpha$ (in EDX spectrum) on copper-coated $Y_3Ga_5O_{12}$ at 10 kV and 50 nA, using different beam modes

Beam mode	Magnification	kV	nA	Cu Kα present?
Scanning	60 x	10.00	18.67	YES
Scanning	100 x	9.68	16.00	YES
Scanning	200 x	8.44	11.67	NO
Scanning	600 x	7.86	12.00	NO
Scanning	2000 x	7.64	14.33	NO
Scanning	6000 x	6.68	17.00	NO
SPOT	———	6.48	18.00	NO

our experiences such a coating must meet a number of minimum specifications; the most important one being that the short-wavelength cut-off is restored to the original nominal voltage. This means that matters such as the thickness and the nature of the coating are in themselves irrelevant: the only important requirement is that the voltage is brought back to the nominal one. A too thin and/or bad quality coating, even if it is copper, may not be successful and in some cases we observed a deterioration of the results rather than an improvement, resulting in the disappearance of the Y-peak due to a drop in voltage below 2 keV.

The value of the short-wavelength cut-off is the result of a number of complex factors, among which the conductivity of the specimen, the quality of the coating and the electrical load imposed by the electron beam on the specimen are the most important. The electrical load itself can be varied by varying the area on the specimen which is scanned by the electron beam, the worst load being in the spot mode. A number of experiments to this effect are represented in table 1 for $Y_3Ga_5O_{12}$ with a too thin (or bad quality) copper film at 10 kV and 50 nA beam current. As these results indicate the voltage drop is largest when the beam is in the spot mode which, of course, imposes the largest electrical load on the specimen. A very interesting feature in table 1 is that the voltage drop is evidently continued into the copper coating because as soon as the actual voltage drops below 8.98 keV (threshold for Cu $K\alpha$) the Cu $K\alpha$ line is no longer visible. The absorbed current turns out to be a poor measure of success for the coating because there appears to be no direct correlation between absorbed current and actual voltage.

The only criterion should be whether or not the voltage is restored to its original nominal value and the check on the short-wavelength cut-off is a simple and straightforward test for the quality of the coating. It is logical to expect that the voltage drop has the largest impact on the results for relatively heavy elements because the nominal overvoltage used is relatively low in such cases. How seriously the voltage drop can affect the k-ratios in cases like this is illustrated in figure 2 where the k-ratios for Si $K\alpha$, relative to pure Si, as measured in uncoated Si_3N_4 specimens, are represented for a number of nominal voltages. The expected k-ratio for each voltage is indicated by arrows.

For this particular experiment a number of different Si_3N_4 specimens have been used, each exhibiting a different (range in) electrical conductivity: first a massive Si_3N_4 specimen, produced by chemical vapor deposition, exhibiting the lowest conductivity; then two layers of CVD-deposited Si_3N_4 on a Si-substrate (1 and 2.5 μm thick, respectively), showing better conductivity than the massive specimen. The thinnest layer gave a much better conductivity than the thicker one. Next, a hot isostatically pressed Si_3N_4 specimen and finally a Si-specimen arc-melted under reduced nitrogen pressure. The latter specimen contained

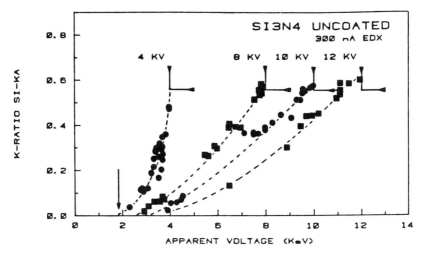

Figure 2. k-ratio for Si $K\alpha$, relative to pure Si, measured in uncoated Si_3N_4 specimens at nominal voltages of 4, 8, 10, and 12 keV. The k-ratio is plotted vs. the observed apparent voltage, measured with the EDX system. The arrows indicate the expected k-ratios for each nominal voltage. The vertical arrow indicates the critical excitation voltage for Si $K\alpha$ (1.8 keV).

numerous finely dispersed Si_3N_4 particles with a wide range in particle size and consequently a wide range in conductivity. When the results for a nominal accelerating voltage of 4 kV are examined it becomes clear that the lowest actual voltage recorded is approximately 2.3 kV. The corresponding k-ratio for Si $K\alpha$ is $\simeq0.03$, i.e., Si $K\alpha$ is hardly detected at all. The highest k-ratio recorded is approximately 0.5 which is rather close to the expected value of $\simeq0.56$. At the same time the actual voltage is very close to the nominal one. In fact, the results indicate a very strong correlation between k-ratio and actual voltage; they can more or less be represented as the branch of a parabola, intersecting the horizontal axis at a voltage of $\simeq1.8$ kV, which is exactly the excitation threshold for Si $K\alpha$. This conclusion is confirmed by the measurements at 8, 10, and 12 kV; only the curvature of the parabola becomes less pronounced.

The correlation between k-ratios and absorbed current, on the other hand, was extremely weak: thus showing again that the absorbed current is a poor indicator for deviations from the nominal voltage. It almost appears in figure 2 that the true k-ratio for Si $K\alpha$ could be obtained, once a number of measurements are available, by a process of extrapolation towards the nominal voltage. For the light-element nitrogen in Si_3N_4 similar measurements were carried out and the qualitative results were comparable: a too low actual voltage simply gave too low k-ratios (relative to Cr_2N), although the deviations were less dramatic than with Si $K\alpha$, due to the much lower excitation threshold for N $K\alpha$ (0.4 keV). For the 1 μm Si_3N_4 layer on Si at 4 kV an actual voltage of 3.5 is found and a correspondingly too low k-ratio (2 as compared to the expected value of approximately 2.5). The values for the 2.5 μm layer were lower still while the massive specimen gave the worst results.

The voltage drop can have a significantly different effect on heavy-element radiations as compared to light-element radiations. In the former case the relative emitted intensity as a function of voltage is usually only a simple increasing function in the usual voltage range

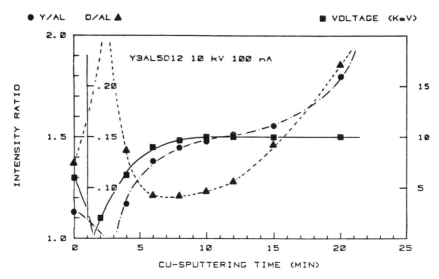

Figure 3. Apparent voltage (right-hand scale, squares), intensity ratio Y/Al (outside left-hand scale, dots) and intensity ratio O/Al (inside left-hand scale, triangles) observed in a $Y_3Al_5O_{12}$ specimen at 10 kV and 50 nA as a function of the copper sputtering time.

(0 to $\sim$40 kV) not unlike the ones shown in figure 2. Hence, any drop in voltage will always result in a lower intensity. Light-element radiations, on the other hand, usually exhibit a maximum at a much lower voltage in the relative intensity-vs.-voltage curve. Depending on the magnitude of the mass absorption coefficient this maximum is usually between 4 and 15 kV. As a result there are two possibilities: either the actual voltage is below the one giving the maximum intensity, in which case a drop in voltage will always yield lower intensities, or it is beyond this value in which case a drop in voltage could yield higher intensities. These remarks apply only, of course, under the assumption that a voltage drop is the only disturbing effect.

After having established the adverse effects of the voltage drop in nonconductors we will now take a closer look at the influence of a conductive coating on the inter-elemental intensity ratios, an aspect which is of the utmost importance for quantitative analysis.

B. THE INFLUENCE OF A COATING ON INTER-ELEMENTAL INTENSITY RATIOS

In order to study this effect a coating of copper was applied to a YAG specimen by DC sputtering at a low argon pressure. Successively increasing layer thicknesses were applied and the actual voltage together with the emitted x-ray signals of Cu $L\alpha$, Cu $K\alpha$, Cu $K\beta$, Al $K\alpha$, Y $L\alpha$ and O $K\alpha$ were measured after each step at 10 kV and 50 nA. The copper layer thickness was not measured; the sputtering time can be used as a gross relative measure for the thickness. In figure 3 the intensity ratios Y/Al and O/Al have been displayed, together with the observed actual voltage, as a function of the sputtering time. The results clearly show that under no circumstances the copper layer can produce constant intensity ratios for the elements in question; the results will always be strongly dependent on the particular copper layer thickness and thus rather arbitrary. A surprising feature is the increasing O/Al ratio with growing copper layer thickness which strongly suggests that the

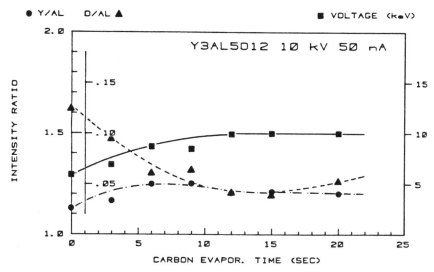

Figure 4. Apparent voltage (right-hand scale, squares), intensity ratio Y/Al (outside left-hand scale, circles) and intensity ratio O/Al (inside left-hand scale, triangles) observed in a $Y_3Al_5O_{12}$ specimen at 10 kV and 50 nA as a function of the carbon evaporation time. Values at 20 s have been obtained by burning a hole in the carbon coating.

$O\,K\alpha$ signal suffers much less from the increasing copper layer thickness than the Al (or Y) signal, which is in itself amazing. We shall come back to this detail later because, as we shall see, this is not a coincidental phenomenon. The peculiar voltage drop and associated discontinuities for the Y/Al and O/Al ratios after two minutes sputtering are most probably caused by impurities from the cathode which are deposited onto the specimen in the initial stages of the process. This phenomenon was avoided in later experiments by first sputtering onto a dummy specimen in order to clean the cathode before coating the specimen.

By and large figure 3 shows the competitive effects of the increase in the actual voltage and the accompanying effect of stronger absorption by the growing copper layer. The former effect leads to the production of more x rays which are in turn more strongly absorbed by the increasing copper layer which is necessary to produce this higher actual voltage. A disadvantage for the Al $K\alpha$ x rays is their stronger absorption in copper as compared to Y $L\alpha$ radiation; this is the reason for the increase in the Y/Al intensity ratio.

A similar test was performed on the same YAG specimen, but now applying an evaporated carbon coating with varying thickness (fig. 4). It is evident that in this case more or less constant Y/Al intensity ratios can be obtained, especially after the longer evaporation times. However, with the increase in this time the O/Al ratio decreases rapidly, due to the extremely high absorption of O $K\alpha$ in carbon. Besides, it will be noted that the Y/Al ratio has hardly ever the same value in figure 4 as it has in figure 3. This illustrates how difficult it is to get meaningful results from coated specimens.

Finally, an experiment was performed by using an air-jet to burn a hole in the carbon coating (after 15 s evaporation time). It was found that when the hole was complete, which takes something between 2 and 5 minutes at 10 kV and can be followed by monitoring the O $K\alpha$ signal, the actual voltage was still 10 kV. The emitted signal ratios obtained in this case are represented (for an evaporation time of 20 s) in figure 4. A separate test on a well-conducting oxide such as Fe_2O_3 showed that the O $K\alpha$ intensity emitted from such a hole (size 6–8 μm cross section) is exactly the same as from an uncoated specimen. This procedure offers the interesting possibility to do measurements with the beneficial effects of a coating, in that it restores the nominal voltage, without suffering from its disadvantages, such as causing a strong absorption for O $K\alpha$ x rays or interfering with the inter-elemental intensity ratios. The conclusion we can draw from this section is that a conductive coating will inevitably alter the inter-elemental intensity ratios and as such one should be extremely careful with its application.

So far, we have only been concerned with the intensity ratios for the elements contained within one specimen. Of interest to quantitative analysis is the question of what the influence of a coating is on the ratio of a specific signal emitted from a specimen as compared to that emitted from a standard when both have the same coating. This will be discussed in the next section.

C.   INFLUENCE OF A COATING ON QUANTITATIVE RESULTS

It has already been mentioned in the introduction that in the past copper coatings have been used [1–3] for the quantitative EPMA of oxygen in (mostly) nonconducting oxides, mainly in order to avoid the excessive absorption of O $K\alpha$ x rays suffered in a carbon coating. The oxygen standards used were either SiO_2 [1–2] or Al_2O_3 [3], both notorious insulators. Weisweiler [1] came to the conclusion that the mere presence of a copper coating (200–400 Å thick) could, in heavily absorbing oxides, lead to deviations in the k-ratios of 15 percent. The other authors (Love et al. [3]) did not even mention the possibility of such an effect. In the latter case a copper layer of less than 100 Å was evaporated onto the specimens. It must be stressed that none of the authors mentioned here made any observation regarding the actual voltage prevailing at the specimen surface. It must be suspected, however, that especially the extremely thin copper layers (< 100 Å) used by Love et al. will probably not have been thick enough to restore the actual voltage to its nominal value.

Apart from the voltage drop at the specimen surface, which can be measured as we have seen, there is still the other vital but unanswered question whether or not the $\phi(\rho z)$ curves are distorted in an insulating material due to the build-up of an electrical field inside the specimen. This latter effect is much more difficult to deal with experimentally because such phenomena are not directly accessible by experimental techniques. The usual measurements in a microprobe are intensity ratios in which all possible disturbing effects are mixed together, especially when nonconductive specimens are measured relative to a nonconductive standard and on top of that a conductive coating is applied, the influence of which on the measurements itself is not known a priori.

We have, therefore, decided to proceed along the following line of reasoning:

– First, it is important to establish how our correction program performs for the analysis of oxygen. In order to exclude as yet unknown influences of nonconductivity on the measurements this test was carried out on sufficiently conductive oxides, relative to a conductive oxygen standard: Fe_2O_3 (hematite).

More than 30 oxides, forming a representative cross section through the periodic system could be found which satisfied our criteria. The lightest oxide studied in this series was TiO_2 (rutile), sufficiently conductive from 4 kV on and higher; the heaviest one was Bi_2O_3 which became sufficiently conductive from 10 kV onwards.

A total number of 333 analyses, in the range between 4 and 40 kV were collected and the results with Henke et al. macs [5], modified in places where necessary (Ti,

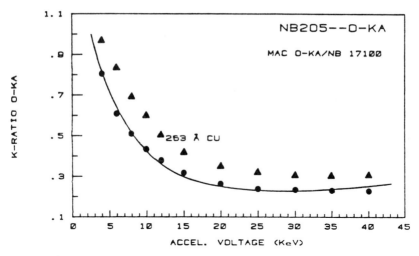

Figure 5. Integral k-ratios for O $K\alpha$ in Nb_2O_5, relative to Fe_2O_3, as a function of accelerating voltage. Solid line represents calculated k-ratios based on a mac O $K\alpha$/Nb of 17100. Full circles represent measurements on uncoated Nb_2O_5 relative to uncoated Fe_2O_3; triangles those on copper-coated (263 Å) Nb_2O_5 relative to copper-coated Fe_2O_3 with identical copper layer thickness. Note the large increase in k-ratio due to the influence of the copper layer.

Zr, Nb, Mo; i.e., close to absorption edges), were excellent: an average k'/k ratio of 0.998 was obtained, together with an r.m.s. value of 2.4 percent. Such results were to be expected considering the excellent results obtained previously for boron and carbon measurements. Apparently our program works fine for conducting oxides.

It seems very likely, therefore, that if unsatisfactory results are obtained for notorious nonconductors, the reason must rather be sought in the lack of conductivity or influence of coating than in the performance of the program.

The next step was to select a few conductive oxides, apply a copper coating on specimen and Fe_2O_3 standard and to compare the results to those obtained before on uncoated specimens.

A nice example of such an experiment is the analysis of Nb_2O_5 which appears to be sufficiently conductive for microanalysis. Figure 5 shows the results obtained for O $K\alpha$ in comparison to the calculated results (solid line) based on a mac of 17100. The values quoted in literature are 17850 (Henke and Ebisu [6]) and 15300 (Henke et al. [5]); the latter value is too low, in our opinion. When a conductive layer of copper with a thickness of 263 Å is applied on specimen and Fe_2O_3 standard then k-ratios are obtained which are 20 percent too high at 4 kV and almost 40 percent too high between 8 and 12 kV.

A second example is the oxygen analysis in Zr(Y)O_2 (0.5968 Zr, 0.1526 Y, 0.2506 O), which can be carried out successfully by burning a hole in a carbon coating, like we described before. As figure 6 shows, these measurements agree very well with calculations based on a mac of 16200 for O $K\alpha$ in Zr and 15100 in Y. For comparison: Henke's older values (1978) are 16140 and 15140, respectively; his newer ones are 14800 and 15100, respectively. The application of a 171 Å thick copper layer on specimen and standard produces the same effect as in Nb_2O_5: a considerable increase in the k-ratios over the full range in accelerating voltages of 20–30 percent. An increase in the copper layer thickness

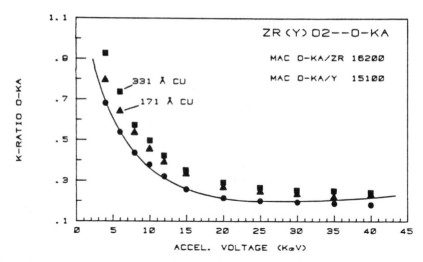

Figure 6. Integral k-ratios for O $K\alpha$ in $Zr(Y)O_2$ (0.5968 Zr, 0.1526 Y, 0.2506 O), relative to Fe_2O_3, as a function of accelerating voltage. Solid line represents calculated k-ratios for macs O $K\alpha/Zr$ of 16200 and O $K\alpha/Y$ 15100. Dots represent measurements on uncoated $Zr(Y)O_2$ (through a hole in the carbon coating, relative to uncoated Fe_2O_3). Triangles were obtained after coating standard and specimen with 171 Å copper. Squares with a 331 Å copper layer. Note the large increase of k-ratios with increasing copper layer thickness.

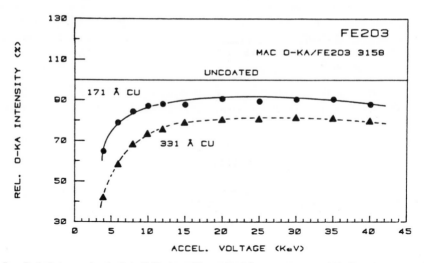

Figure 7a. Ratio between the absolute O $K\alpha$ intensities emitted from copper-coated Fe_2O_3 and uncoated Fe_2O_3 (reference, indicated by horizontal line). Full circles refer to a copper layer of 171 Å, triangles to a layer of 331 Å.

to 331 Å leads to a further increase of this deviation; thus demonstrating that almost arbitrary results can be obtained simply because of the influence of the copper layer.

Once we have seen the dramatic increase in k-ratios, the next interesting question is how the absolute intensities in specimen and standard vary separately with the thickness of the copper layer and some quite amazing observations were made in this respect. Let us first examine the absolute O $K\alpha$ intensity emitted from the coated Fe_2O_3 standard. In figure

7a we have plotted the ratio between the absolute intensity emitted from the Fe_2O_3 standard covered with copper layers of 171 and 331 Å, respectively, and that from the uncoated (conductive) standard. It is evident that the presence of a 171 Å copper layer already reduces the intensity by 10 percent for voltages between 10 and 30 kV; even more substantial reductions are observed at lower voltages. This is the result of the fact that at low voltages the thickness of the copper layer reduces considerably the useful excitation volume for the generation of O $K\alpha$ x rays. The thicker layer of 331 Å has an even more dramatic effect at low voltages.

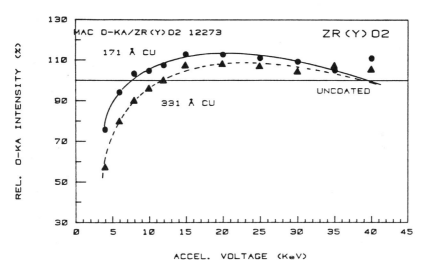

Figure 7b. As a; this time for the $Zr(Y)O_2$ specimen. Full circles refer to a copper layer thickness of 171 Å; triangles to a layer of 331 Å. The reference line corresponds to the absolute intensity emitted from the specimen through a hole burned in a carbon coating.

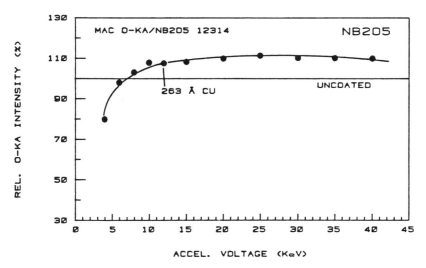

Figure 7c. As a; now for a Nb_2O_5 specimen with a copper layer of 263 Å.

Similar measurements were carried out for $Zr(Y)O_2$, $ZnGa_2O_4$, Co_2TiO_4, Mn_3O_4 and Nb_2O_5, with astonishing results. For $Zr(Y)O_2$ we also observed (fig. 7b) the expected reduction in intensity at low voltages, although it was significantly less severe than for Fe_2O_3. Beyond 10 kV, however, the absolute intensities were even higher than those emitted from the uncoated specimen: up to 12 percent higher. The most amazing thing is perhaps that these effects are more pronounced for the thinnest layer. This might have been caused by the fact that the actual voltage was perhaps lower than the nominal one, which could be favorable for oxygen in a heavily absorbing system, once the maximum in the intensity-vs.-voltage curve is passed. The actual voltage was, however, within experimental error equal to the nominal one. That these results are not a mere coincidence is illustrated for Nb_2O_5 in figure 7c, where the intensity from a coated (263 Å copper) specimen exceeds that from an uncoated one already beyond 7 kV. These observations fully explain the peculiar increase in k-ratios in figure 5 and 6: a considerable reduction in standard intensity takes place while the intensity from the specimen can even be enhanced substantially. To a somewhat lesser extent similar results were obtained for Co_2TiO_4 where a small range (15–25) in accelerating voltage was found with a higher absolute intensity than from an uncoated specimen for the 171 Å copper layer. (This same effect is presumably responsible for the increasing O/Al ratio in figure 3 for sputtering times exceeding 6 min.)

When these results are analyzed in more detail there appears to be a not yet understood correlation between the extent of the observed effects and the mac for oxygen in the compound in question: the most drastic reduction in intensity was found in Mn_3O_4 (mac 2835), followed by Fe_2O_3 (mac 3158), $ZnGa_2O_4$ (mac 5508), Co_2TiO_4 (mac 6792), $Zr(Y)O_2$ (mac 12273) and finally Nb_2O_5 (mac 12314). In the last three cases higher intensities can actually be observed from coated than from uncoated specimens. This observation is very strange in itself, because one would expect the largest interference of the coating in those cases with the shallowest emission volumes for O $K\alpha$ x rays; i.e., in compounds with the heaviest absorption. In fact, calculations about the influence of the coating on the k-ratios for oxygen in the case of, e.g., Nb_2O_5 show that increases of 5–7 percent are to be expected by coating with copper. So it seems that the sense of the effect is correct; only its magnitude turns out to be unexpectedly larger.

Even if the reasons for the peculiar phenomena observed are not completely understood, the results can and must be taken as a serious warning against the use of metallic coatings on (non)conducting specimens because their use can obviously lead to gross errors in the intensity measurements which cannot be corrected for because, in our opinion, there are simply no models capable of quantitatively describing the amazing effects found.

Of course, the problem remains what we must do with notorious insulators like MgO, $MgAl_2O_4$, Al_2O_3, and SiO_2 now that we have seen that a copper coating cannot be trusted. All our measurements on these specimens, when coated with 171 Å of copper, were not only too high (up to 15%) over the full range in accelerating voltages: it was observed at the same time that the variation of k-ratio with voltage was quite different from the calculations: the largest deviations were observed at the lowest voltages while the discrepancies gradually decreased with increasing voltage. Such results can never be explained by a wrong value of the mass absorption coefficient: the variation in deviations should in this case be in the opposite direction and the largest deviation should be found at the highest voltage. Even burning a hole in a carbon coating on these notorious insulators will yield 10–15 percent too high k-ratios.

The theoretical possibility that the correction program would suddenly predict deviating k-ratios after producing excellent results from Bi_2O_3 down to TiO_2 is in our opinion, highly unlikely. It is more plausible, therefore, to suppose that in such insulators a distortion in the $\phi(\rho z)$ curves takes place, leading to a much smaller average depth of x-ray emission. If such a distortion, due to the build-up of an electrical field, indeed takes place this is certainly not taken into account by any existing correction program. As long as the parameters governing such a distortion are not quantitatively understood they are not likely to be incorporated into any program either. Besides, it will always be difficult to specify "the conductivity" of a specimen because the conductivity can strongly be influenced by small amounts of impurities, the presence of which is not a priori known. On top of that a certain degree of conductivity can be induced by a high-energy electron beam capable of creating lattice defects in the irradiated material. Our experimental evidence shows that there is a group of oxides which cannot be measured below 10 kV without a coating (e.g., MoO_3, $BaTiO_3$, Ta_2O_5); beyond 10 kV this may become perfectly possible. The range between 10 and 20 kV is probably optimal and will not present any problems at all for a correction program.

Summarizing this brief discussion, we would like to repeat our warning against the use of metallic coatings and, where possible, advise to try to analyze either without a coating at higher voltages or to burn a hole in a carbon coating. A closed carbon coating is probably too dangerous because small differences in coating thickness would produce large differences in oxygen count rates due to the excessive absorption of O $K\alpha$ x rays in carbon.

III. References

[1] Weisweiler, W. (1974), Arch. Eisenhüttenwesen **45**, 287.
[2] Weisweiler, W. (1978), Arch. Eisenhüttenwesen **49**, 555.
[3] Love, G., Cox, M. G. C., and Scott, V. D. (1974), J. Phys. D: Appl. Phys. **7**, 2131.
[4] Cazaux, J. (1986), J. Appl. Phys. **59**, 1418.
[5] Henke, B. L., et al. (1982), Atomic Data and Nuclear Data Tables **27**, 1–144.
[6] Henke, B. L. and Ebisu, E. S. (1974), Adv. in X-Ray Analysis **17**, 639.

THE R FACTOR: THE X-RAY LOSS DUE TO ELECTRON BACKSCATTER

R. L. MYKLEBUST AND D. E. NEWBURY

National Institute of Standards and Technology
Gaithersburg, MD 20899

I. Introduction

With the evolution of electron microprobes that employ more stable electronics which permit high precision measurements of x-ray intensity, a new interest has arisen in the accuracy of methods and parameters used to compute the matrix corrections for quantitative electron probe microanalysis. One of the two components that contribute to the "atomic number correction," part of the ZAF matrix correction procedure, is the R-factor. The R-factor is an estimate of the ratio of the total inner shell ionization which is actually produced in the specimen over that which would occur in the absence of electron backscattering. There are several formulations of the R-factor in use today. They are based on either empirical fits to experimental data on electron backscattering and backscattered electron energy distributions [1] or on data generated by a Monte Carlo simulation for electron scattering in solids [2]. In this publication we will examine several of these formulations and compare the results obtained by each method.

II. Theory

Duncumb and Reed [1], using Bishop's data [3] for the energy distribution of backscattered electrons, determined the R-factors using the following equations:

$$R = 1 - \frac{\int_{w_K}^{1} \eta(w) \frac{Q}{S} \, dw}{\int_{E_K}^{E_0} \frac{Q}{S} \, dE} \tag{1}$$

$$S = c \left[\frac{Z}{A}\right] \left[\frac{1}{E}\right] \ln \left[\frac{1166 \, E}{J}\right] \tag{2}$$

where w is E_K/E_0 and $\eta(w)$ is the integral form of the backscattered electron energy distribution [$\eta(w)$ is the number of backscattered electrons with energy greater than $w E_0$]. E_K is the critical excitation energy for the K-shell of the atom, and E_0 is the incident beam energy. The formula for the ionization cross section, Q, was given by Webster et al. [4] and the equation for the stopping power, S, was from Bethe [5]. Duncumb and Reed assumed that the value of the mean ionization potential, J, was given by $J = 11.5Z$. The values obtained for R are affected by the values selected for J, as seen in eqs (1) and (2). For many years, because of deficiencies in the expression $J = 11.5Z$, quantitative analysis programs have used values of J computed with an expression arrived at by Berger and Seltzer [6]:

$$\frac{J}{Z} = 9.76 + 58.8 Z^{-1.19} . \tag{3}$$

The values of R were then determined with the expression of Duncumb and Reed or by

Electron Probe Quantitation, Edited by K.F.J. Heinrich and
D. E. Newbury, Plenum Press, New York, 1991

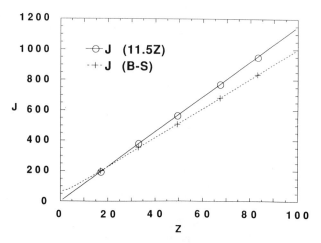

Figure 1. A comparison of two expressions for mean ionization potential, J. Circles are from $J = 11.5Z$ and crosses are from eq (3).

various expressions that were fitted to their data. In this work we have computed the R-factors for several elements with the aid of a modified version of the Microanalysis Monte Carlo simulation program [7]. In addition we have recalculated the R-factors from Bishop's data to determine the error involved in computing R with different values of S. Figure 1 shows a plot of J versus Z for the two expressions listed above. There is little difference in the J values for low atomic numbers; however, the J values computed by the two expressions differ widely for high atomic number. Duncumb [8] has fitted a polynomial to the R-factors as a function of Z and w.

$$
\begin{aligned}
R = {}& 1 + (-0.581 + 2.162\, w_K - 5.137\, w_K{}^2 + 9.213\, w_K{}^3 \\
& - 8.619\, w_K{}^4 + 2.962\, w_K{}^5) \times 10^{-2}\, Z + (-1.609 - 8.298\, w_K \\
& + 28.791\, w_K{}^2 - 47.744\, w_K{}^3 + 46.540\, w_K{}^4 - 17.676\, w_K{}^5) \\
& \times 10^{-4}\, Z^2 + (5.400 + 19.184\, w_K - 75.733\, w_K{}^2 + 120.050\, w_K{}^3 \\
& + 110.700\, w_K{}^4 + 41.792\, w_K{}^5) \times 10^{-6}\, Z^3 + (-5.725 - 21.645\, w_K \\
& + 88.128\, w_K{}^2 - 136.060\, w_K{}^3 + 117.750\, w_K{}^4 - 42.445\, w_K{}^5) \\
& \times 10^{-8}\, Z^4 + (2.095 + 8.947\, w_K - 36.510\, w_K{}^2 + 55.694\, w_K{}^3 \\
& - 46.079\, w_K{}^4 + 15.851\, w_K{}^5) \times 10^{-10}\, Z^5 \; .
\end{aligned}
\tag{4}
$$

This expression has been and continues to be used in many computer programs for quantitative electron probe microanalysis. Since this and other expressions in this paper were derived from mathematical fits, rounding the coefficients to fewer significant figures may cause erroneous results. The function produces values that generally agree with the computations by other formulas for R except for extreme cases (high Z with either very high or very low overvoltage). Yakowitz et al. [9] produced a much simpler expression for R which was used in the FRAME computer programs.

$$
R = A - B \ln (DZ + 25)
\tag{5}
$$

where

$$A = 0.00873\ U^3 - 0.1669\ U^2 + 0.9662\ U + 0.4523$$

$$B = 0.002703\ U^3 - 0.05182\ U^2 + 0.302\ U - 0.1836$$

$$D = 0.887 - \frac{3.44}{U} + \frac{9.33}{U^2} - \frac{6.43}{U^3}$$

and U is the overvoltage E_0/E_K. This function did produce values similar to those of Duncumb's above; however, Labar [10] has observed that this expression does not produce a continuous curve as a function of Z. Myklebust [11] has proposed an even simpler expression which is used in the current version of the FRAME programs.

$$R = 1 - 0.0081512\ Z + 3.613 \times 10^{-5}\ Z^2 + 0.009582\ Z \exp(-U) + 0.00114\ E_0. \tag{6}$$

This expression was fitted from values of R computed by the Microanalysis Monte Carlo program. While it agrees with most of the other expressions at low atomic numbers, it produces the lowest R-factors of any of the expressions at higher atomic number.

Love and Scott [12] have employed an atomic number correction in their $\phi(\rho z)$ method for quantitative analysis with the following expression for R:

$$R = 1 - \eta\ [I(U_0) + \eta G(U_0)]^{1.67} \tag{7}$$

where

$$I(U_0) = 0.33148 V + 0.05596 V^2 - 0.06339 V^3 + 0.00947 V^4$$

$$G(U_0) = \frac{1}{U_0} [2.87898 V - 1.51307 V^2 + 0.81312 V^3 - 0.08241 V^4]$$

in which

$$V = \ln U_0$$

and the backscatter coefficient is given by

$$\eta = \eta_{20} \left[1 + \frac{H(Z)}{\eta_{20}} \ln \left(\frac{E_0}{20}\right)\right] \tag{8}$$

where

$$\eta_{20} = (-52.3791 + 150.48371 Z - 1.67373 Z^2 + 0.00716 Z^3) \times 10^{-4}$$

$$\frac{H(Z)}{\eta_{20}} = (-1112.8 + 30.29 Z - 0.15498 Z^2) \times 10^{-4}.$$

This expression produces R-factors that are smaller than Duncumb's at low atomic numbers and larger than Duncumb's at higher atomic numbers. Pouchou and Pichoir [13] have also introduced a backscatter correction factor into their $\phi(\rho z)$ method for quantitative analysis.

$$R = 1 - \bar{\eta}\ \bar{w}\ [1 - G(U_0)] \tag{9}$$

where

$$\bar{\eta} = 1.75 \times 10^{-3} Z_p + 0.37\ [1 - \exp(-0.015 Z_p^{1.3})]$$

$$Z_p = (C_i\ \sqrt{Z_i})^2$$

$$\bar{w} = \frac{E_{av}}{E_0} = 0.595 + \frac{\bar{\eta}}{3.7} + \bar{\eta}^{4.55}$$

$$G(U_0) = \frac{U_0 - 1 - \left[\dfrac{1 - 1/U_0^{(a+1)}}{1 + \alpha}\right]}{(2 + \alpha)J(U_0)}$$

$$J(U_0) = 1 + U_0(\ln U_0 - 1)$$

$$\alpha = \frac{(2\bar{w} - 1)}{(1 - \bar{w})}.$$

Again, the agreement with other methods is good for low atomic numbers but R diverges for higher atomic numbers where this expression produces R-factors larger than Duncumb's. August et al. [14] have proposed another analytical equation based on a large number of R-factors computed from an expression for the energy distribution of backscattered electrons used by Czyzewski and Szymanski [15] (see below).

$$R = 1 + \sum_{j=1}^{5} \sum_{u=1}^{j} a_{i,j-i+1} \left(\frac{1}{U_0} - 1\right)^i \cdot Z^{j-i+1} \tag{10}$$

where

$$a_{1,3} = -0.5531081141 \times 10^{-5}$$

$$a_{1,4} = 0.5955796251 \times 10^{-7}$$

$$a_{1,5} = -0.3210316856 \times 10^{-9}$$

$$a_{2,1} = 0.3401533559 \times 10^{-1}$$

$$a_{2,2} = -0.1601761397 \times 10^{-3}$$

$$a_{2,3} = 0.2473523226 \times 10^{-5}$$

$$a_{2,4} = -0.3020861042 \times 10^{-7}$$

$$a_{3,1} = 0.9916651666 \times 10^{-1}$$

$$a_{3,2} = -0.4615018255 \times 10^{-3}$$

$$a_{3,3} = -0.4332933627 \times 10^{-6}$$

$$a_{4,1} = 1.0300997920 \times 10^{-1}$$

$$a_{4,2} = -0.3113053618 \times 10^{-3}$$

$$a_{5,1} = 0.3630169747 \times 10^{-1}.$$

The values resulting from this expression agree closely with those of Pouchou and Pichoir [13] and the double summation employed is easy to program. Czyzewski [16], using the differential expression for the energy distribution of backscattered electrons proposed by Czyzewski and Szymanski [15], has proposed an integral function for the computation of R.

$$R = 1 - \frac{1}{(U_0 - 1)^{1.67}} \int_1^{U_0} \frac{d\eta}{dU} (U - 1)^{1.67} \, dU \tag{11}$$

where the energy distribution of the backscattered electrons is given by

$$\frac{d\eta}{dU} = kps\eta \frac{U^{(k-1)}/U_0^{(k-1)}}{U_0(1 - U^k/U_0^k)^{(1+p)} \, [1 - s + s/(1 - U^k/U_0^k)^p]^2} \tag{12}$$

and

$$k = 1.6$$

$$p = (0.8 + 2\eta) \ln\left(\frac{1}{\eta}\right)$$

$$s = 0.1054 \, [2^p - 1 + 1/(2.25 \cos \alpha)^p] \, .$$

If $d\eta/dU$ is plotted vs. $1/U$, this function reproduces the plots of the energy distribution of backscattered electrons as seen in figure 2, where the results are compared to Bishop's data [3] and to the energy distribution obtained from the Monte Carlo simulation program. This expression does not directly use the atomic number but does require that the backscatter coefficient η be entered. The function is designed to work for inclined specimens as well as

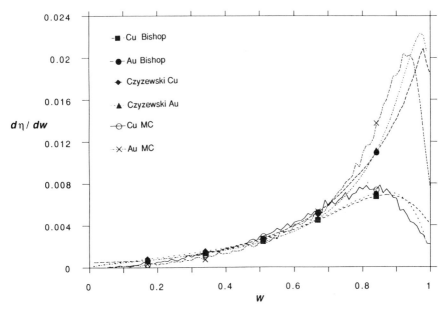

Figure 2. A comparison of the energy distribution of backscattered electrons from gold and copper at 20 keV. Bishop's data are compared with the Microanalysis Monte Carlo results and with the expression of Czyzewski and Szymanski [eq (14)].

specimens normal to the electron beam, with the introduction of the specimen tilt angle (α). The expression may be easily integrated numerically on a computer and agrees reasonably well with Pouchou and Pichoir's [13] results. Figure 3 is a comparison of all of the above computations for R.

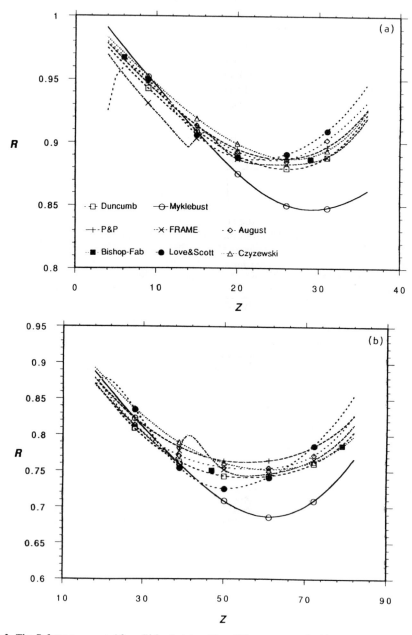

Figure 3. The R-factors computed from Bishop's data with eq (18) are compared with seven published expressions for R for K (a), L (b), and M (c) x-ray lines. The identifying legend is the same in (b) and (c) as is listed for (a).

III. Experimental

The loss of x-ray generation due to backscatter of primary electrons has been re-evaluated with the aid of a Monte Carlo simulation of electron scattering in solids. The Monte Carlo simulation we have employed is a single scattering model using a screened Rutherford elastic scattering cross section, the Bethe continuous energy loss expression, and the Bethe inner shell ionization cross section as modified by Mott and Massey [17]. The x rays generated by electrons are computed as long as the electrons stay within the specimen. If the electron backscatters, the x rays that would have been generated if the electron had remained in the specimen are computed separately. The sum of these two numbers is equal to the total generated x rays and the R-factor is the ratio of the x rays generated in the specimen to the total generated x rays. In addition, the program produces a histogram of the energies of the backscattered electrons. R may then be computed from this histogram using the equation of Webster et al. [4].

$$R = 1 - \frac{\int_{w_K}^{1} \frac{d\eta}{dw} \int_{E_K}^{wE_0} \frac{Q}{S} \, dE \, dw}{\int_{E_K}^{E_0} \frac{Q}{S} \, dE} \qquad (13)$$

where Q is the x-ray cross section, S is the stopping power from eqs (2) and (3), and $d\eta/dU$ is the energy distribution of backscattered electrons. We have performed the integration in eq (13) with backscattered electron energy distributions from Bishop [3], from the Monte Carlo simulation, and from eq (12) written as a function of W instead of U:

$$\frac{d\eta}{dw} = kps\eta \frac{w^{(k-1)}}{(1-w^k)^{(1+p)} [1-s+s/(1-w^k)p]^2} \qquad (14)$$

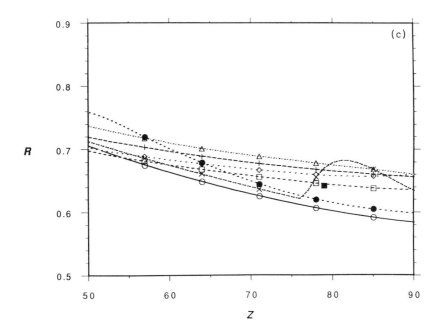

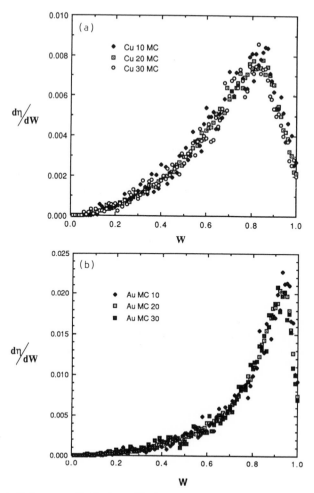

Figure 4. The Monte Carlo energy distributions of backscattered electrons for copper (a) and gold (b) are compared for 10, 20, and 30 keV beam voltages.

where all of the variables are defined the same as in eq (12) and the backscatter coefficient was computed with eq (8). It should be noted that the primary beam voltage (E_0) does not occur in this expression except in the computation of the backscatter coefficient. The energy distribution of backscattered electrons is therefore relatively free of any voltage dependence, as is shown in figure 4 which compares energy distribution plots at 10, 20, and 30 keV as generated by the Monte Carlo simulation. All of the integrations were done with a form of the Bethe cross section from Powell [18]:

$$Q = C \cdot \frac{\ln(c_K U)}{U E_K^{\,2}} \qquad (15)$$

where

$$c_K = 2.42 .$$

The expression to be integrated then becomes

$$R = 1 - \frac{\displaystyle\int_{w_K}^{1} \frac{d\eta}{dw} \int_{E_K}^{WE_0} \frac{\ln\left(\frac{c_K}{E_K}E\right)}{\ln\left(\frac{1.166}{J}E\right)} dE \cdot dW}{\displaystyle\int_{E_K}^{E_0} \frac{\ln\left(\frac{c_K}{E_K}E\right)}{\ln\left(\frac{1.166}{J}E\right)} dE} . \tag{16}$$

The integrations were repeated with a cross section from Fabre de la Ripelle [19]

$$Q = C \cdot \frac{\ln(U)}{(U+1.32)E_K^2} \tag{17}$$

and the expression to be integrated becomes

$$R = 1 - \frac{\displaystyle\int_{w_K}^{1} \frac{d\eta}{dw} \int_{E_K}^{WE_0} \frac{\ln\left(\frac{1}{E_K}E\right)}{\left(\frac{1}{E_K}+\frac{1.32}{E}\right)\ln\left(\frac{1.166}{J}E\right)} dE \cdot dW}{\displaystyle\int_{E_K}^{E_0} \frac{\ln\left(\frac{1}{E_K}E\right)}{\left(\frac{1}{E_K}+\frac{1.32}{E}\right)\ln\left(\frac{1.166}{J}E\right)} dE} . \tag{18}$$

We have performed Monte Carlo simulations on a wide variety of pure elements at many different beam voltages. Each of these simulations produces a variety of data including the R-factor for the x-ray line of the element. This data is distributed over a range of atomic numbers and overvoltages (U) and it is therefore difficult to plot R as a function of Z with U constant. An additional set of simulations was performed on binary alloys in which the mass fraction of the matrix element was 0.999 and the mass fraction of the element whose line we were calculating was 0.001. In this manner we were able to obtain the R-factors for any x-ray line in any matrix. The matrix elements selected were Be(4), Al(13), Ca(20), Zn(30), Zr(40), Sn(50), Nd(60), Yb(70), Au(79), and Th(90). The x-ray lines calculated were Na $K\alpha$, S $K\alpha$, Ti $K\alpha$, Ga $K\alpha$, Br $K\alpha$, and Nb $K\alpha$. All of these Monte Carlo simulations were done with a beam voltage of 20 keV; therefore, the range of overvoltages used was $U = 1.05$ to $U = 18.66$.

IV. Results

Changing the expression for computing J produced negligible differences in the values obtained for R. Therefore, even a simple expression like $J = 11.5Z$ should suffice; however, since most programs would have already computed J using the expression of Berger and Seltzer [6], it would be preferable to use this latter value.

The integration used in eq (16) was tested by comparing the results with the R-values computed with the Monte Carlo simulation program. The comparison is shown in figure 5

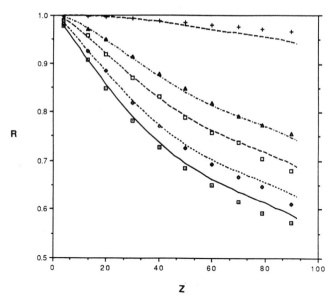

Figure 5. The R-factors computed with the Microanalysis Monte Carlo simulation program are compared with the R-factors computed by integrating the energy distribution of the backscattered electrons [eq (16)]. The energy distributions were also computed with the Monte Carlo simulation program.

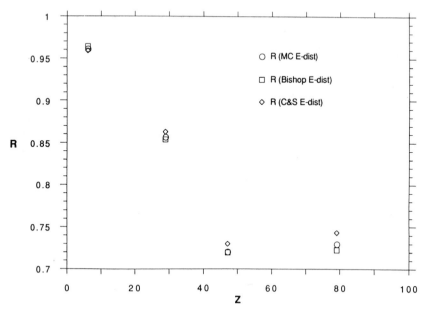

Figure 6. R-factors for carbon, copper, silver, and gold computed by integrating the respective energy distributions of the backscattered electrons. The energy distributions are from the Monte Carlo simulation program, R(MC E-dist), from Bishop's data, R(Bishop E-dist), and from eq (14), R(C&S E-dist).

for 20 keV electrons. The results are almost identical, as they should be, since the energy distribution of the backscattered electrons for the integration was obtained from the Monte Carlo simulation. The integrations of the energy distributions of the backscattered electrons from Bishop [3], Czyzewski and Szymanski [15], and the Monte Carlo simulation all produced similar results, as seen in figure 6. This is not surprising since the energy distributions are similar. The only real differences between the energy distributions occur for values of w near 1.0, which could result in discrepancies in the R-values if very low overvoltages are used ($w_0 > 0.95$). Since overvoltages this low are not recommended for good quantitative analysis methods, it would appear that any of the energy distributions are satisfactory. We have decided on the expression of Czyzewski and Szymanski [15] for the remainder of the integrations since this is the only available analytical expression.

The R-values obtained from the integrations of eqs (16) and (18) are compared with the values computed with eqs (4), (9) and (11) in figure 7. It is evident that of all the parameters we have examined, the cross section has the greatest effect on R. Use of the cross section of Fabre de la Ripelle [19] produces results that are in better agreement with the majority of the methods presented. In addition, this cross section is more suitable for use at low overvoltages than that of Powell [18]. Even with the rather large difference between these two functions, the effect on the results of a quantitative analysis are fairly small. For example, table 1 contains the computed compositions for lead and silicon in a lead silicate glass analyzed at 20 keV. The results were determined using both eqs (16) and (18). The standards were silicon dioxide and a lead selenide compound.

The results obtained with eq (18) are only slightly better than those obtained with eq (16). Neither accounts for the unusually high values calculated for silicon in this material, an anomalous result which is currently under study.

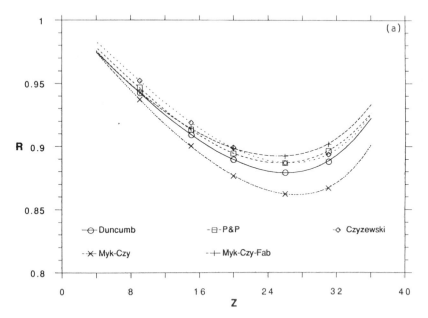

Figure 7. The R-factors computed with the integrations in eqs (16) and (18) are compared with three of the published expressions for R. (a) K-lines, (b) L-lines, and (c) M-lines. The identifying legend is the same in (b) and (c) as is listed for (a).

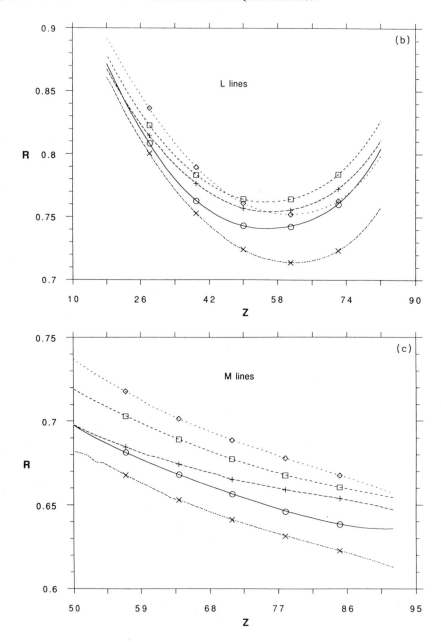

We have also tested the method suggested by Duncumb and Reed [1], that the R-factor for compounds could be determined by computing the mass concentration average of the R-factors for pure elements. Four R-factors were determined by Monte Carlo simulation and by two fitted expressions for Cu $K\alpha$ and for Au $L\alpha$ in pure gold and pure copper (table 2). The mass concentration average for a 10%Cu-90%Au alloy was then calculated (table 3).

Table 1.

Element	True conc.	eq (16)	eq (18)
Silicon	0.1402	0.1558	0.1529
Lead	.6498	.6389	.6448

Table 2. R-factors

Specimen	Line	Monte Carlo	Duncumb	Myklebust
Cu	Cu $K\alpha$	0.8548	0.8821	0.8468
Au	Cu $K\alpha$	.6834	.7366	.6860
Cu	Au $L\alpha$	.8939	.9139	.8687
Au	Au $L\alpha$	.7423	.7874	.7458

Table 3. Mass concentration averages of R for a 10% Cu-90% Au Alloy

Line	Monte Carlo	Duncumb	Myklebust
Cu $K\alpha$	0.7005	0.7512	0.7021
Au $L\alpha$	.7575	.8001	.7581

The values obtained can be compared to those obtained from a Monte Carlo simulation of the alloy which are for Cu $K\alpha$, $R = 0.7046$ and for Au $L\alpha$, $R = 0.7509$. They agree very well with the mass concentration values calculated from the Monte Carlo R-factors from table 2. Mass concentration averaging does appear to be valid for computing the R-factors of alloys.

Regardless of the differences in the computations of R, there seems to be only a small effect on the quantitative analysis results. There may be some effects in extreme cases such as large atomic number differences together with either very high or very low overvoltages, however, in these cases, the absorption correction may cause a more serious error than the atomic number correction. We do recommend that the computation be done with eq (18), as it should produce the best results even for the extreme cases.

V. References

[1] Duncumb, P. and Reed, S. J. B., The Calculation of Stopping Power and Backscatter Effects in Electron Probe Microanalysis, NBS Spec. Publ. 298, K. F. J. Heinrich, ed., 1968, 133.

[2] Myklebust, R. L., An evaluation of x-ray loss due to electron backscatter, J. de Physique **45**, 1984, C2.

[3] Bishop, H. E., Some Electron Backscattering Measurements for Solid Targets, Fourth Int. Cong. X-Ray Opt. Mic., Orsay, Hermann, Paris, 1966, 153.

[4] Webster, D. L., et al., Phys. Rev. **37**, 1931, 115.

[5] Bethe, H. A., Theory of the transmission of corpuscular radiation through matter, Ann. Phys. (Leipzig), **5**, 1930, 325.

[6] Berger, M. J. and Seltzer, S. M., Tables of Energy Losses and Ranges of Electrons and Positrons, Report number NASA SP-3012, Washington, DC, 1964.

[7] Myklebust, R. L., Newbury, D. E., and Yakowitz, H., NBS Monte Carlo Electron Trajectory Calculation Program, NBS Spec. Publ. 460, K. F. J. Heinrich, D. E. Newbury, and H. Yakowitz, eds., 1975, 105.

[8] Duncumb, P., in Electron Beam X-Ray Microanalysis, K. F. J. Heinrich, ed., Van Nostrand Reinhold, New York, 1981, 249.

[9] Yakowitz, H., Myklebust, R. L., and Heinrich, K. F. J., FRAME: An On-Line Correction Procedure for Quantitative Electron Probe Microanalysis, NBS Tech. Note 796, 1973.

[10] Labar, J. L., Comparison of backscatter loss calculations in electron probe microanalysis, J. Phys. D **11**, 11079, 721.

[11] Myklebust, R. L., An evaluation of x-ray loss due to electron backscatter, J. de Physique **45**, supplement 2, 1984, C2-41.

[12] Love, G. and Scott, V. D., Evaluation of a new correction procedure for quantitative electron probe microanalysis, J. Phys. D **11**, 1978, 1369.

[13] Pouchou, J-L. and Pichoir, F., Basic expressions for PAP computation for quantitative EPMA, Proc. 11th Int. Cong. X-Ray Optics and Microanalysis, London, Ontario, J. D. Brown and R. H. Packwood, eds., 1986, 249.

[14] August, H.-J., Razka, R., and Wernisch, J., Calculation and comparison of the backscattering factor R for characteristic x-ray emission, Scanning **10**, 1988, 107.

[15] Czyzewski, Z. and Szymanski, H., Proc. 10th Int. Cong. Electron Microscopy, Vol. 1, Offizon Paul Hartung, Hamburg, 1982, 261.

[16] Czyzewski, Z., Backscattering factor of the ZAF-correction procedure in quantitative electron-probe microanalysis, Phys. Stat. Sol. A, 92, 1985, 563.

[17] Mott, N. F. and Massey, H. S. W., The Theory of Atomic Collisions, 2nd Edition, Oxford University Press, London, 1949, 243.

[18] Powell, C. J., Evaluation of Formulas for Inner-Shell Ionization Cross Sections, NBS Spec. Publ. 460, K. F. J. Heinrich, D. E. Newbury, and H. Yakowitz, eds., 1975, 97.

[19] Fabre de la Ripelle, M., J. Phys. (Paris) **10**, 1949, 319.

THE USE OF TRACER EXPERIMENTS AND MONTE CARLO CALCULATIONS IN THE $\phi(\rho z)$ DETERMINATION FOR ELECTRON PROBE MICROANALYSIS

PETER KARDUCK AND WERNER REHBACH

Gemeinschaftslabor für Elektronenmikroskopie der RWTH
D5100 Aachen, Federal Republic of Germany

I. Introduction

Electron Probe Microanalysis (EPMA) is a well-established technique for the quantitative elemental analysis of solid materials on a microscopic scale. For a reliable quantification of the characteristic x-ray measurements an appropriate set of standards would be necessary, to cover all possible elements and compounds. In practice only a limited set of pure elements or compounds is available. Therefore for a general use of the technique the relation between the emitted x-ray intensity and the weight fraction c_i of the emitting element in the sample should be known for all elements and their possible combinations in compounds. Following the work of Castaing [1] one can write the general form for the emitted primary intensity of element i-radiation:

$$I_i^e = \text{const.} \cdot c_i \cdot \omega_i \cdot Q_i(E_0) \int_0^\infty \phi_i(\rho z) \exp(-\chi_i \rho z) d\rho z \tag{1}$$

Q_i is the ionization cross section for the ith level, c_i the weight fraction of element i, ω_i the fluorescent yield of the ith level, ρz the mass depth and $\chi_i = (\mu/\rho) \cos \theta$ with the absorption coefficient μ/ρ and the take-off angle θ. ϕ_i is the absolute distribution of the characteristic x radiation excited by the electrons in a depth ρz of the target.

By referring this intensity to that of a standard of known compositions the following expression is obtained:

$$k = \frac{I_i^u}{I_i^s} = \frac{c_i^u}{c_i^s} \cdot \frac{\int_0^\infty \phi_i^u(\rho z) \exp(-\chi_i^u \rho z) d\rho z}{\int_0^\infty \phi_i^s(\rho z) \exp(-\chi_i^s \rho z) d\rho z} \tag{2}$$

where u and s denote the unknown sample and the standard, respectively. Possible contributions of secondary excitation such as characteristic and continuum fluorescence are not considered in this expression. Equation (2) contains both the absorption and the so-called atomic number correction. For the case that specimen and standard have the same composition, the quotient of integrals becomes one and the relation between k and c_i^u is linear.

Assuming that $\phi(\rho z)$ is known, eq (2) represents a full and exact correction of matrix effects in electron probe analysis, disregarding fluorescence. Therefore, the knowledge of a generalized formula for $\phi(\rho z)$ is a basic requirement for an exact matrix correction. $\phi(\rho z)$ itself depends on the primary electron energy E_0, the atomic number Z, the sample composition c_i and the excitation energy E_c.

Electron Probe Quantitation, Edited by K.F.J. Heinrich and
D. E. Newbury, Plenum Press, New York, 1991

Physically these dependences are functions of the energy and angular distributions of the primary electrons in each element of depth ρz, and of the energy-dependent ionization cross section $Q_i(E)$ of the corresponding energy level. Since the estimates of distributions of the electron energies and their effective paths $d\rho z/\cos\theta$ in each element of depth $d\rho z$ are very complex many-particle problems, it is not possible to set up a general theoretical expression for the depth distribution $\phi(\rho z)$. To bypass these problems, in the past three different approaches have been used to obtain $\phi(\rho z)$ data:

—Experimental determination of $\phi(\rho z)$ by appropriate tracer techniques,

—Simulation of individual electron trajectories by Monte Carlo techniques and subsequent application of an appropriate ionization cross section for the calculation of the x-ray production,

—Description of the flux of electrons in a sample as well as their scattering and energy losses as a function of depth by a distribution function, which was a numerical solution of the Boltzmann transport equation (Brown et al.) [3].

According to the literature, experimental techniques and Monte Carlo simulations are mostly used to determine $\phi(\rho z)$ systematically. Therefore in the following communication both techniques will be shortly reviewed. Although the use of the Boltzmann equation and its solution for electron transport gave reasonable results for x-ray depth distributions and backscattering [4], because of its complexity this method has not been used widely.

II. Experimental Techniques for the $\phi(\rho z)$ Determination

A. THE TRACER TECHNIQUE

The problem of determining the production of characteristic x rays in EPMA as a function of depth was first treated experimentally by Castaing and Descamps [2]. They used the so-called sandwich sample technique which is schematically described in figure 1. A

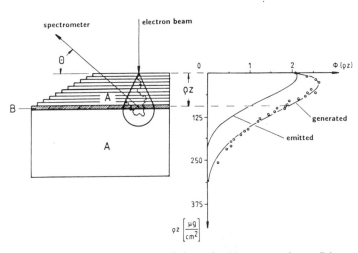

Figure 1. Depth distribution of generated x radiation emitted by a tracer element B in a matrix A.

thin layer of a pure element B is deposited on a polished block of element A. The elements should match closely in atomic number Z and the system should not be affected by characteristic fluorescence. The tracer B is covered again by successively thicker layers of element A, using a technique to obtain a suitable staircase-shaped specimen of different layer thicknesses for each step. The characteristic x radiation of element B emerging from the different depths of the tracer is measured and corrected with exp-$(\chi\rho z)$ for the absorption in element A. By relating these intensities to the intensity emitted from an isolated layer of the same thickness under identical electron bombardment conditions one obtains the absolute depth distribution of x-ray production. Thus the function $\phi(\rho z)\cdot d\rho z$ is defined as the intensity generated in a depth element $\rho z + d\rho z$ divided by the intensity emerging from a self-supporting layer of the same thickness $d\rho z$. It follows that $\phi(0)$, the so-called surface ionization contains the contribution of the backscattered electrons penetrating again through the upper layer of thickness $d\rho z$, added to the intensity $I_0 \propto Q_i(E_0)\,d\rho z$, that is generated by the impinging primary electrons only. As a result $\phi(0)$ is always >1. The tracer layer must be so thin that no elastic scattering of the primary electrons can occur in it. Castaing and Descamps already pointed out that for a thick tracer the path of the electrons through the upper layer increases $d\rho z$ by a factor $1/\cos\theta$ due to scattering. This would result in an increase of I_0 and would affect the whole $\phi(\rho z)$ curve.

From their experimental $\phi(\rho z)$ curves Castaing and Descamps calculated the absorption factors

$$f(\chi)=\frac{F(\chi)}{F(0)}=\frac{\displaystyle\int_0^\infty \phi(\rho z)\exp(-\chi\rho z)d\rho z}{\displaystyle\int_0^\infty \phi(\rho z)d\rho z} \tag{3}$$

for Al, Cu, and Au-matrices and carried out the first absorption correction in EPMA.

Subsequently the sandwich sample technique was adopted by other authors (table 1). Vignes and Dez [5] chose tracer thicknesses for the $\phi(\rho z)$ determination in lead and titanium which increased with higher depth of the tracer. By this they overcame the problem of insufficient count rates due to absorption of the x radiation emerging out of great depths. Some of their curves seem to decrease too slowly with increasing ρz, which can be attributed to the effects of the very thick tracer at great depth.

The tracer technique was also applied, with closely matching tracer and matrix elements, by Castaing and Henoc [6] and by Shimizu et al. [7]. The latter compared the results of Zn $K\alpha$ in a copper matrix measured by the tracer technique with depth distributions obtained by means of the analysis of the angular distribution of x rays emitted from a bulk target. This technique gives the Laplace transform of $\phi(\rho z)$ for variable take-off angles θ_i:

$$F(\chi_i)=\int_0^\infty \phi(\rho z)\exp\{-\chi_i\rho z\}dz. \tag{4}$$

As a solution of eq (4) the depth distribution was computed for angular distributions at 25, 30, and 35 keV by a slope correction method [7]. The authors preferred the analysis of angular distributions to the tracer method mainly because the angular distribution technique avoids experimental difficulties in preparing tracers with a homogeneous thickness. On the other hand a complex numerical evaluation is necessary to determine $\phi(\rho z)$ from the $F(\chi)$ distribution. Nowadays this technique should be given more attention since it enables the measurement of $\phi(\rho z)$ for compounds including light and ultra light elements, without the

Table 1. Summary of measured $\phi(\rho z)$ curves

Authors	Electron energy (keV)	Matrix	Tracer, radiation	Reference
Castaing & Descamps	29 29 29	Al Cu Au	Cu $K\alpha$ Zn $K\alpha$ Cr $K\alpha$ Bi $K\alpha$ Zn $K\alpha$ Bi $L\alpha$	[2]
Castaing & Henoc	10,15,20,25,29	Al	Mg $K\alpha$	[6]
Brown	13.4,18.2,23.1,27.6	Cu	Zn $K\alpha$ Ni $K\alpha$ Co $K\alpha$ Fe $K\alpha$	[9]
Shimizu et al.	14.2–37	Cu	Zn $K\alpha$	[7]
Vignes & Dez	17,20,24,29,35 17,20,24,29,35 29,33	Ti Ni Pb	V $K\alpha$ Cu $K\alpha$ Bi $L\alpha$	[5]
Schmitz et al.	20,25,5,30	Ni	Cu $K\alpha$	[12]
Büchner	10–50	Al Co Ni Pd	Ag $K\alpha$ Ag $L\alpha$ Cu $K\alpha$ Ni $K\alpha$ Cu $K\alpha$ Ag $K\alpha$ Ag $L\alpha$	[13]
Brown & Parobek	15,20,25,30	Al Cu Ag Au	Si $K\alpha$ Cd $L\alpha$ Zn $K\alpha$ Bi $M\alpha$ Bi $L\alpha$	[10]
Parobek & Brown	6,8,10 12,15	Al Ni Ag Au Al Ni Ag	Si $K\alpha$ Cu $K\alpha$	[11]
Sewell et al.	10,15,20,25 20,25,29 5,10,15	Al Au C	Si $K\alpha$ Bi $L\alpha$ Al $K\alpha$	[14]
Rehbach	5,7,10,12,15 5,7,10,12,15 4,5,7,10,12,15 4,5,7,10,12,15 4,5,7,10,12,15 7,10,12 5,7,10,12,15 5,7,10,12,15	Ti Ag Sb Al Cu Au Cr Ni	B $K\alpha$ C $K\alpha$ Cu $L\alpha$ B $K\alpha$ C $K\alpha$ Cu $L\alpha$ B $K\alpha$ C $K\alpha$ Cr $L\alpha$ Cu $L\alpha$ Cr $L\alpha$ Cr $L\alpha$ Cu $L\alpha$ Cu $L\alpha$	[35]

experimental difficulties related to specimen preparation. In this context it should be noted that a new $F(\chi)$-apparatus is introduced by Small et al. in this volume [8].

The applicability of the tracer technique was further extended by Brown [9], Brown and Parobek [10], and Parobek and Brown [11] to the measurement of the atomic number effect by combining elements with widely varying atomic number Z for the tracer and the matrix, respectively (table 1). If tracers of identical thickness of one element Z are combined with matrices having different atomic numbers, the measured $\phi(\rho z)$ curves show differences in the shape of the $\phi(\rho z)$ curves and also in the magnitude of the integral over $\phi(\rho z)$ (total intensity generated) for different matrix elements. These changes could be attributed qualitatively to differences in backscattering and stopping power. The tracer

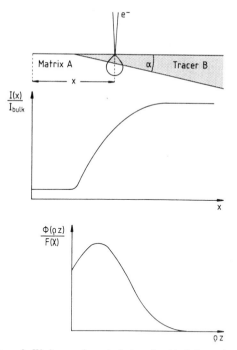

Figure 2. Wedge specimen technique for $\phi(\rho z)$ determination.

experiments simulate the depth distribution of x rays emitted by an element B (tracer), which is diluted to a small concentration in an element A (matrix). Recent attempts to determine $\phi(\rho z)$ for an additional set of tracer and matrix elements including carbon, aluminum, and gold were carried out again with the tracer technique by Sewell et al. [14]. Their results can be considered reliable, as indicated by previously published results of Castaing and Descamps [2] and Castaing and Henoc [6]. But a comparison of the experimental results of Sewell et al. [14] with Monte Carlo calculations of the same $\phi(\rho z)$ curves by Love et al. [17] and Myklebust et al. [18] pointed out large discrepancies in the absolute magnitude of $\phi(\rho z)$ and as well as in the shape.

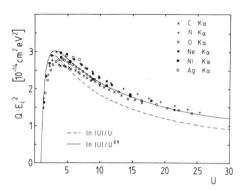

Figure 3. Ionization cross section.

B. THE WEDGE SPECIMEN TECHNIQUE

A slightly different approach was proposed by Schmitz et al. [12] for copper. They used a shallow wedge-shaped tracer ($a \approx 3$–$5°$), deposited on a matrix of an atomic number lower than that of the tracer ($\Delta Z = 1$) to avoid characteristic fluorescence. The characteristic x-ray intensity of the tracer element is measured along a line across the wedge as a function of distance x, giving the integral of $\phi(\rho z)$ as a function of the wedge thickness ρz. Differentiation of the integral curve yields a normalized depth distribution $\phi(\rho z)/F(\chi)$, as illustrated schematically in figure 2. The method has also been applied by Büchner [13].

The advantages of the wedge method are:

—to get the possible use of alloys as tracer,

—the fairly good determination of the tracer thickness by means of the precisely known wedge angle.

The major disadvantages are the restriction to tracer and matrix elements of similar Z, and the difficulties in the accurate determination of $\phi(0)$. This becomes obvious for low primary energies and high atomic number of the material, where the maximum of $\phi(\rho z)$ can be in depths of about 50 μg/cm^2, within a region of uncertain thickness especially at the wedge tip.

C. TRACER EXPERIMENTS FOR $\phi(\rho z)$ OF SOFT X RAYS

The tracer experiments reported so far have Si $K\alpha$ as the lowest x-ray energy, and the minimum primary energy E_0 was 6 keV (table 1). For an investigation of the depth distribution of soft x rays the $K\alpha$ lines of boron to oxygen or the L lines of titanium to copper must be considered as tracer radiations. Due to the high absorption of these lines in almost all matrix materials and the low x-ray yield of soft radiation and the rather poor efficiency of spectrometers, accurate intensity measurements can only succeed for absorption coefficients of the radiation in the matrix of much less than 10,000 cm^2/g. The maximum depth from which the x rays are detectable is limited in these cases to 200–300 μg/cm^2. As a consequence only incomplete $\phi(\rho z)$ curves can be measured for E_0 greater than 15 keV since the depth of emergence increases beyond the maximum depth of detectability.

All these negative factors require a very accurate specimen preparation and microprobe measurements with long counting times.

Experimental

Karduck and Rehbach recently reported new tracer experiments with soft $K\alpha$ and $L\alpha$ lines as tracer radiations [41,44]. In their work a single polished block of matrix material with an evaporated tracer layer on it (thickness 2.4–5.8 μg/cm^2) was covered by 48 matrix layers of different mass thicknesses within 13 individual evaporations. Details of the technique are described in ref. [35]. The layer thicknesses were controlled during the evaporation by a quartz crystal device. At the same time thin glasses of pre-determined area and weight were evaporated with the total of all layers. Their thickness $D = \Sigma d_i$ could be determined by weighing and also by atomic absorption spectrometry. Simultaneously the single layers were deposited side by side on a sheet of vitreous carbon. Their relative thicknesses could be measured by the electron microprobe, which gives several equations $d_i/d_{i+1} = C_j$. This system of equations can be solved easily and yields the individual thick-

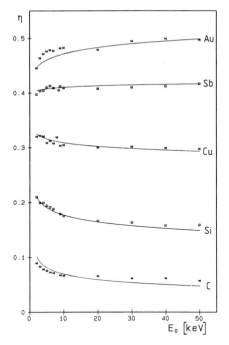

Figure 4. Backscatter coefficients by Monte Carlo calculation, (—) experimental results by Hunger.

nesses d_i. Simultaneously the results are examined by the values obtained by the quartz crystal device. Thus the thicknesses could be determined within deviations of less than 5%. Besides the several methods to determine and to control the layer thicknesses the measured $\phi(\rho z)$ curve itself is a sensitive indicator of errors in the thickness determination. As an example, figure 6 gives $\phi(\rho z)$ curves for Cu $L\alpha$ in silver with values for the generated x rays. The experimental points are coupled in groups of seven points, which belong to one specific column of layers on the sandwich sample. Actually these fluctuations are very smooth in the measured curve, i.e., the errors in the thickness determination are small but systematic. After the absorption correction they become more obvious. Results of these measurements of Rehbach [35] are discussed later in this paper.

III. $\phi(\rho z)$ Determination by the Monte Carlo Simulation of Electron Trajectories

The Monte Carlo method simulates theoretically three-dimensional electron trajectories of parallel incident primary electrons in a sample. The solid target is treated as a body with a density ρ and randomly distributed atoms. Each electron trajectory itself is composed of single straight line segments, i.e., the mean free paths of an electron between two collisions. The orientation of each segment is determined by using random numbers to select the particular scattering angle for each segment (or the energy loss after an inelastic collision) according to equations describing the appropriate physical processes. Since experimental measurements mostly yield quantities that represent an integral value over many single events, in a Monte Carlo simulation many individual trajectories must be calculated to obtain reasonable results with statistical reliability.

There exists a large body of literature dealing with Monte Carlo simulation of electron scattering in solids for several purposes including x-ray generation. It is not the intention of this paper to review the methodology in that field; rather we refer to the special literature in which the development of the last 20 years is summarized [18,20].

Surprisingly the examination of the literature shows that just a few investigations were aimed at the systematic collection of x-ray data for EPMA. Most of the research work reported in literature deals with the details of approximation to the physical scattering processes. The results are usually examined by comparison with experimental electron backscatter or transmission data, with $f(\chi)$ data and to a minor extent with $\phi(\rho z)$ data.

A systematic investigation of $\phi(\rho z)$ depth distributions was carried out by Love et al. [17] using a simple Monte Carlo method according to Curgenven and Duncumb [21]. The program contains an adjustable maximum impact parameter for the scattering processes, which is adjusted to yield correct backscatter coefficients. The model was used to generate $\phi(\rho z)$ curves for a wide range of experimental conditions. From these data the authors deduced an empirical expression for the mean depth of x-ray generation ρz, which they then used for their absorption correction models [22,15]. The absolute $\phi(\rho z)$ data presented by Love et al. [17] and by Sewell et al. [14] in most cases are in unsatisfactory agreement with experimental results. A systematic comparison with $\phi(\rho z)$ available at that time was not carried out by the authors.

A more sophisticated Monte Carlo method has been proposed by Shimizu et al. [24] for calculating correction factors for the absorption and atomic number effects. A single scattering model with a screened Rutherford cross section is used. The results were not very satisfactory, which led the authors to an improvement of their Monte Carlo approach by adjusting the screening parameter and the mean ionization potential so as to fit experimental data. In a recent more accurate approach Murata et al. [23] used Mott cross sections instead of the screened Rutherford cross section for elastic scattering and replaced the Bethe energy loss equation for electron energies $E < 6.338\ J$ by a modification of Rao Sahib and Wittry [31] which will be discussed later in detail. With this modification reasonable $\phi(\rho z)$ results, even for low E_0 and high atomic number Z (gold), were obtained. Although the approach is more accurate in describing the physical processes of scattering and energy loss there are still discrepancies between the experiments and the Monte Carlo results which could not be explained convincingly by the authors due to the lack of a more extended set of Monte Carlo data.

Reimer and Krefting [25] proposed several extensions of the techniques described so far. Their program has been adopted and slightly modified by Karduck and Rehbach [26] to investigate the $\phi(\rho z)$ depth distribution of x-ray energies of less than 1 keV. A recent version of the Reimer-Krefting program has been published by Reimer and Stelter [27]. The program takes into account the following:

—For single elastic electron scattering with angles $> 10°$ exact Mott cross sections are used according to Reimer and Lödding [24a]. Reimer and Krefting demonstrated the importance of exact cross sections for elastic scattering by their accurate calculations of energy and angular distributions of backscatter electrons and of the backscatter coefficient η with increasing film thickness.

—For angles $< 10°$ multiple elastic small angle scattering according to Lewis [28] is applied with a mean angular deviation after each half mean free path.

—Single inelastic electron-electron scattering is considered according to the classical theory of Gryzinski [29], including the production of fast secondary electrons (FSE) with energies above 200 eV. The FSE are also taken into account for the calculation of characteristic x rays. The FSE are simulated up to the 3rd generation.

—The energy loss is calculated by the Bethe formula [30] from which the losses by single inelastic scattering are subtracted:

$$\frac{dE}{dS} = \left(\frac{dE}{dS}\right)_{Bethe} - \left(\frac{dE}{dS}\right)_{Single}. \tag{5}$$

—In calculating the mean free path and for the decision whether an elastic or inelastic scattering takes place random numbers are also used.

—For low energies ($E < 6.338\ J$) the energy loss is calculated following Rao-Sahib and Wittry [31].

$$\frac{dE}{dS} = -\frac{62360 \cdot Z \cdot \rho}{A \cdot \sqrt{E} \cdot \sqrt{J}} \tag{6}$$

to avoid an "energy gain" for $E \to 0$ as would result from the Bethe formula.

—A very important aspect of the calculation of the x-ray production is the choice of an appropriate ionization cross section. For the program the authors prefer the expression of Bethe [30] modified by Hutchins [32]:

$$Q\,E_c^2 \propto \frac{\ln U}{U^m}. \tag{7}$$

The exponent m is added by Hutchins to give a better fit to experimental values. Following a comparison with experimental ionization cross sections for $K\alpha$ radiation collected by Powell [33,34] m is set at 0.9 (fig. 3).

To prove the reliability of the Monte Carlo program, two tests were made. First, backscatter coefficients η were calculated for several E_0 and Z. Very good agreement with experimental values of Hunger [39] (fig. 4) was found, even in the range of very low energies. Second, the energy dependence of emitted x-ray intensities on the primary electron energy was determined by Monte Carlo calculations as well as by experiments. The relative intensities calculated by Monte Carlo are in very good agreement with the measured ones for a large range of E_0, E_c, and Z (figs. 5a–d). It should be mentioned that the agreement in the backscatter coefficients as well as the intensities, was achieved without adjusting any parameter in the Monte Carlo program.

IV. Recent Results

A. INFLUENCE OF THE TRACER THICKNESS IN EXPERIMENTAL $\phi(\rho z)$ CURVES

The tracer thickness can strongly influence the shape of $\phi(\rho z)$ curves as Castaing and Descamps [2] already pointed out in their early work. This is demonstrated quantitatively in figure 7 with chromium tracers of 4 and 12 $\mu g/cm^2$ thickness respectively, in an aluminum matrix at 10 keV and 12 keV. For the thick tracer the value of the maximum ϕ is reduced by nearly 10 percent compared to the curve obtained with a thin tracer, whereas the value of $\phi(0)$ stays constant or even increases. The reason for that effect is that owing to electron

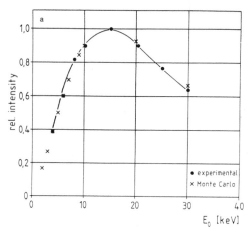

Figure 5a. Relative C $K\alpha$-intensity emitted from pure vitreous carbon, (●) experimental results, (×) Monte Carlo results.

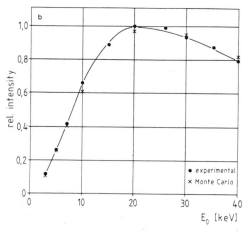

Figure 5b. Relative Cu $L\alpha$-intensity emitted from pure copper, (●) experimental results, (×) Monte Carlo results.

scattering in the tracer the intensity of the isolated tracer layer increases the value $I_0 \propto Q$ (E_0) dρz by a factor of $1/<\cos \omega>$, with $<\cos \omega>$ being the mean angular deflection of the incident electrons in the tracer. By relating all other intensities from greater depth to that intensity, the curve is compressed to lower values. A thick tracer also complicates the accurate definition of the position $\rho z = 0$, which for infinitesimally thin tracers can be set to half the thickness of the tracer.

B.   THE SURFACE IONIZATION $\phi(0)$

For the correction of EPMA measurements in highly absorbing systems especially the surface ionization $\phi(0)$ and part of the $\phi(\rho z)$ curves close to the surface are very important and therefore must be described exactly. The commonly used empirical expressions for $\phi(0)$ are proposed by Love et al. [37] and Reuter [38], respectively. Love et al. used their own

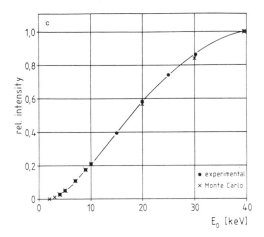

Figure 5c. Relative Si $K\alpha$-intensity emitted from silicon, ($\bullet$) experimental results, ($\times$) Monte Carlo results.

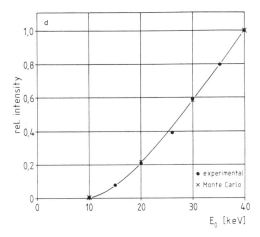

Figure 5d. Relative Cu $K\alpha$-intensity emitted from pure copper, ($\bullet$) experimental results, ($\times$) Monte Carlo results.

Monte Carlo values [22] as a basis for their approach. Reuter obtained his expression by fitting experimental $\phi(0)$ data which were obtained by measuring x-ray intensities of a tracer on several substrates with different atomic numbers. The intensity of the isolated layer equals the extrapolated intensity at $Z=0$. Both empirical equations are based on $\phi(0)$ values for x-ray energies above 1 keV.

Rehbach and Karduck [36] recently presented a new empirical approach which has been fitted to experimental $\phi(0)$ values for soft x rays and also to Monte Carlo results over the total range of x rays used in EPMA. The major aspects of the new approach and significant differences from the other models can be summarized as follows (figs. 8a–d):

In its dependence on the overvoltage ratio $U_0=E_0/E_c$, $\phi(0)$ increases less rapidly for $U_0<20$ than predicted. But in the range $U_0>20$ the increase of $\phi(0)$ continues slightly and is not independent of U_0.

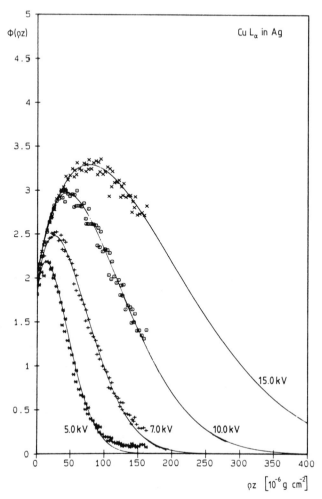

Figure 6. Influence of inaccuracy in thickness determination on experimental $\phi(\rho z)$, Cu $L\alpha$ in silver.

Also, the curvature and the absolute values of $\phi(0)$ depend explicitly on the excitation energy E_c. This becomes evident particularly for B $K\alpha$ and high atomic numbers of the matrix in figure 8a. The very high surface ionization of soft x rays can be explained physically by an increasing contribution of fast secondary electrons (FSE) to the total amount of ionization with decreasing E_c. By means of a detailed observation of the FSE in the Monte Carlo program up to the 3rd generation this contribution could be separated from the ionization by primary electrons. Figure 9 gives an example for B $K\alpha$ in gold which shows clearly that the fraction exceeding the Reuter curve can be attributed to the FSE. The new approach is based on the Reuter equation

$$\phi(0) = 1 + (1 - 0.9/U_0) \cdot 2.8 \, \eta \tag{8}$$

in good agreement with data for $E_c > 1$ keV and medium overvoltages. To consider the changed curvature and the separate E_c dependence, $\phi(0)$ is assumed to be

$$\phi(0)=1+(1-1/\sqrt{U_0})^a \cdot b. \tag{9}$$

The exponent a considers the E_c and a slight Z dependence and can be described by

$$a=\left(1+0.005 \cdot \frac{Z}{E_c}\right) \cdot (0.68+3.7/Z), \ E_c \ \text{in keV}. \tag{10}$$

The parameter b contains the backscatter properties and the x-ray production by FSE:

$$b=(1+0.05/E_c) \cdot b_\infty, \qquad E_c \ \text{in keV} \tag{11}$$

with $b_\infty=-0.01+0.04805 \cdot Z-0.51599 \cdot 10^{-3}Z^2+0.20802 \cdot 10^{-5}Z^3$.

The excellent fit of the experimental and MC data is shown in figures 8a–d by the fully drawn curves. For all cases of ultra soft and soft x rays in low-Z or high-Z material, respectively, the data are fitted better by the new approach than by those proposed by Reuter or by Lowe et al. For x rays of higher energy, e.g., Cu $K\alpha$, the established expressions differ only slightly and the values of the new formula are within that range [35].

The results also show the good agreement between the experimental data and the Monte Carlo values over a wide range of boundary conditions. This is attained without manipulating the Monte Carlo results by any adjustable parameter.

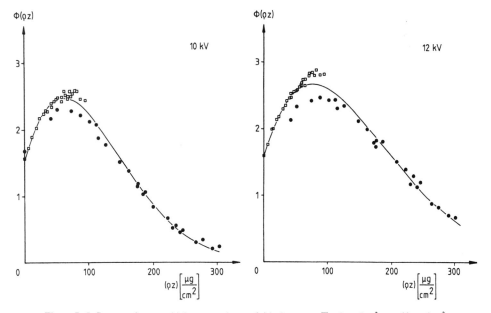

Figure 7. Influence of tracer thickness on shape of $\phi(\rho z)$ curves. $\square=4$ $\mu g/cm^2$, $\bullet=12$ $\mu g/cm^2$.

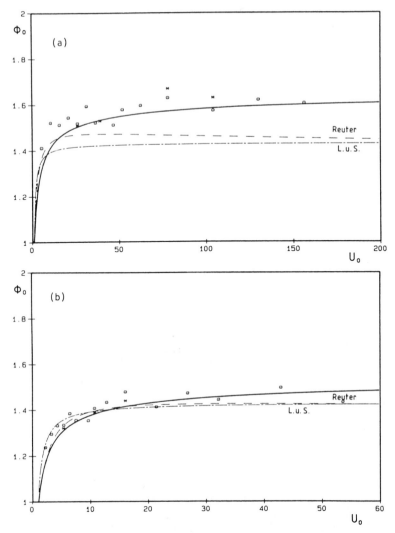

Figure 8a, b. Comparison of the new formula for $\phi(0)$ with the formulation by Reuter and by Love and Scott respectively. Experimental values (*) and MC values (○). (a) B $K\alpha$ and (b) Cu $L\alpha$ in aluminum.

C. THE DEPTH DISTRIBUTION OF X-RAY GENERATION $\phi(\rho z)$

Recent Models for $\phi(\rho z)$ in Correction Procedures

In the recent development of the correction procedure for quantitative EPMA two different ways of curve fitting procedures have been proposed which are especially conceived for light element analysis. The first approach was presented by Packwood and Brown [40], who fitted a modified Gaussian expression for $\phi(\rho z)$ to a series of experimental $\phi(\rho z)$ curves. The same authors proved the following general expression to describe actual $\phi(\rho z)$ curves accurately over a wide range of boundary conditions:

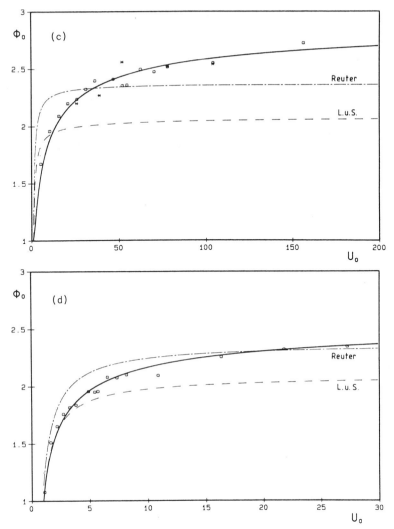

Figure 8c, d. Comparison of the new formula for $\phi(0)$ with the formulation by Reuter and by Love and Scott respectively. Experimental values (*) and MC values (o). (c) B $K\alpha$ and (d) Si $K\alpha$ in gold.

$$\phi(\rho z) = \gamma_0 \left[1 - \frac{\gamma_0 - \phi_0}{\gamma_0} \exp\{-\beta\rho z\} \right] \exp\{-\alpha^2(\rho z)^2\}. \tag{12}$$

For the parameters γ_0, β, and α, Packwood and Brown suggested expressions which include the dependence on E_0, E_c, Z and the ionization potential J, and which were deduced by simple theoretical considerations and by fitting them to the experimental curve parameters. Equation (12) describes the $\phi(\rho z)$ curve by a Gaussian profile centered at the target surface and modifies the initial shape of the Gaussian near the surface by a transient exponential function. This function is able to approach the actual shape of $\phi(\rho z)$ curves very accurately for a wide range of E_0, E_c and target compositions [40,41,44].

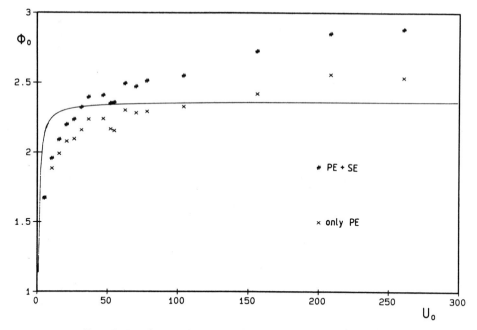

Figure 9. Contribution of FSE to $\phi(0)$, B $K\alpha$ in gold, (—) $\phi(0)$ by Reuter.

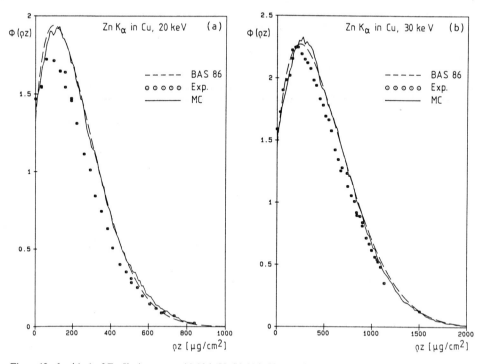

Figure 10a, b. $\phi(\rho z)$ of Zn $K\alpha$ in copper, (a) 20 keV, (b) 30 keV, experimental values of Brown and Parobek [10].

The second correction model conceived to describe the real shape and magnitude of the $\phi(\rho z)$ curve as accurately as possible was proposed by Pouchou and Pichoir [42,43] and called PAP. The authors used a simple and efficient mathematical description of the $\phi(\rho z)$ curve by two parabolic branches that are continuous and differentiable. For a reliable atomic number correction in PAP, the integral of $\phi(\rho z)$ is set equal to the real intensity, i.e., $I/Q(E_0)$. The shape of the $\phi(\rho z)$ function is fixed by the surface ionization $\phi(0)$, the location of the maximum ϕ_{max}, and the ultimate depth of ionization. All parameters of the PAP approach are determined by fitting to an extended set of intensity measurements at varying E_0, and by a comparison with Monte Carlo calculations to form the shape of the curves [45].

The Gaussian $\phi(\rho z)$ method of Packwood and Brown is based on measured $\phi(\rho z)$ curves for $E_0 > 6$ keV. Thus for the correction of light element or soft x-ray analysis the model refers to extrapolations from its basic assumptions. Bastin et al. [46,47] have taken into account this disadvantage of the Gaussian $\phi(\rho z)$ model and have optimized the parameterization of Packwood and Brown by changing the parameters α, β, and γ_0 in eq (12) by a trial and error procedure to obtain an optimum correction of k values from a set of binary carbides and borides measured by the authors themselves. Nevertheless this procedure gives no clear indication either about how accurately and reliably the real $\phi(\rho z)$ distributions are described by this model, especially for soft and ultra soft x rays.

Experimental and Monte Carlo Results $\phi(\rho z)$ for Soft X Rays

The authors of the present paper have carried out systematic experimental and Monte Carlo investigations of $\phi(\rho z)$ depth distributions of x rays with energies below 1 keV to

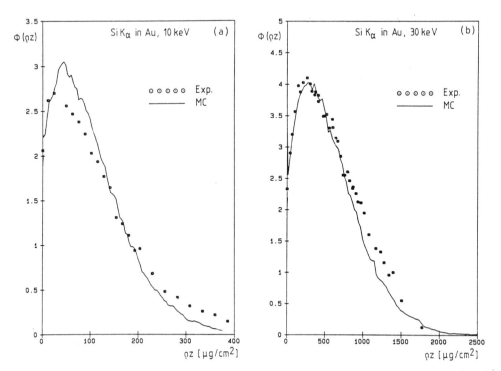

Figure 11a, b. $\phi(\rho z)$ of Si $K\alpha$ in gold, (a) 10 keV, (b) 30 keV, experimental values of Parobek and Brown [10,11].

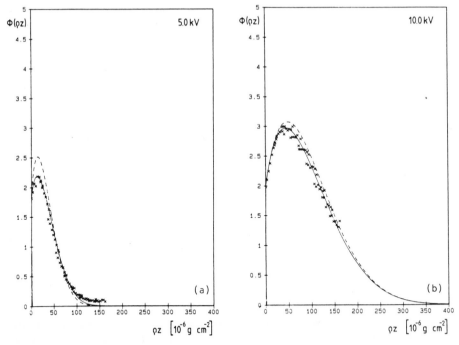

Figure 12a, b. $\phi(\rho z)$ of Cu $L\alpha$ in silver ($\times$) experimental, (– – –) Monte Carlo, (a) 5 keV, (b) 10 keV.

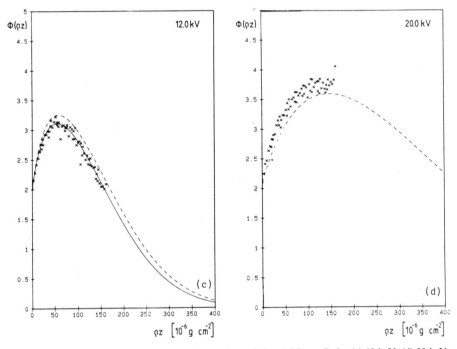

Figure 12c, d. $\phi(\rho z)$ of Cu $L\alpha$ in silver ($\times$) experimental, (– – –) Monte Carlo, (c) 12 keV, (d) 20 keV.

elucidate this very important range of EPMA application further [26,35,41,44]. It could be demonstrated that the experimental $\phi(\rho z)$ curves of soft x rays describe the expected physical behavior of the x-ray production in depth as a function of the parameters E_0, E_c, and Z [44]. Since both experimental and Monte Carlo data are available the Monte Carlo data can be proved concerning their agreement with the experimental $\phi(\rho z)$ curves. As a result the Monte Carlo program described in chapter III is able to calculate $\phi(\rho z)$ curves with an excellent agreement to experimental data over the whole range of EPMA application [26]. This can be demonstrated by some examples in figures 10–13. In all curves the depth of maximum ionization and also the absolute values of $\phi(\rho z)$ are described by the MC data with very good agreement to the experimental ones. Exceptions to the agreement are observed when relative low primary electron energies are applied; i.e., in figures 10a, 11a, 12a, and 13a the MC curves are systematically higher than the corresponding experimental ones. Recently similar results have been reported by Sewell et al. [14], who attributed this effect to an overestimation by their simplified Monte Carlo program. However, the result of the observed tracer influence on the shape of $\phi(\rho z)$ curves (ch. IVA) suggests a severe error in experimental curves at low E_0, when the tracer thickness is too high compared to the depth of maximum x-ray generation. Indeed the experimental curves in figures 10a and 11a have been obtained with a tracer thickness of 20–30 $\mu g/cm^2$, but the maxima of the curves are at 24 and 80 $\mu g/cm^2$, respectively. This confirms that significant scattering of the electrons takes place in the tracer, which is contrary to the definition of the appropriate tracer thickness. The same holds for the soft x-ray experiments of the present work with very thin tracers but low primary energy E_0 (figs. 12a and 13a). As a result, in all cases of thick tracers and/or low E_0 severe distortions of the experimental $\phi(\rho z)$ curves are ob-

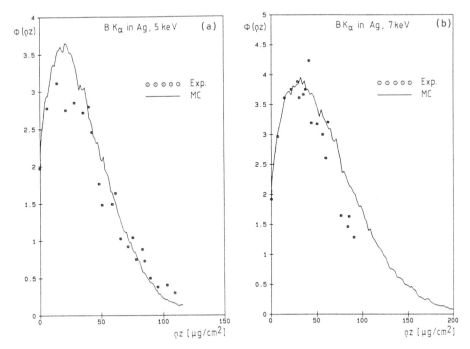

Figure 13a, b. $\phi(\rho z)$ of B $K\alpha$ in silver, (a) 5 keV, (b) 7 keV.

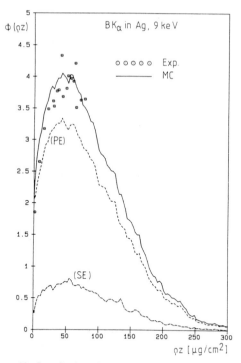

Figure 14. Contribution of FSE to $\phi(\rho z)$, B $K\alpha$ in silver, 9 keV.

served. For that reason, even for low E_0 the MC data can be regarded as reliable, particularly since the exact Mott cross sections for elastic electron scattering used in the program make it also valid in this energy region.

A further reason for the good performance of the MC model in predicting $\phi(\rho z)$ curves of soft x rays is the consideration of fast secondary electrons (FSE), which contribute a noticeable amount of ionizations to the total $\phi(\rho z)$ especially for low E_c. An example is given in figure 14 for B $K\alpha$ in silver at 9 keV. The FSE generate about 19 percent of the totally generated x-ray intensity. Without the consideration of these FSE contributions the MC model would fail for excitation energies below 1 keV. This effect is usually neglected in the literature and has been first discussed by Reimer and Krefting [25].

Limits of the Gaussian $\phi(\rho z)$ Approach

Several authors have reported an excellent fit of the Gaussian approach given in eq (12) to experimental $\phi(\rho z)$ data. This holds also for nearly all $\phi(\rho z)$ curves of soft x rays [41,44]. Examples are given in figures 6, 7, and 12 by the solid lines, which are a least squares fit of eq (12) to the experimental and Monte Carlo data, respectively.

But Packwood and Brown [40] have already pointed out a few exceptions in an aluminum matrix at high overvoltage ratios, where for eq (12) anomalous values for γ_0 and β are produced. Indeed there exists a limit of validity of the Gaussian approach to describe $\phi(\rho z)$ curves for matrix materials with a low atomic number. The reason for the invalidity

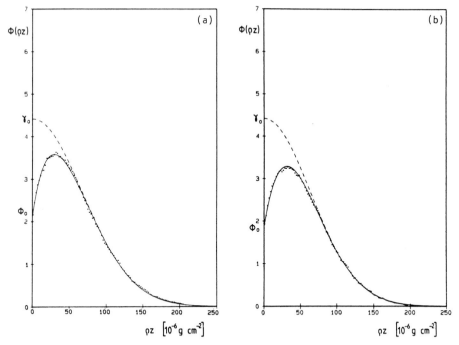

Figure 15a, b. $\phi(\rho z)$ of C $K\alpha$ by Monte Carlo in (a) silver, (b) copper, (– – –) pure Gaussian.

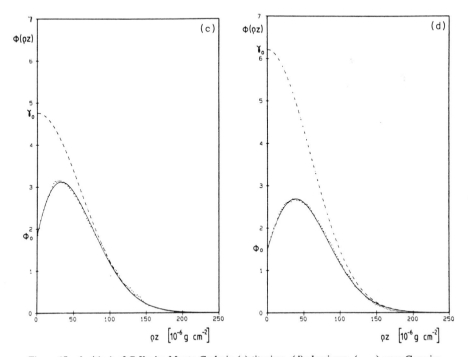

Figure 15c, d. $\phi(\rho z)$ of C $K\alpha$ by Monte Carlo in (c) titanium, (d) aluminum, (– – –) pure Gaussian.

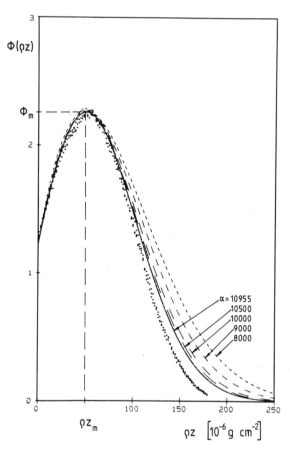

Figure 16. Depth distribution of C $K\alpha$ in carbon; $E_0=7$ keV; MC values (points) and Gaussian $\phi(\rho z)$ curves for different values of α.

can be explained qualitatively by figures 15a–d: the solid lines represent the results of the fitted Gaussian expressions. The dashed lines show the pure Gaussian curves, that describe the depth distributions for the case that the electrons follow a random walk beginning at the sample surface. As can be seen curves (a) and (b) behave according to the theoretical predictions, that near the surface the electron randomization is controlled by the exponential with β and that the tail of the $\phi(\rho z)$ curves is completely described by the pure Gaussian ("random walk") [40]. For titanium in curve (c) the randomization is reached in the lower part of the curve, but for aluminum in curve (d) a range of complete randomization is not reached in depth important for the emitted x rays. Although this curve can also be described fairly well by the Gaussian approach the values for γ_0 and β become unrealistic for the aluminum matrices. For a carbon matrix a realistic value for α could not be found. Figure 16 shows that by increasing the value for α and keeping the maximum ϕ_{max} of $\phi(\rho z)$ fixed, one can only approximate the measured depth distribution by the calculated curve. But with $\alpha=10955$ an upper limit is reached, where γ_0 tends to infinity and β becomes zero. This problem becomes more evident with figure 17. For a fixed value of α, evaluated from the slope of $\ln(\phi)$ versus $(\rho z)^2$ for large ρz, the positions of ϕ_{max} for several values of γ_0 and

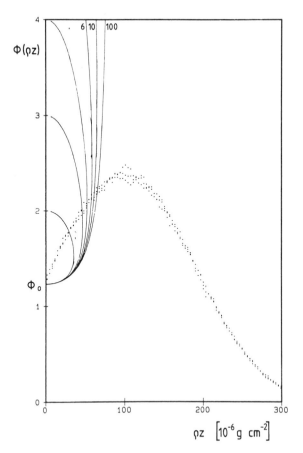

Figure 17a. $\phi(\rho z)$ of C $K\alpha$ in carbon, 10 keV, by Monte Carlo calculation (points), (—) possible positions of ϕ_{max} by variation of γ_0 and β.

varying β are calculated. As a result the $\phi(\rho z)$ curve of C $K\alpha$ in carbon at 10 keV can not be described properly by the Gaussian $\phi(\rho z)$ approach. As a consequence the range of invalidity is given by $Z \leqslant 13$ and $E_c < 1$ keV. From these facts it can easily be concluded, that the discrepancy between the actual $\phi(\rho z)$ curve and the appropriate Gaussian curve in figure 17 will become more serious when the primary electron energy is increased further to 15 or 20 keV, which is quite in the range of practical light element EPMA work.

Comparison with the PAP Model

Pouchou and Pichoir claimed that their $\phi(\rho z)$ model describes the shape and the magnitude of $\phi(\rho z)$ accurately [42,43]. Furthermore their mathematical approach of $\phi(\rho z)$ by two parabolic branches is very flexible in describing also the extreme depth distribution of soft x rays in light element matrices, where the positions of ϕ_{max} are at relatively large depth compared to the total x-ray range. For comparison a $\phi(\rho z)$ curve of C $K\alpha$ in carbon at 10 keV, calculated by the PAP equations, is plotted in figure 17b together with the present Monte Carlo results. Indeed the PAP approach is in very good agreement with the Monte

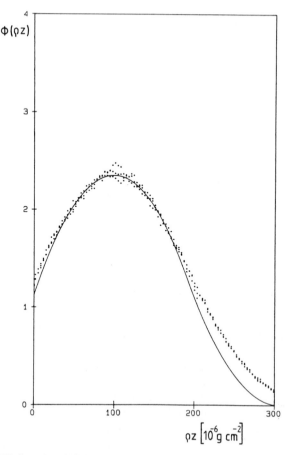

Figure 17b. $\phi(\rho z)$ of C $K\alpha$ in carbon, 10 keV by Monte Carlo calculation (points), (- - -) $\phi(\rho z)$ according to PAP.

Carlo result of the present work concerning the shape as well as the absolute values of $\phi(\rho z)$. So for the range of light matrix materials and high primary electron energies it can be concluded that the PAP model is fully capable of describing $\phi(\rho z)$.

V. Conclusions and Future Prospects

Tracer experiments were of great importance in the development of modern matrix correction procedures, although the experimental techniques are very complicated and susceptible to errors. That is why a long period of time has passed since the first application of the technique to its extension to the research of soft x-ray depth distributions. Most of the correction procedures frequently used in practice are so-called curved fitting methods. These methods require a complete set of reliable $\phi(\rho z)$ data for an accurate generalization of the expression for $\phi(\rho z)$. Experimental $\phi(\rho z)$ curves produced up to the seventies were the basis for establishing the Gaussian $\phi(\rho z)$ approach, which its authors claimed describes the real $\phi(\rho z)$ curves over the whole range of EPMA application [40]. The remaining difficulty was that up to this time no data for soft x rays were available; thus the model represented an extrapolation for the range of soft x ray analyses. Within the framework of the present investigation it has been attempted to measure $\phi(\rho z)$ for soft x ray lines with energies below 1 keV by the tracer technique. The investigation showed that the tracer technique is suitable for investigating $\phi(\rho z)$ of soft x rays. But there are several severe restrictions to the applicability, mostly in highly absorbing matrix materials, difficulties in the thickness determination of the very thin layers necessary and due to other experimental problems. Furthermore the technique suffers from still unreliable mass absorption coefficients for soft x rays needed to transform the measured into generated intensities. Last but not least the measurement of $\phi(\rho z)$ data for alloys remains a legitimate but at the moment unrealizable requirement. Thus, at the moment an improvement of correction procedures by means of further experimental $\phi(\rho z)$ measurement is not to be expected. Nevertheless the present results served as a basis for the assessment of a new Monte Carlo approach in calculating $\phi(\rho z)$ for soft x rays [25,26]. This Monte Carlo method provides $\phi(\rho z)$ curves in excellent agreement with the experimental curves. A detailed comparison of experimental $\phi(\rho z)$ curves with the Monte Carlo results led to the conclusion that several experimental curves of previous work and also of the present investigation are distorted by the use of too thick tracers. Curves measured at low electron energies and with thick tracers are not suitable for optimization purposes. The Monte Carlo and tracer investigations reported here make possible the proposal of a new consistent expression for the surface ionization $\phi(0)$, which is valid for the whole range of x-ray lines and overvoltage ratios [36]. The expression is more accurate than the previous formulations of Reuter [38] and Love and Scott [37].

An attempt also to improve the Gaussian $\phi(\rho z)$ expression of Packwood and Brown [40] by means of the new $\phi(\rho z)$ data was made by Rehbach and Karduck [36] but above all it revealed serious limitations of the validity of this approach for matrix materials with atomic numbers lower than 13. It could be demonstrated for light matrices with $Z \leqslant 13$, that a complete randomization of the electron diffusion as postulated by Packwood and Brown is not probable and the resulting Gaussian shape of at least the tail of the curves is not observed for soft x radiation. This result can have severe consequences for the correction of quantitative analysis of technically important compounds such as borides, carbides, nitrides and oxides of elements with $Z \leqslant 22$. Nevertheless an improved $\phi(\rho z)$ approach by Bastin et al. [46,47] provides good accuracy for the quantitative analysis of light elements. The reason is that $\phi(\rho z)$ is described by an optimized compromise. It seems that the deviation of Bastin's approach from the real $\phi(\rho z)$ curves does not affect the result of the correction severely since it is well known that $f(\chi)$ is rather insensitive to slight changes in $\phi(\rho z)$. A greater influence of deviations in the $\phi(\rho z)$ description is to be expected when a deviating parameterization is used to determine thin layer thicknesses or compositions. But precisely this application of $\phi(\rho z)$ approaches is developed and propagated in recent publications [48,50].

A further development and improvement of correction procedures by $\phi(\rho z)$ investigations can be attained by models that describe the shape and the integral of $\phi(\rho z)$ as accurately as possible by a very flexible mathematical expression. The expression should also be based on a reliable set of $\phi(\rho z)$ curves for soft x rays as it has been presented in the present work. In this context the model of Pouchou and Pichoir [42] can be regarded as a promising approach, even though this model seems to run into trouble in the case of high-Z and low-overvoltage ratio [43].

Very much attention should be paid in the future to the improvement of $\phi(\rho z)$ correction methods for the analysis of thin layer compositions and thickness. This application is of increasing interest in materials research. To generalize $\phi(\rho z)$ descriptions also for layer-substrate systems will raise some problems, especially when the atomic numbers of layer and substrate are largely different. Under these conditions the $\phi(\rho z)$ curves will become discontinuous near the interface. This could be demonstrated by the Monte Carlo calculations of Desalvo and Rosa [49]. In these cases a completely different method for quantifying the measurements from layered samples would consist of establishing calibration curves by a set of Monte Carlo simulations with varying layer thickness and layer composition. This method has been suggested by Kyser and Murata [51].

Both methods will require an extensive set of reliable $\phi(\rho z)$ curves to study all influences produced by the discontinuity of the scattering behavior at both sides of the interface between the layer and the substrate. A solution could be provided by the use of a complex Monte Carlo program with Mott cross sections and single inelastic scattering as a routine analysis tool.

VI. References

[1] Castaing, R. (1951), O.N.E.R.A.—Publ. No. 55.
[2] Castaing, R. and Descamps, J. (1955), J. Phys. et le Radium **16**, 304.
[3] Brown, D. B. and Ogilvie, R. E. (1966), J. Appl. Phys. **37**, 4429.
[4] Ogilvie, R. E. and Brown, D. B. (1966), 4th ICXOM, Castaing, R., Deschamps, P., and Philibert, J., eds., Paris, p. 139.
[5] Vignes, A. and Dez, G. (1968), Brit. J. Appl. Phys., ser. 2, 1, 1309.
[6] Castaing, R. and Henoc, J., in Ref. 4, p. 120.
[7] Shimizu, R., Murata, K., and Shinoda, G., in Ref. 4, p. 127.
[8] Small, J., Myklebust, R. L., Newbury, D. E., and Heinrich, K. F. J., this volume.
[9] Brown, J. D. (1966), Ph.D. Thesis, Univ. of Maryland.
[10] Brown, J. D. and Parobek, L. (1972), 6th ICXOM, Shinoda, G., Ohra, K., and Ichinokawa, T., eds., Univ. of Tokyo Press 1972, p. 163.
[11] Parobek, L. and Brown, J. D. (1978), X-Ray Spectrom. **7**, 26.
[12] Schmitz, U., Ryder, P. L., and Pitsch, W. (1969), 5th ICXOM, Möllenstedt, G. and Gaukler, K. H., eds., Springer, Berlin 1969, p. 104.
[13] Büchner, A. R. (1971), Doctoral Thesis, Aachen.
[14] Sewell, D. A., Love, G., and Scott, V. D. (1985), J. Phys. D: Appl. Phys. **18**, 1233.
[15] Ref. 14, p. 1245.
[16] Ref. 14, p. 1269.
[17] Love, G., Cox, M. G. C., and Scott, V. D. (1977), J. Phys. D: Appl. Phys. **10**, 7.
[18] Use of Monte Carlo Calculations in Electron-Probe Microanalysis and Scanning Electron Microscopy, Heinrich, K. F. J., Newbury, D. E., and Yakowitz, H., eds., NBS Spec. Publ. 460, 1976.
[19] Myklebust, R. L., Newbury, D. E., and Yakowitz, H. (1976), in Ref. 18, p. 105.
[20] Electron Beam Interaction with Solids for Microscopy, Microanalysis and Microlithography, Kyser, D. F., Niedrig, H., Newbury, D. E., and Shimizu, R., eds., Chicago, SEM, Inc., AMF O'Hare, 1982.
[21] Curgenven, L. and Duncumb, P., Tube Investments Res. Rep. No. 303.
[22] Love, G., Cox, M. G. C., and Scott, V. D. (1976), J. Phys. D: Appl. Phys. **9**, 7.
[23] Murata, K., Kotera, H., and Nagami, K. (1983), J. Appl. Phys. **54**, 1110.
[24] Shimizu, R., Nishigori, N., and Murata, K. (1972), 6th ICXOM, Shinoda, G., Ohra, K., and Ichinokawa, T., eds., Univ. of Tokyo Press 1972, p. 95.
[24a] Reimer, L., and Lödding, B. (1984), Scanning **6**, 128.
[25] Reimer, L. and Krefting, E. R., in Ref. 18, p. 45.
[26] Karduck, P. and Rehbach, W. (1988), Microbeam Analysis, 227.
[27] Reimer, L. and Stelter, D. (1986), Scanning **8**, 265.
[28] Lewis, H. W. (1950), Phys. Rev. **78**, 526.
[29] Gryzinski, M. (1965), Phys. Rev. **138**, 2A, 337.
[30] Bethe, H. (1930), Ann. Phys. **5**, 325.
[31] Rao Sahib, T. S. and Wittry, D. W. (1974), J. Appl. Phys. **45**, 5060.
[32] Hutchins, G. A. (1974), Electron Probe Microanalysis, in Characterization of Solid Surfaces, Kane, P. F. and Larrabee, G. B., eds., New York, p. 441.
[33] Powell, C. J., in Ref. 18, p. 97.
[34] Powell, C. J., in Ref. 20, p. 19.

[35] Rehbach, W., Doctoral Thesis, Aachen 1988.
[36] Rehbach, W. and Karduck, P. (1988), Microbeam Analysis 1988, 285.
[37] Love, G., Cox, M. G., and Scott, V. D. (1978), J. Phys. D: Appl. Phys. **11**, 23.
[38] Reuter, W., in Ref. 10, p. 121.
[39] Hunger, H.-J. and Küchler, L. (1979), Phys. Status Solidi A: **56**, K45.
[40] Packwood, R. H. and Brown, J. D. (1981), X-Ray Spectrom. **10**, 138.
[41] Karduck, P. and Rehbach, W. (1985), Mikrochim. Acta, Suppl. **11**, 289.
[42] Pouchou, J. L. and Pichoir, F. (1984), Rech. Aerosp. **3**.
[43] Pouchou, J. L. and Pichoir, F. (1987), Proc. 11th ICXOM, Packwood, R. H. and Brown, J. D., eds., London, Ontario, 1987, p. 249.
[44] Karduck, P. and Rehbach, W., in Ref. 43, p. 244.
[45] Pouchou, J. L., Pichoir, F., and Girard, F. (1980), J. Microsc. Spectrosc. Electr. **5**, 425.
[46] Bastin, G. F., van Loo, F. J. J., and Heijligers, H. J. M. (1984), X-Ray Spectrom. **12**, 91.
[47] Bastin, G. F., Heijligers, H. J. M., and van Loo, F. J. J. (1986), Scanning **8**, 45.
[48] Willich, P. and Obertrop, D. (1985), Mikrochim. Acta, Suppl. **11**, 299.
[49] Desalvo, A. and Rosa, R. (1979), Materials Chemistry **4**, 495.
[50] Pouchou, J. L. and Pichoir, F. (1984–5), Rech. Aerosp. **5**, 349.
[51] Kyser, D. F. and Murata, K. (1976), NBS Spec. Publ. 460, 129.

EFFECT OF COSTER-KRONIG TRANSITIONS ON
X-RAY GENERATION

JÁNOS L. LÁBÁR

Research Institute for Technical Physics of the Hungarian Academy of Sciences
P.O. Box 76
H-1325 Budapest, Hungary

I. Introduction

The first of the physical processes leading to x-ray generation is the ionization of an atom in one of the inner shells, caused by energetic particles (e.g., electrons, protons, or x rays). Different radiative and nonradiative transitions present concurrent channels for the atom to return to the ground state. As is known, the most common nonradiative transitions are the Auger transitions, in which the vacancy of the singly ionized atom is transferred from one of the inner shells to another and the difference in the energy is released by the emission of one of the outer shell electrons. A special case of these two-electron processes is that of the Coster-Kronig transitions in which the two inner-shell electrons are situated on two different subshells of the same inner shell (e.g., L_1 and L_3). Although these transitions do not produce x rays directly, they change the distribution of the primary vacancies between the different subshells of the same shell. As a consequence they alter the absolute intensity of the analytical line, the relative intensities of the different lines in the same line family and the depth distribution of the generated x rays. The magnitude of the change caused is a function of the atomic number and of the excitation. Although proton-induced x-ray emission (PIXE) and x-ray fluorescence analysis (XRF) are also affected in a similar way and some of these effects can be examined in the transmission electron microscope (TEM) too, we concentrate on the effects experienced in electron probe x-ray microanalysis of bulk samples (i.e., the microprobe).

II. Shape of the $\phi(\rho z)$ Curve: Absorption Correction

Because of the Coster-Kronig transitions the ionizations of the different subshells of the same (e.g., the L) shell are not independent of each other. Although the $L\alpha1$ (i.e., the analytical) line originates from a vacancy in the L_3 subshell only it is not sufficient to calculate the depth distribution of the primary L_3 vacancies if we want to determine the distribution of the $L\alpha1$ photons. Part of the vacancies generated on the L_2 and L_1 subshells is transferred to the L_3 as well. The probability of the transfer is characterized by the $f_{i,j}$ Coster-Kronig transition rates. Since the excitation energies of the three subshells differ from one another, the depth distributions of the vacancies in the three subshells slightly differ from one another as well. The difference is a function of the primary beam energy. (The closer it is to the excitation energy the larger the effect.) Figure 1a demonstrates this difference for the three L subshells of gold at 15 keV beam energy. The total amount (and distribution) of the L_3 vacancies is determined by a weighted sum of the three distributions, because

$$V_{L3} = P_{L3} + f_{2,3}P_{L2} + (f_{1,3} + f_{1,2} \cdot f_{2,3})P_{L1}$$

$$\phi(\rho z)_{L3}^{\text{total}} + \phi(\rho z)_{L3} + f_{2,3}\,\phi(\rho z)_{L3} + (\phi_{1,3} + \phi_{1,2}\phi_{2,3})P_{L1} \tag{1}$$

Electron Probe Quantitation, Edited by K.F.J. Heinrich and
D. E. Newbury, Plenum Press, New York, 1991

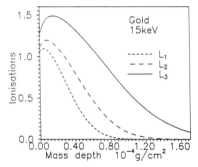

Figure 1a. Differences in the depth distributions of ionizations in the three L subshells of gold at 15 keV.

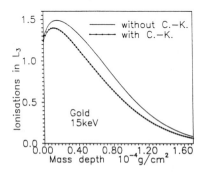

Figure 1b. Change in the shape of the $\phi(\rho z)$ due to the Coster-Kronig transitions.

where P_{L_i} is the number of primary vacancies in the L_i subshell. Since the same processes occur in the thin film too, the distribution must be normalized with respect to the increased intensity in the thin layer, described in

$$\text{Norm}=\frac{Q(U_{L3})+f_{2,3}Q(U_{L2})+f_{1,3}+(f_{1,2}+f_{2,3})\,Q(U_{L1})}{Q(U_{L3})}$$

and

$$\phi(\rho)_{L3}^{\text{final}}=\frac{1}{\text{Norm}}\,\phi(\rho z)_{L3}^{\text{total}}\ . \tag{2}$$

Q is the ionization efficiency for the given subshell, taken at the primary beam energy. Figure 1b shows $\phi(\rho z)_{L3}$ compared with $\phi(\rho z)_{L3}^{\text{final}}$. It can be seen that the shift in the peak position and the change in the peak height and shape are noticeable. Figure 2 demonstrates the same effect for silver at 4 keV primary beam energy. Although the difference between the two curves (i.e., $\phi(\rho z)$ with and without Coster-Kronig transitions) is significant, its practical importance is limited because on the one hand the small range of x-ray generation makes direct $\phi(\rho z)$ measurement almost impossible for these primary beam energies and on the other hand, absorption corrections calculated from the two distributions differ less than one percent. For larger primary beam energies (hence larger penetration depths and as a

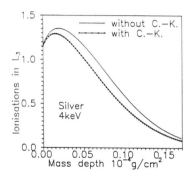

Figure 2. Change in the shape of the $\phi(\rho z)$ due to the Coster-Kronig transitions.

consequence larger absorptions) the two distributions approach each other, so the difference is low for most situations. This fact explains why the effect of Coster-Kronig transitions has not been noticed for decades in conventional microanalysis. The absolute magnitude of the generated intensity, on the contrary to the absorption, is altered noticeably because of the presence of the Coster-Kronig transitions.

III. Generated Intensity, Atomic Number (Z) Correction

Because of the Coster-Kronig transitions, the number of vacancies generated in the L_3 subshell is enhanced from P_{L3} to V_{L3} as calculated in (1). The enhancement is up to 30 percent depending mainly on the atomic number and to a lesser extent on the primary beam energy. So high a difference is notable and appears in measurements of the absolute intensities of, e.g., the $L\alpha1$ lines. An example of such a comparative measurement is given in the first part of Lábár's paper [1]. Measured intensities of 22 elements were compared with predictions of calculations with and without the effect of the Coster-Kronig transitions. As a consequence of the incorporation of the Coster-Kronig transitions, a significant improvement in the prediction of the intensities is demonstrated in figure 1 of the above paper. Unfortunately, the last section of the cited paper was based on a publication of Pouchou and Pichoir [2] which contained a misprint.[1] Results calculated with the corrected formula corroborate our previous conclusion that incorporation of the Coster-Kronig transitions into the calculations significantly improves the match between measured and calculated intensities of the $L\alpha1$ lines in single element samples. The average error of these 22 intensities was reduced from 10.5 to 5.4 percent relative. The generation factor was calculated with Pouchou's corrected formula, absorption with Bastin's method [3] and fluorescence and Coster-Kronig yields were taken from Krause [4] and the weight of the $L\alpha1$ line within the lines originating from the L_3 subshell was calculated from Salem et al. data [5].

The effect of the Coster-Kronig transitions on the relative intensities of the minor L lines was examined by Schreiber and Wims [6]. Its consequences for the overlap correction of energy dispersive spectra and for standardless analysis were investigated by Lábár [7]. These effects were found significant.

As far as the atomic number correction in conventional microanalysis is concerned, V_{L3} is calculated for the sample and for the standard in a similar way. Their ratio (the Z-correc-

---

[1] I was informed of the misprint later, at the MAS'88 conference, where Pouchou presented an erratum. That misprint caused the ambiguous result in my above cited paper, when the effect on nonradiative transitions could not be demonstrated and the same misprint was responsible for the large errors reported in it.

tion factor with the effect of the Coster-Kronig transitions) is not too far from the ratio of P_{L3} for the sample and for the standard (i.e., the Z-factor without Coster-Kronig transitions) for most of the cases. (A difference of less than 1% was found for a 10%Ag–90%Au sample and pure silver standard system at 5 keV.)

IV. Conclusions

The effect of the Coster-Kronig transitions on the $\phi(\rho z)$ function can only be demonstrated for low overvoltages when the range of x-ray production is small.

Changes in the shape of the $\phi(\rho z)$ function are mainly canceled during the integration and division, so that the effect of Coster-Kronig transitions on the A and Z correction factors is generally negligible in the microprobe analysis of bulk samples if standards are used.

Absolute intensities are significantly affected by the enhanced ionization in the L_3 subshell because of the presence of the Coster-Kronig transitions. Both the absolute intensities of the analytical ($L\alpha1$) lines and the relative intensities of the minor lines are noticeably affected. Differences of up to 30 percent can be observed in the intensities of the $L\alpha1$ lines. Standardless methods which are sensitive to deficiencies in the physical models, are greatly affected by that fact.

V. References

[1] Lábár, J. L. (1988), Microbeam Analysis–1988, Newbury, D. E., ed., San Francisco Press, 253–257.
[2] Pouchou, J. L. and Pichoir, F. (1986), 11th International Congress of X-Ray Optics and Microanalysis, London (Canada), 249–253.
[3] Bastin, G. F., et al., (1986), Scanning 8, 45 pp.
[4] Krause, M. O. (1979), J. Phys. Chem. Ref. Data 8, 307–327.
[5] Salem, S. I., et al., (1974), Atomic Data and Nuclear Data Tables 14, 91–109.
[6] Schreiber, T. P. and Wims, A. M. (1982), Microbeam Analysis–1982, Heinrich, K. F. J., ed., San Francisco Press, 161–166.
[7] Lábár, J. L. (1987), X-Ray Spectrometry 16, 33–36.

UNCERTAINTIES IN THE ANALYSIS OF *M* X-RAY LINES
OF THE RARE-EARTH ELEMENTS

J. L. LÁBÁR AND C. J. SALTER

Research Institute for Technical Physics of the Hungarian Academy of Sciences
P.O. Box 76*
H-1325 Budapest

Oxford University, Department of Metallurgy and Science of Materials
Oxford OX1 3PH

I. Introduction

Electron probe x-ray microanalysis by *M* lines poses some inherent problems similar in a certain respect to those arising during the analysis of light elements. The energy of the radiation is rather low, so that the measurement is more sensitive to absorption and contamination effects and the closeness of other lines is more of a problem. The requirements for the quality of the surface are therefore stricter than usual. Because of the important role of the outer shells in the x-ray generation process, chemical effects can also be pronounced. To make matters worse, the lack of reliable atomic constants (mass absorption coefficients (macs), fluorescence and Coster-Kronig yields, relative line intensities etc.) for the *M* lines is often conspicuous.

The situation seems to be worst for the *M* lines of the rare-earth elements. Fisher and Baun [1] proved that an anomalous line type absorption is present in the *M* spectra of rare-earth elements, and they listed a new set of absorption edge and line energies. The discrepancies between these new values and the original values of Bearden [2] and Bearden and Burr [3] are small but they are of great importance and consequences, because they sometimes cause the reversal of the energies of the absorption edge and the emission line. According to Fischer and Baun [1], this physically surprising situation exists because of the unfilled inner shells taking part in the x-ray generation. If we use one of the common computer programs for the calculation of the mass absorption coefficients, a line energy on the wrong side of the absorption edge may result in a radically erroneous absorption coefficient. Examination of the compilation of mass absorption coefficients by Saloman and Hubbell [4] shows that hardly any experimental values can be found in the energy range where the *M* lines of rare-earth elements are situated. The latest formulas by Heinrich [5] for the extrapolation of mass absorption coefficients also contain warnings that calculations are not reliable when close to the absorption edges, especially below the $M5$ edge of elements with $Z < 70$. All mac values for $M\alpha$ and $M\beta$ of rare-earth elements are marked with a question mark in his list to emphasize this fact.

Wendt and Christ [6] pointed out that relative intensities of the $M\zeta$ to the $M\alpha$ lines differ from the predictions of White and Johnson [7] by more than three orders of magnitude for the rare-earth elements. Their energy-dispersive (EDS) measurements were not able to resolve the $M\alpha$ and $M\beta$ lines, but they suggested that the relative intensities of other *M* lines of the same elements also can differ significantly from expectations (Wendt [8]).

*The present work was completed at Oxford University in the framework of a postdoctoral scholarship sponsored by the Soros Foundation. Valuable consultations with Dr. John Jakubovics are gratefully acknowledged.

Electron Probe Quantitation, Edited by K.F.J. Heinrich and
D. E. Newbury, Plenum Press, New York, 1991

The present paper summarizes the results of a quantitative study on the M lines of the rare-earth elements. Experimental data were collected to discover the absorption properties of these lines, and to determine the atomic data necessary for overlap and absorption corrections. A new absorption model is proposed to explain the experimentally observed variation of relative intensities with primary beam energy. Without the new model, it is not possible to account for the experimental data. The new model is an extension of the previous models of normal absorption, created to incorporate the description of the anomalous line type absorption phenomenon. Relative intensities of $M\zeta$, $M\beta$, $M\gamma$, and $M2N4$ lines and mass absorption coefficient of anomalous absorption for the $M\alpha$ and $M\beta$ lines are simultaneously determined from the measurements. A new parameter, the width of anomalous absorption, is determined for $M\alpha$ and $M\beta$ lines during the same data reduction. This parameter is unique to the anomalous process. Further research is necessary to clarify the measure of chemical effects to test the new model as a part of the ZAF correction calculation in analytical situations.

II. Experimental

Intensity as a function of wavelength was recorded for different rare-earth elements in wavelength ranges covering both the full width of a line and some part of the background on both sides of it. The wavelength intervals examined involved the $M\alpha$ and $M\beta$ and the $M\zeta$, $M\gamma$, and $M2N4$ lines. The intensities were calculated by integrating over the whole peak width above the background. The background was determined by linear interpolation based on fitting to the measured values on both sides of the peak. By relative intensity of a line we mean the ratio of the integral peak intensity of that line to the integral intensity of the $M\alpha$ line. Relative intensities of the above mentioned lines were measured as a function of primary beam energy. The variations of relative intensity with the primary beam energy were used to determine both the absorption properties and the atomic parameters of the lines under examination, as described in more detail in the next section.

In ambiguous situations peaks were identified under conditions leading to low absorption and the limits of integration were set based on this identification and retained for high absorption spectra as well. Overlap correction calculations were omitted. That implies the assumption that the $M\alpha$ and $M\beta$ peaks have approximately the same size and that their position is approximately symmetrical with respect to the limit of integration. (We shall see that these are acceptable assumptions for the examined elements.)

The results of the wavelength scans were digitally recorded using a CAMECA microprobe under computer control. One of two identical spectrometers (with TAP crystals) was used for the measurements described, while the other was kept in an unaltered position to check the stability of the system. Wavelength scans recorded under unstable conditions (instability more than 5%) were neglected and repeated. The contamination was reduced by a trap cooled with liquid nitrogen and by a slow lateral movement of the samples. The applied primary beam energies were 2.5, 5, 10, and 15 keV. Count rates of less than 10 kc/s made dead-time correction unimportant. The effects of multiple Bragg reflections were eliminated by careful selection of pulse height analyzer settings.

Rare-earth metal (La, Ce, Pr, Nd, Sm, Gd, Tb, Dy, Ho, and Er) samples[1] were freshly polished on each occasion and instantly placed into the vacuum system for measurement in order to minimize oxidation. In some cases oxide (CeO_2) and fluoride (SmF_3, HoF_3, ErF_2, and YbF_3) samples were analyzed as well, but problems due to charge build-up under the surface made these results ambiguous, lacking the required reproducibility.

[1] The authors thank Dr. D. Fort at the University of Birmingham for supplying the samples.

Intensity vs. wavelength diagrams on pure boron and pure titanium samples were recorded to determine the response function of the experimental setup from comparison of theoretically predicted and measured intensities of bremsstrahlung radiation.

III. Data Evaluation

The basic assumptions implicit in our data evaluation are:

1.   The intensity of the $M\zeta$ line relative to the $M\alpha$ line is independent of the primary beam energy at the moment of generation. The observed dependence of the relative intensity on the primary beam energy originates from the difference in the absorptions of the two lines.

That assumption is justified because both $M\zeta$ and $M\alpha$ originate from the same (M_5) subshell. So on the one hand, factors of vacancy generation are the same for both of them and cancel in the ratioing of intensities. On the other hand, the rates of different radiative (and nonradiative) transitions are independent of the way the atom is excited; they only depend on the excited state itself, but not on the energy of the exciting radiation. The response function of the spectrometer is, of course, independent of the excitation mechanism of the sample. Hence, the self-absorption of the lines within the sample is the only factor which depends on the primary beam energy in a different way for the $M\zeta$ than for the $M\alpha$ lines. This difference in self-absorption is obviously due to the different absorption coefficients of the two lines and the path length within the sample varying with primary beam energy:

$$R_{\zeta/\alpha}=(I_\zeta/I_\alpha)=R_{\zeta/\alpha}^{0,\mathrm{Meas}}\ (f_\zeta/f_\alpha) \tag{1}$$

where $R_{\zeta/\alpha}^{0,\mathrm{Meas}}$ designates the limiting (constant) value of the measured relative intensities of the $M\zeta$ line (i.e., what would be the value measured in the absence of absorption). The absorption correction function "f" is defined as usual; it gives the emitted fraction of the generated intensity.

2.   The effects of self-absorption within the sample and of the response function of the experimental setup can be separated, because the first is a function of the primary beam energy (E_0) and the second is independent of it. This separability implies that the absorption properties can be deduced from the measured data without a knowledge of the response function (T^{Resp}). The latter is only necessary for the deduction of the real relative line intensity ($R^{0,\mathrm{True}}$) from the measured $R^{0,\mathrm{Meas}}$ value in eq (1):

$$R^{0,\mathrm{True}}=R^{0,\mathrm{Meas}}\ T^{\mathrm{Resp}} \ . \tag{2}$$

3.   The absorption properties ($f[\mu/\rho]$) of the $M\zeta$, $M\gamma$, and $M2N4$ lines are known. That means that for them the normal absorption is present, with a known mac (μ/ρ). The absence of the anomalous absorption can be seen in the experimental wavelength scans below. The assumption for the mac originates from the fact that these lines are far from the absorption edges, so that their absorption coefficients are available in the literature, with the usual reliability.

Based on these assumptions the data evaluation proceeds in four separate steps:

A.   The absorption properties of the $M\alpha$ line (i.e., parameters in f_α) and the relative intensity of the $M\zeta$ line ($R_{\zeta/\alpha}^{0,\mathrm{Meas}}$) are determined from eq (1). Using the known f_ζ, a pre-assumed functional form of f_α (see below) is fitted to the measured values of $R_{\zeta/\alpha}$ (determined at 2.5, 5, 10, and 15 keV). $R^{0,\mathrm{Meas}}$ is determined from the requirement that f_α converges to 1 when the absorption becomes negligible.

B.   Applying the absorption of the $M\alpha$ determined above, a similar fit to the measured $R_{\beta/\alpha}$ values can be performed to deduce the absorption and relative intensity data for the $M\beta$ line. An additional assumption is involved in this procedure, namely, that the ratio of the vacancy generation in the M_5 subshell to that in the M_4 subshell is independent of the primary beam energy. This assumption is not exactly valid, but it seems to be a good approximation for these two subshells (it gave a small difference for holmium, the only element for which the correction could be performed).

C.   The excitation of the M_3 and M_2 subshells varies in a different way from that of the M_5 subshell, because of their different excitation potentials. The correction for the differences in the generation term during the calculation of relative intensities of the M_γ and $M2N4$ lines is not negligible. Unfortunately, it can only be calculated using the values of the Coster-Kronig transition rates. We have found in the literature these Coster-Kronig yield data for holmium and ytterbium only. Based on the calculations for these two elements we can state that the ratio of the generation terms for the M_3 to M_5 and the M_2 to M_5 subshells almost goes to saturation (i.e., goes toward a limiting constant value) in the 10–15 keV energy range. This is so because the overvoltages for the different subshells become so high for high E_0 that their differences become negligible in the slowly varying generation function. This saturation enables us to use the $R_{\gamma/\alpha}$ values measured at 10 and 15 keV together with the absorption of the α line (determined above) and the known absorption of the γ line to deduce $R_{\gamma/\alpha}^{0,\text{Meas}}$ similarly as $R_{\beta/\alpha}^{0,\text{Meas}}$ was calculated. $R_{M_2N}^{0,\text{Meas}}$ can be determined in the same way from $R_{M_2N_4}$ measured at 10 and 15 keV.

D.   The experimentally determined T^{Resp} can be substituted into eq (2) to calculate $R_i^{0,\text{True}}$ for $i = \zeta$, β, γ, and $M2N4$. T^{Resp} is a quotient of two values of the response function taken at the energies of the two lines. It is therefore possible to determine the response function without the exact knowledge of a constant multiplicative factor, because this factor cancels during the division. Calculations by Small et al. [9] were applied in order to deduce the response function of the system. A multiplication with energy (E) was necessary to convert their formula for use with the wavelength dispersive spectrometer (WDS). (They developed their calculations for an energy dispersive spectrometer, i.e., for a fixed energy interval ΔE. However, the characteristic of a WDS is described with $\Delta\lambda/\lambda = \text{const}$. An energy window of $\Delta E = \text{const}\, E$ is the result of this property of the WDS.)

IV.   A New Model for Anomalous Absorption

The presence of unfilled inner electron shells in the atom causes remarkable distortions in the M x-ray spectra. The affected lines are broader and because of vacancies in the inner subshell involved, some of the electron transitions become partially reversible. Hence, x rays having the energy which is required to initiate that transition, are strongly absorbed in the material. However, this process is different from the usual x-ray absorption. The latter is caused by transitions from an inner shell to the continuum state which means that all x rays with energies above a certain threshold can cause the transition. The result is the known edge type absorption. The anomalous absorption caused by an unfilled inner shell significantly differs from that process. X rays are only absorbed if their energy is exactly that required to move one electron from one inner shell to another, unfilled, inner shell. The consequence of that strict prerequisite is a line type absorption. The two types of absorption are completely independent processes, enabling two concurrent modes of electron transition. Studies of Fischer and Baun [1] showed that absorption spectra exhibit some fine structure too, but the main component is a single absorption line accompanying a main emission line. Experimental results demonstrate that the whole width of the emission line is not affected homogeneously (see figs. 1–10 in this paper, and figs. 6–9 in [1]) but the width of absorption is smaller than that of the emission line. This limited absorption width is also an unusual characteristic of the anomalous absorption phenomenon, which should be reflected in the new model.

In order to express the independence of the normal and anomalous absorption the new model is built up in the form of a product of the two functions:

$$f^{\text{new}}[(\mu/\rho)^n, (\mu/\rho)^a, x] = f^n[(\mu/\rho)^n] f^a[(\mu/\rho)^a, x] \qquad (3)$$

where the superscript "n" denotes "normal" and "a" stands for "anomalous." The functional form (f^n) and mac [$(\mu/\rho)^n$] of normal absorption can be taken from the literature. Because every absorption correction (f) approaches 1 in the limiting case of negligible absorption, our new model is simplified to the usual absorption correction in the absence of anomalous absorption ($(\mu/\rho)^a = 0$ or $x = 0$). (For explanation of width of absorption "x" see below.)

To obtain the functional form of the anomalous absorption part (f^a), we use the idea of limited absorption width introduced above. This part of the model is semiempirical; first physical principles are not used to calculate the fraction (x) of the radiation which is affected by the anomalous absorption. Experimental data are used instead to deduce parameters of the model [x and $(\mu/\rho)^a$]. The method is based on the empirical fact that only a fraction x of the emitted line is affected by the line type absorption [characterized by $(\mu/\rho)^a$]. The usual definition of absorption correction (f) is that it gives the emitted fraction of the generated intensity. Let us denote the absorbed fraction by B.

$$B = 1 - f . \qquad (4)$$

If an only fraction x is affected by the absorption process, the absorbed fraction becomes xB, so the emitted fraction after the anomalous absorption is

$$f^a[(\mu/\rho)^a, x] = 1 - xB = 1 - x + x f^n[(\mu/\rho)^a] . \qquad (5)$$

In the last step we substituted (4). Because the anomalous properties of the absorption are incorporated into the x and $(\mu/\rho)^a$ parameters and because eq (4) does not make use of any properties of the absorption, the usual functional form of absorption (f^n) can be used on the right-hand side of eq (5). Substituting eq (5) into (3) we get the final form of the new model.

$$f^{\text{new}}[(\mu/\rho)^n, (\mu/\rho)^a, x] = f^n[(\mu/\rho)^n]\{1 - x + x f^n[(\mu/\rho)^a]\} . \qquad (6)$$

The term of f^n is taken from Bastin et al. [10] and $(\mu/\rho)^n$ is calculated with the program of Heinrich [5]. Atomic data x and $(\mu/\rho)^a$ are determined from fits in the present paper. We expect that these constants together with eq (6) can be used in the future as part of a ZAF correction calculation in analytical situations.

V. Comparison of Models; Fitting

Experimental $R_{l/a}$ vs. E_0 and $R_{\beta/a}$ vs. E_0 were compared with the predictions of three different models.

A. The new absorption model described above is applied in eq (1) for theoretical prediction of relative intensities at four different E_0. Parameters in eq (1) were determined from a fitting procedure. Values of $(\mu/\rho)^a$ were iterated and $R_{l/a}^{0,\text{Meas}}$ and x were determined from linear least-squares fitting in each step of the iteration. The final selection of a set of the three parameters was controlled by the requirement that root mean square of error (rms) between measured and computed values of $R_{l/a}$ should be a minimum.

B.   It is necessary but not sufficient to prove that the new model is appropriate for the description of the experimental data. It is equally important to show that the introduction of the new model is inevitable, i.e., to demonstrate that the established models cannot account for the experimental findings. That is why, as a second approach, normal absorption only was taken into consideration. Bastin et al. [10] absorption correction and Heinrich's [5] mac were applied to all lines. We are aware of their warnings concerning the usage of their values. The shortcomings of this approach (i.e., that we try and apply them to a situation they were not intended to be used for) do not imply the failure of the above authors' model and data. We only wanted to demonstrate that the complex situation with the M_α and M_β lines of rare-earth elements cannot even be described qualitatively with the established microanalytical models and data.

C.   In an attempt to prove that there is no absorption coefficient that could restore the match between the predictions of the established model and experiment, we tried a third model as well. The effect of anomalous absorption was incorporated into an "effective" mass absorption coefficient. The functional form of the normal, established microanalytical model was retained as above, but the mac was treated as a parameter to be fitted in it. This is a similar approach that Pouchou successfully applied to determine the absorption coefficients for the L_α lines of the transition metals (Pouchou and Pichoir [14]). Iterated values of $(\mu/\rho)^{\text{eff}}$ were substituted into f_α in eq (1) and $R_{\zeta/\alpha}^{0,\text{Meas}}$ was determined by least-squares fitting. Final parameters were selected on the basis that they should result in a minimum rms between the measured and calculated relative intensities. We show that no single value of mac can restore the accord between theory and experiment.

VI.   Some Remarks to this Procedure

–   The $M\zeta$ line is not only used to determine its relative intensity but also to establish the quantitative model of the absorption of the usual analytical line (the $M\alpha$). This fact increases the importance of the $M\zeta$ line which is generally not used in analysis (save its role in overlap correction).

–   Although the use of the electron microprobe may not be the best way to obtain mass absorption coefficients, it did provide a reasonably rapid means of acquiring the required data. The precision of the parameters obtained in this way could have been increased by using a larger number of E_0 values. However, the selection of the number of primary beam energies was determined by the time and hardware available.

–   An examination of chemical effects cannot be avoided. Preliminary results hint that the main properties of the experimental data sets remain the same if we turn from one chemical form of the examined element to another (i.e., the same model can be used for all forms of the material). Fine details of the model (i.e., its parameters) can change to a certain extent. The usefulness of the model in an analytical application will be finally determined by the extent of this change.

– $(\mu/\rho)^{\text{a}}$ should be averaged for composite materials in the same way as $(\mu/\rho)^{\text{n}}$, i.e., by weight fractions.

VII.   Results

A.   INDIVIDUAL DATA

Figures 1–10 present the measured line scans. Vertical bars designate discontinuities on the abscissa (i.e., limits of sections covering single or double lines). Sections from left to right cover the $M2N4$, $M\gamma$, $M\beta - M\alpha$, and $M\zeta$ lines.

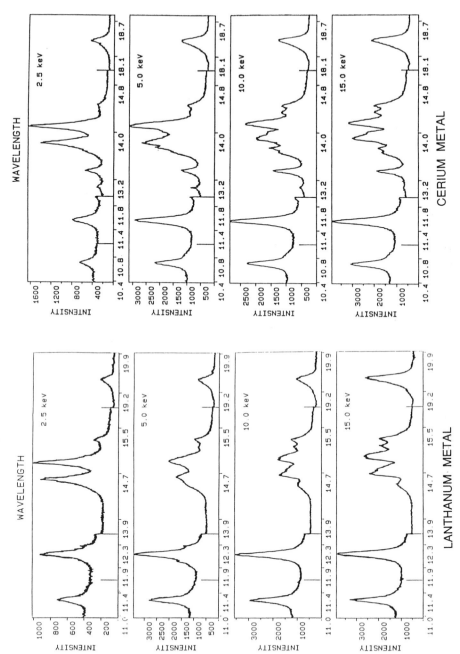

Figure 2. Measured intensities of x rays as a function of wavelength at four primary beam energies for elementary cerium.

Figure 1. Measured intensities of x rays as a function of wavelength at four primary beam energies for elementary lanthanum.

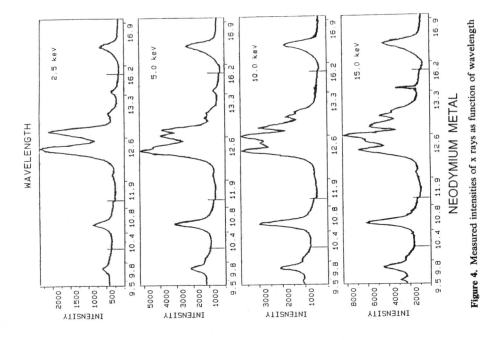

Figure 4. Measured intensities of x rays as function of wavelength at four primary beam energies for elementary neodymium.

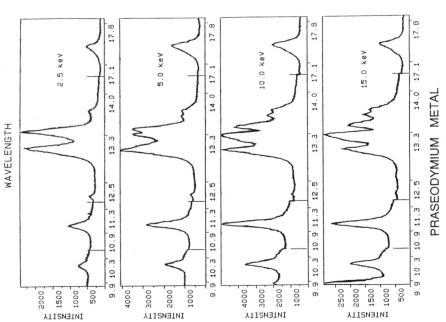

Figure 3. Measured intensities of x rays as a function of wavelength at four primary beam energies for elementary praesodymium.

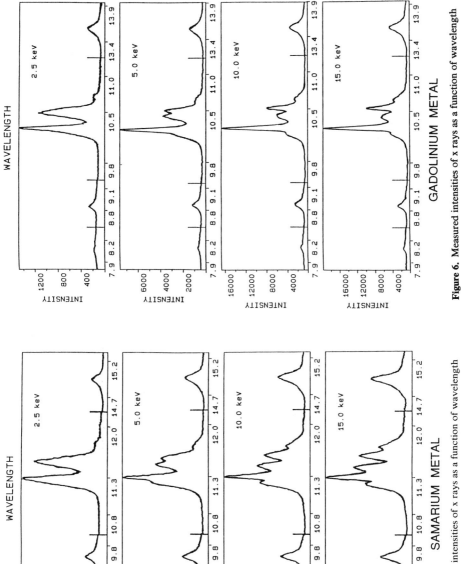

Figure 6. Measured intensities of x rays as a function of wavelength at four primary beam energies for elementary gadolinium.

Figure 5. Measured intensities of x rays as a function of wavelength at four primary beam energies for elementary samarium.

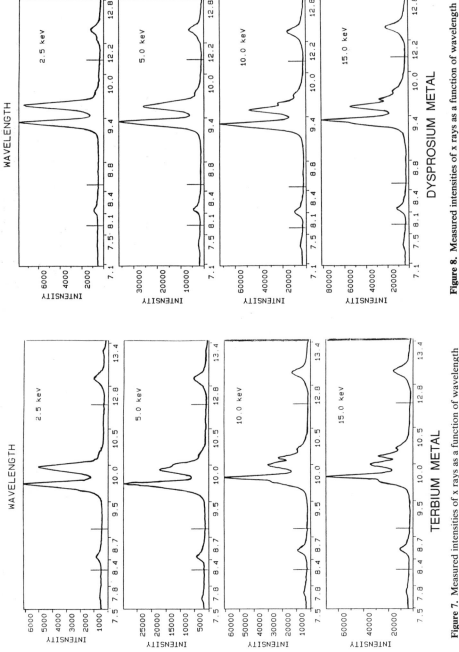

Figure 8. Measured intensities of x rays as a function of wavelength at four primary beam energies for elementary dysprosium.

Figure 7. Measured intensities of x rays as a function of wavelength at four primary beam energies for elementary terbium.

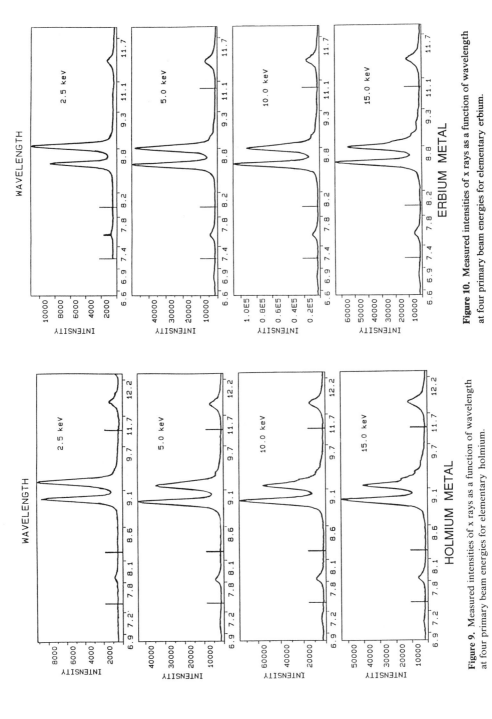

Figure 10. Measured intensities of x rays as a function of wavelength at four primary beam energies for elementary erbium.

Figure 9. Measured intensities of x rays as a function of wavelength at four primary beam energies for elementary holmium.

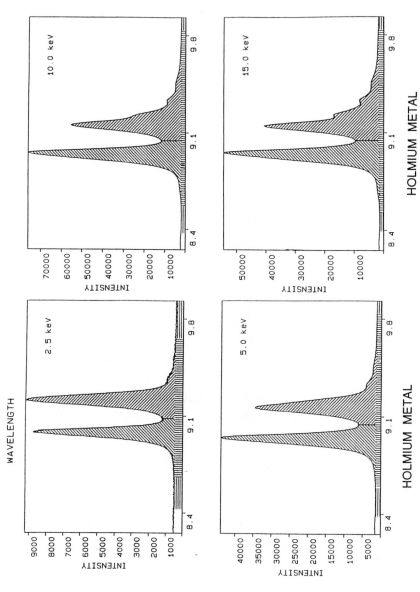

Figure 11. Background and overlap correction for the $M\beta$ and $M\alpha$ peaks of elementary holmium at four different primary beam energies.

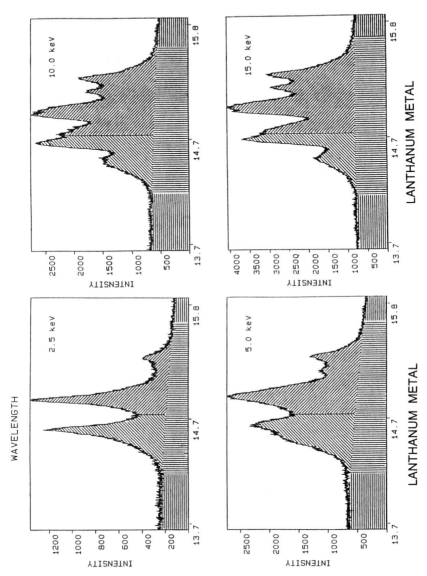

Figure 12. Background and overlap correction for the $M\beta$ and $M\alpha$ peaks of elementary lanthanum at four different primary beam energies.

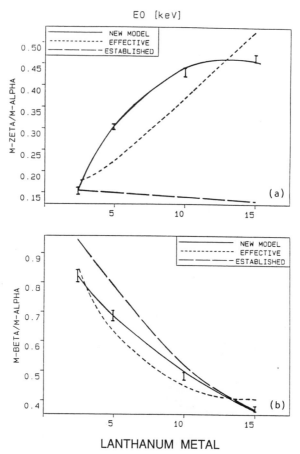

Figure 13a-b. Dependence of relative intensities of the $M\zeta$ and of the $M\beta$ lines on the primary electron beam energy for elementary lanthanum.

Figures 11–12 give examples of a relatively straightforward and a rather complicated situation of a peak identification and separation for the $M\beta$ and $M\alpha$ lines.

The dependence of the measured relative intensity of the $M\zeta$ lines on primary electron beam energy (E_0) can be seen in figures 13a–22a. The figures contain a comparison of measured and calculated values as described above. All of them have the same general characteristics:

– The curves start as increasing function of E_0. This contradicts the prediction of a combination of the conventional model and established mac values.

– The steepest part of the measured $R_{\zeta/\alpha}$ vs. E_0 function is at the lowest energies. The slope of the curve seems to decrease with E_0 and the increasing curve can turn into a decreasing one. (This feature can be even more emphasized in connection with the β line; see below.) Models of normal absorption cannot even qualitatively describe such a behavior. They would predict a monotonous change with E_0 and a slope close to zero at the lowest penetration depth of electrons (i.e., at the lowest E_0). The effective mac model can serve as a better approximation in a restricted E_0 range but it cannot account for the two special features mentioned in the paragraph. The new model of anomalous absorption is

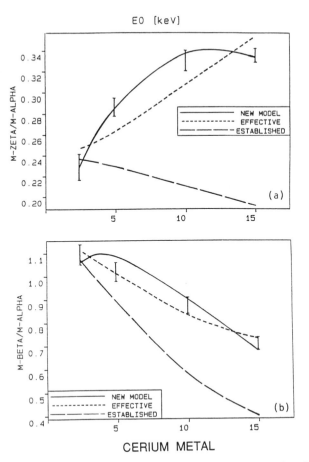

Figure 14a-b. Dependence of relative intensities of the $M\zeta$ and of the $M\beta$ lines on the primary electron beam energy for elementary cerium.

able to describe these very strange characteristics of the experimental data. The fit not only seems to be good in a qualitative sense but also quantitatively. The "turn-back" feature may be a little overemphasized by the new model. (A similar "turn-back" can be observed in fig. 3 of Wendt and Christ [6], but they did not attribute importance to it. It was explained as a saturation in their paper. For a comparison of the numerical values of their measurement to the values in this paper, see below.)

Figures 13b–22b present the measured values of $R_{\beta/\alpha}$ and their comparison with predictions of the three models. Because the relative intensity of the β line is affected by two anomalous absorptions (that of the α and of the β), the resulting dependence on E_0 can even be stranger than usual. The same remarks apply to the evaluation of the $R_{\zeta/\alpha}$ data: the experimental data can only be properly interpreted with the help of the new model of anomalous absorption. The effective mac model only seems to be a tradeoff within a restricted E_0 range. Although this approximation is rough it can be acceptable in certain situations (see conclusions). The difference between the absorption behavior of the α and of the β line is obvious. A rather complicated change of observed intensity with energy can be anticipated in an EDS as a consequence, since the technique cannot resolve these lines.

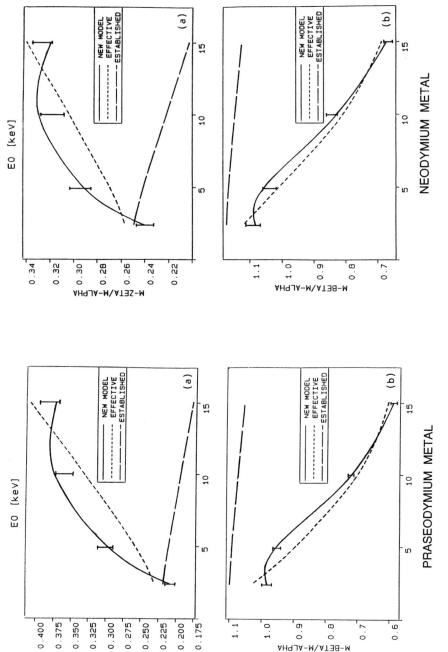

Figure 15a-b. Dependence of relative intensities of the $M\zeta$ and of the $M\beta$ lines on the primary electron beam energy for elementary praseodymium.

Figure 16a-b. Dependence of relative intensities of the $M\zeta$ and of the $M\beta$ lines on the primary electron beam energy for elementary neodymium.

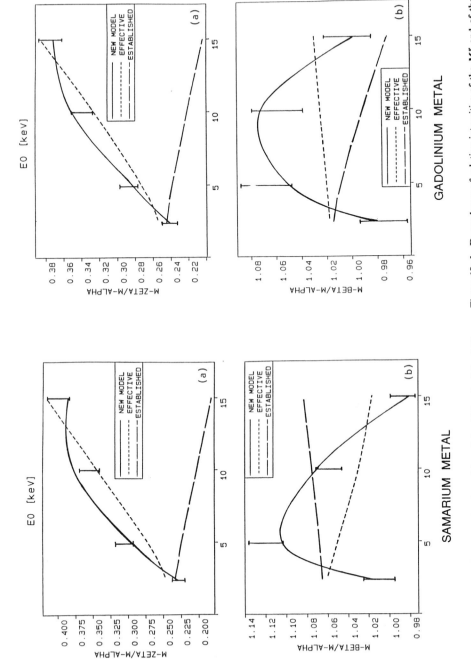

Figure 18a-b. Dependence of relative intensities of the $M\zeta$ and of the $M\beta$ lines on the primary electron beam energy for elementary gadolinium.

Figure 17a-b. Dependence of relative intensities of the $M\zeta$ and of the $M\beta$ lines on the primary electron beam energy for elementary samarium.

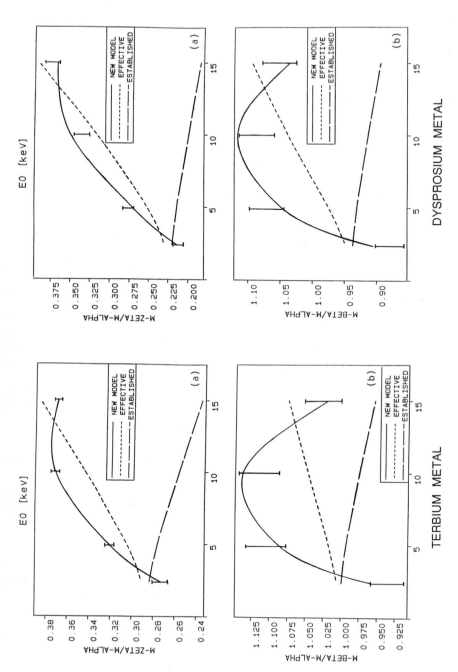

Figure 19a-b. Dependence of relative intensities of the $M\zeta$ and of the $M\beta$ lines on the primary electron beam energy for elementary terbium.

Figure 20a-b. Dependence of relative intensities of the $M\zeta$ and of the $M\beta$ lines on the primary electron beam energy for elementary dysprosium.

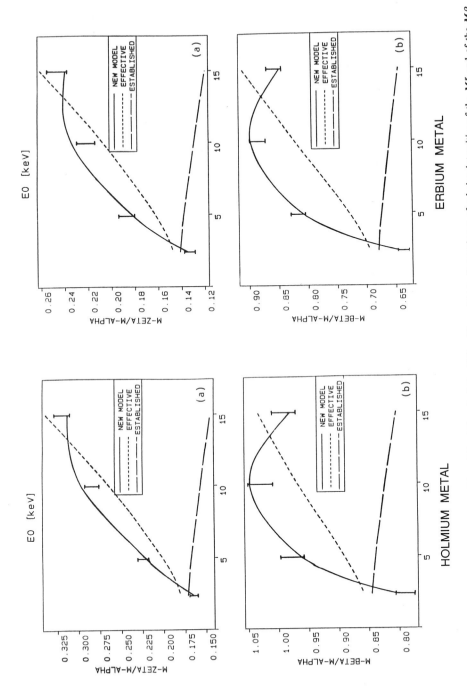

Figure 21a-b. Dependence of relative intensities of the $M\zeta$ and of the $M\beta$ lines on the primary electron beam energy for elementary holmium.

Figure 22a-b. Dependence of relative intensities of the $M\zeta$ and of the $M\beta$ lines on the primary electron beam energy for elementary erbium.

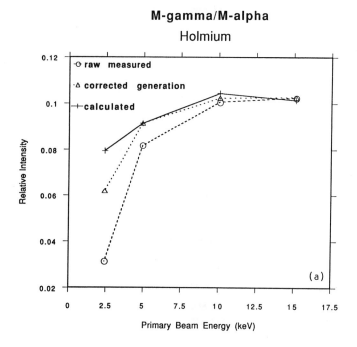

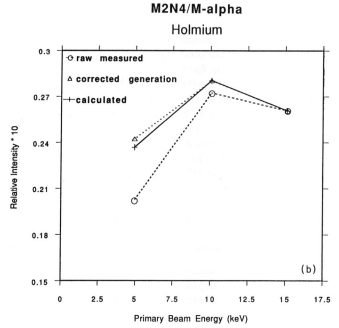

Figure 23a-b. Effect of differences in the efficiencies of vacancy generation in different *M* subshells on relative intensities of *M*γ and *M*2*N*4 lines for elementary holmium.

M-alpha Lines

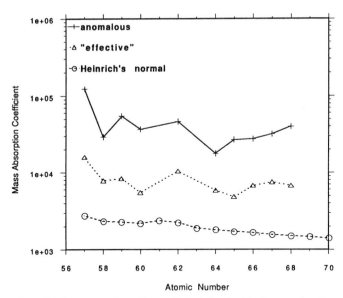

Figure 24. Comparison of different mass absorption coefficients for the $M\alpha$ lines as a function of atomic number.

Figures 23a–b demonstrate the magnitude of the necessary correction for the differences between excitations of the different subshells. Holmium has been chosen as an example because atomic data were available for it (McGuire [11], Chen et al. [12,13]). The vacancy generation efficiency was calculated with an extended version of the Pouchou and Pichoir (P&P) model [14]. The primary ionization of the subshells (N_i) was calculated with the same function which was introduced by P&P for total L shell ionization. Taking into account the number of electrons in the subshells and the possible ways of radiationless rearrangement within the shell, the final distribution of vacancies between the subshells was calculated as a function of excitation (E_0). In the latter step the $f_{i,j}$ Coster-Kronig transition rates of McGuire [11] and Chen et al. [12,13] were applied. For example, the ionization of the L_3 subshell is calculated by

$$V_3 = N_3 + N_2 f_{2,3} + N_1 (f_{1,3} + f_{1,2} f_{2,3}) \; . \tag{7}$$

B. ATOMIC NUMBER DEPENDENCE OF DATA

Figures 24 and 25 compare normal mac from the literature (Heinrich [5]) with "effective" mac and mac of the anomalous absorption for the $M\alpha$ and $M\beta$ lines. Some tendencies in the atomic number dependence of the normal mac are more accentuated for the anomalous absorption. It is conspicuous that there is an order of magnitude difference between the mass absorption coefficients of normal absorption phenomena and those which characterize the anomalous absorption process.

Figure 26 comprises the absorption widths of the $M\alpha$ and $M\beta$ lines. The abrupt decrease in $x(M_\beta)$ at $Z = 58$ is not understood and requires further examination.

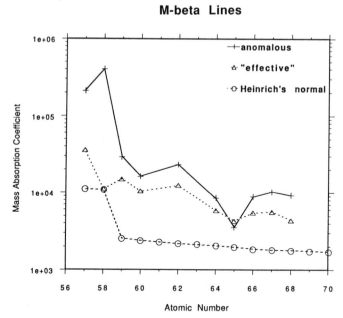

Figure 25. Comparison of different mass absorption coefficients for the $M\beta$ lines as a function of atomic number.

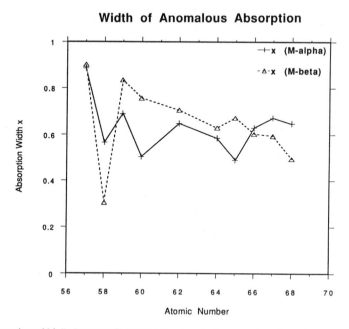

Figure 26. Absorption width "x," an atomic parameter unique to the anomalous absorption process, as a function of atomic number for the $M\alpha$ and $M\beta$ lines.

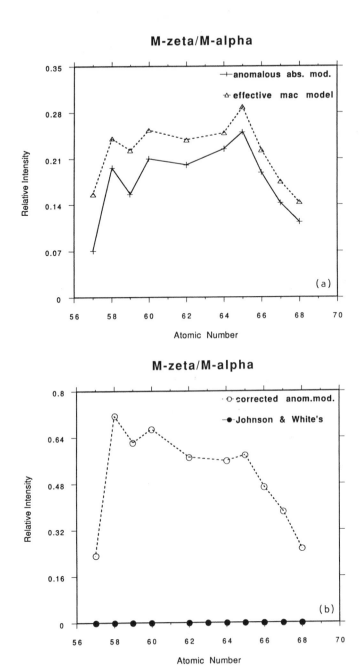

Figure 27a-b. Comparison of measured relative intensities of the M_ζ lines, after correction for self-absorption within the sample and for response function of the spectrometer, with predictions of Johnson and White.

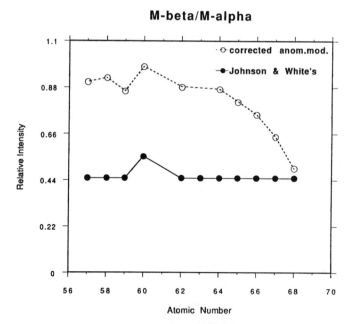

Figure 28. Comparison of measured relative intensities of the M_β lines, after correction for self-absorption within the sample and for response function of the spectrometer, with predictions of Johnson and White.

Measured relative intensitites after absorption correction can be found in figures 27–30. The atomic numbers 58 and 65, where an anomaly exists in the filling up of the atomic subshell $4f$, seem to produce a deviating value on each of the curves.

Figures 27b and 28–29 contain Johnson and White's data compared with our ultimate values of relative intensities, after correction for the response function of the spectrometer. The importance of this correction is obvious from the plots. It must be emphasized that the response function plays a significant role both in ED and WD analysis. Without that correction, relative intensities measured on different types of equipment cannot be compared to each other and to theoretical predictions. For example, if we want to compare the relative intensity of $M\zeta$ line of gadolinium measured in the present investigation to that published by Wendt and Christ [6], the corrected value (56%) must be used. Because these authors compared their $M\zeta$ line to the sum of the $M\alpha$ and $M\beta$ lines, $56/(100+87)=30\%$ is the relevant value from our measurement, after correction for effects of both self-absorption and response function. On first sight it seems to contradict their values of approximately 10 percent which can be read from figure 3 in [6] as a limiting value. The authors assumed that detection efficiency does not play an important role. (On the one hand they considered it too small compared with the discrepancy they observed. On the other hand it is indeed small for platinum which they selected for demonstration of the magnitude of the effect.) But if we calculate the absorption in a beryllium window with a thickness of 13 μm (taken from their paper), we can see that the transparency of the window is only 12 percent at the energy of gadolinium $M\zeta$ and 40 percent for the $M\alpha$. Hence, the $M\zeta$ is absorbed approximately three times more than the $M\alpha$. After this obviously significant correction their value agrees with our relative intensity. Evidently, White and Johnson's [7] prediction (0.01%) compares to the above mentioned 56 percent merely because it is related to the $M\alpha$ line alone. The discrepancy is obvious.

M-gamma/M-alpha

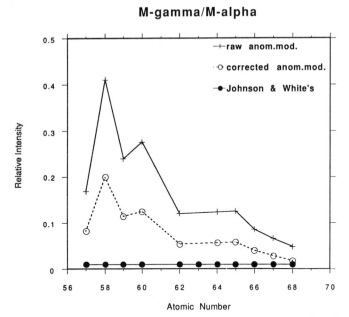

Figure 29. Comparison of measured relative intensities of the M_γ lines, after correction for self-absorption within the sample and for response function of the spectrometer, with predictions of Johnson and White.

M2N4/M-alpha

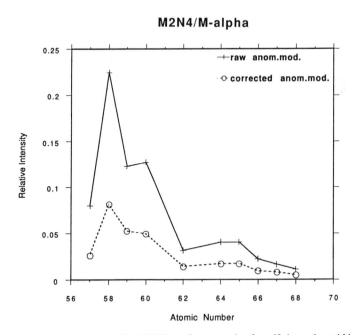

Figure 30. Measured relative intensities of the M_2N_4 lines after correction for self-absorption within the sample and for response function of the spectrometer.

Johnson and White's [7] predictions are also plotted in figures 27b–29 together with our values for all elements and lines. A steady decrease of relative intensities of $M\zeta$, $M\beta$, $M\gamma$, and $M2N4$ lines above $Z = 65$ and a similar decline below $Z = 58$ can be seen on these curves of corrected values. It suggests that anomalously high relative intensities are unique to the rare-earth elements, and the balance between experimental values and predictions tends to hold outside of this region of atomic number. Besides, predicted relative intensities are characteristically lower than measured values after corrections for both anomalous absorption and for response function within this interval of atomic numbers. We conclude that experimental data cannot be interpreted as apparent values only shifted from tabulated relative intensities by the anomalous absorption. On the contrary, the presence of unfilled inner shells must be the common root of both the anomalous absorption and the anomalously high relative intensities.

The relative intensities published in the present paper have several advantages over former collections. The experimental data are corrected for the significant anomalous absorption, $M\alpha$ and $M\beta$ lines were resolved in the measurements and the correction for response function made them universally comparable and applicable.

VIII. Conclusions

(1)   The intensity of M lines of rare-earth elements cannot be measured by their peaks only, due to the serious and excitation-dependent distortion of the peak shape by anomalous absorption. Instead, the area under the peaks must be determined in a manner similar to that used for the analysis of ultra-light elements.

(2)   A new model of absorption has been suggested which accounts for both normal and anomalous absorption phenomena. The latter are unique to the M lines of rare-earth elements, and originate from the partial reversibility of electron transitions because of the presence of unfilled inner shells. The observed dependence of relative line intensities on primary electron beam energy cannot be explained even qualitatively with the conventional approach. The new model correctly describes the experimental data. Even the most unusual characteristics of the measured data are quantitatively explained with the help of the new model. Atomic constants, determined from the behavior of the relative intensity of the $M\zeta$ line, can later be used in the absorption correction of the $M\alpha$ line in electron probe x-ray microanalysis.

(3)   The "effective" mass absorption coefficient approach showed that there is no absorption coefficient which could predict the experimental data within the established absorption models. This fact proves that the introduction of the new model is unavoidable.

(4)   The relative intensities of the $M\zeta$, $M\beta$, $M\gamma$, and $M2N4$ lines were determined. They are compared to the intensity of the $M\alpha$ line. Our results corroborate the discrepancy between measured and tabulated values of the $M\zeta$ lines reported by Wendt and Christ [6] for some elements. $M\beta/M\alpha$ values, which can be deduced from spectra measured under minimum absorption condition in figure 4 of Fischer and Baun [1], are not far from our data. Presented values are corrected for self-absorption in the sample and for response function of the spectrometer. Their anomalously high values, as compared to predictions of White and Johnson [7], must originate from the presence of unfilled inner shells. In line overlap correction procedures these relative intensities must be applied together with corrections for self-absorption in the sample and for response function of the spectrometer.

(5)   Further research is required to determine the extent of chemical effects in the values of parameters of the model presented above.

IX. References

[1] Fischer, D. W. and Baun, W. L. (1967), J. Appl. Phys. **38**, 4830–4836.

[2] Bearden, J. A. (1967), Rev. Mod. Phys. **39**, 78.

[3] Bearden, J. A. and Burr, A. F. (1967), Rev. Mod. Phys. **39**, 125.

[4] Saloman, E. B. and Hubbell, J. H. (1988), At. Data Nucl. Data Tables.

[5] Heinrich, K. F. J. (1986), 11th Int. Cong. on X-Ray Optics and Microanalysis, London (Canada), 67–119.

[6] Wendt, M. and Christ, B. (1985), Cryst. Res. Technol. **20**, 1443–1449.

[7] White, E. W. and Johnson, G. G. (1970), ASTM Data Series **DS 37A**; Johnson, G. G. and White, E. W. (1970), ASTM Data Series **DS** 46.

[8] Wendt, M. (1987), personal communication.

[9] Small, J. A., et al. (1987), J. Appl. Phys. **61**, 459.

[10] Bastin, G. F., et al. (1986), Scanning **8**, 45.

[11] McGuire, E. J. (1972), Phys. Rev. **A5**, 1043.

[12] Chen, M. H., et al. (1980), Phys. Rev. **A21**, 449.

[13] Chen, M. H., et al. (1983), Phys. Rev. **A27**, 2889.

[14] Pouchou, J. L. and Pichoir, F. (1986), 11th Int. Cong. on X-Ray Optics and Microanalysis, London (Canada), 249–253. Errata was presented in: 1st European Workshop on Modern Developments and Applications in Microbeam Analysis, 1989, Antwerp, p. 146. Concerning the mass absorption determinations, see Appendix 5, p. 59 of Pouchou and Pichoir's paper in this publication.

STANDARDS FOR ELECTRON PROBE MICROANALYSIS

R. B. MARINENKO

National Institute of Standards and Technology
Gaithersburg, MD 20899

I. Introduction

Accurate quantitative electron probe microanalysis requires standards of known, homogeneous compositions. This is true for all data reduction procedures. Many pure elements and stoichiometric compounds that are commercially available can be used as reliable standards. Also, well-characterized naturally occurring minerals are used extensively as standards in the analyses of minerals and related materials.

In spite of these readily available standards, there is still a need for additional types of standards. Some elements cannot be found in the forms cited above. For some analyses, complex standards are desirable. In addition, testing of new theories and algorithms used in quantitative data reduction procedures requires specific types of standards or sets of standards to facilitate their evaluation.

For these reasons, the National Institute of Standards and Technology (formerly the National Bureau of Standards) has been involved in the development, evaluation, and certification of standards for microanalysis for more than 20 years. Twelve different Standard Reference Materials (SRM's) have been issued for use in microanalysis [1,2]. Some of these SRM's include one specimen, while others include as many as six different specimens, for a total of 34 different specimen compositions.

II. Certification of SRM's

SRM certification is administered by the Office of Standard Reference Materials (OSRM) which also oversees the distribution and generation of publicity for these standards.

Scientific evaluation of potential SRM's is made by one or more research scientists at NIST. The certification process for SRM's used in microanalysis, as well as many other SRM's, is rigorous and time-consuming. The quantitative composition of a material being evaluated for certification is determined by one of three ways: (1) by a definitive analytical method [3]; (2) by two or more different, independent, reliable analytical methods; or (3) by a comprehensive study through a network of cooperating laboratories [4]. For all potential SRM's the extent of macrohomogeneity, or interspecimen homogeneity, is evaluated. This assures the purchaser that any particular specimen is representative of the certified lot of SRM specimens. For microanalysis, the homogeneity on the micrometer scale must also be evaluated, and this homogeneity (or extent of heterogeneity) of each specimen must be shown to be consistent with the lot.

Another type of material issued by OSRM is the Research Material (RM). RM's are not as rigorously analyzed as SRM's and are not generally certified for composition or homogeneity. These materials are issued with a Report of Investigation by an NIST staff member who is the sole authority on the information provided. The purpose of an RM is to further scientific research on a particular material.

Homogeneity for SRM certification is evaluated from five or six specimens randomly selected from a lot of 100–500 specimens. These are mounted and polished for electron probe microanalysis (EPMA). The specimens are analyzed in random order using wave-

Electron Probe Quantitation, Edited by K.F.J. Heinrich and
D. E. Newbury, Plenum Press, New York, 1991

length- or energy-dispersive spectrometers to analyze the elements of interest. Duplicate readings are taken at a total of five to six randomly selected locations on the specimen. The resulting data are statistically evaluated to determine the within and between specimen homogeneities [5]. The experimental error (measurement imprecision) is also determined from the data.

In addition to the statistical specimen intercomparison, periodic integrator traces [6] such as those in figures 1 and 2 are prepared. These plots provide a rapid visual display of the microhomogeneity of the specimen being studied. Such traces are prepared by moving the specimen in a stepwise fashion under the electron beam. At each point, x-ray counts for the elements of interest are accumulated for a preselected time period, usually 10 s. While x-ray counts are integrated on the next point for the allotted time period, the total counts accumulated for the previous point are recorded on the continuously moving chart paper. A trace of the experimental error alone, such as the one to the left in each figure, is acquired by defocussing the electron beam onto a 20-μm diameter spot on the specimen. Without moving the specimen under the beam, the integrated number of counts per time period is repeatedly recorded. Sample inhomogeneities are therefore suppressed; signal deviations are attributed to the experimental error alone.

To the right of the periodic integrator trace is a double-headed arrow that defines the $\pm 3s$ limits about the average number of integrated counts/time period for the trace. Deviations that fall outside of this $\pm 3s$ limit are expected to be from sample fluctuations or heterogeneity from point to point. For a specimen that is stable under the electron beam, the trace at the left of the stationary specimen under the defocussed beam should lie within these limits unless a serious experimental error is present.

In figure 1 are periodic integrator traces of iron, chromium, and nickel simultaneously recorded with wavelength spectrometers from a specimen of SRM 479a. The trace of the moving specimen is no different than the trace of the stationary specimen under the defocussed beam, and signal deviations are consistently within the $\pm 3s$ limits. Therefore, this specimen is homogeneous on the micrometer scale. Periodic integrator traces for a specimen that exhibits observable inhomogeneities are shown in figure 2. Although this specimen is an SRM, it was not certified as a standard for microanalysis. Chromium and magnesium show large deviations beyond the $\pm 3s$ limits. Deviations for both elements occur simultaneously, suggesting that such variations are probably due to inclusions that occur just below the sample surface. (The region of the specimen selected for the traverse appeared to be inclusion-free to the naked eye.) Even though the molybdenum trace appears to be homogeneous, i.e., within the $\pm 3s$ limits, a conclusion that this is a homogeneously distributed element would be wrong since the signal for this element above the background is quite low.

A procedure for homogeneity testing that has been used more recently in our laboratory is digital compositional mapping [7,8]. Although the technique was not developed for homogeneity testing, it is useful for this application. In the mapping procedure the electron beam is moved under digital control to collect data at specific points on the specimen in an array pattern, e.g., 64×64, 128×128, or 256×256 pixels. The beam resides on each specimen point for a preselected time period during which counts for each of the elements of interest are accumulated. This can be done using either energy dispersive (EDS) or wavelength dispersive (WDS) detectors or a combination of both. Since there are only three wavelength spectrometers on our instrument, EDS mapping is preferred when the number of elements of interest exceeds three and when the concentration of each is sufficient to provide good count rates at reasonable counting times per point. It would also be preferred for low magnifications where the mapped region is larger than 50 μm $\times$ 50 μm since WDS mapping would require defocussing corrections at these magnifications. But the lower

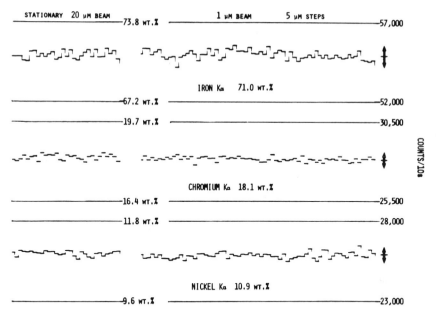

Figure 1. Periodic integrator traces showing homogeneity of iron, chromium, and nickel simultaneously recorded from NBS SRM 479a. (Excitation potential $=20$ kV, specimen current $=4.2\times10^{-8}$ A.) In the traces on the right, the sample was advanced in 5-μm steps under a 1-μm electron beam after each 10-s counting period. To the left, the sample was not moved during repeated 10-s counting periods with a 20-μm$\times$20-μm scanning raster. The double-headed arrows to the right represent a range of ±3 s around the average number of counts per 10 s, $\bar{N}$, for the entire trace.

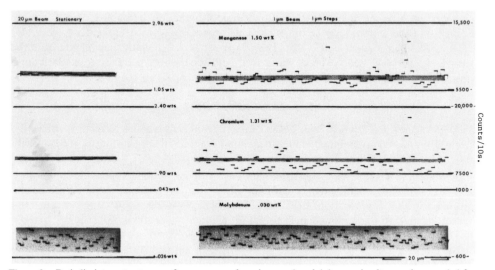

Figure 2. Periodic integrator traces of manganese, chromium, and molybdenum simultaneously recorded from NBS SRM 663 steel. (Excitation potential $=30$ kV, beam current $=1.3\times10^{-8}$ A.) In the traces on the right, the sample was advanced in 1 μm steps under a 1-μm electron beam after each 10-s counting period. To the left, the sample was not moved during repeated 10-s counting periods with a 20-μm$\times$20-μm scanning raster. The shaded regions represent a range of ±3 s around the average number of counts per 10 s, $\bar{N}$, for the entire trace.

Table 1. NIST (NBS)[a] Standard Reference Materials for Microanalysis

SRM NO.	Type	Form	Elements (nominal wt%)	Ref.
SRM 470	Mineral Glasses for Microanalysis	Slices	Two glasses containing SiO_2, FeO, MgO, CaO, and Al_2O_3	[14]
SRM 471	Glasses for Microanalysis	Slices	15 compositions of various oxides	[15]
SRM 478	Cartridge Brass	Cube & cylinder	Cu-73, Zn-27	[11]
SRM 479a	Fe-Cr-Ni Alloy	Wafer	Fe-71, Cr-18, Ni-11	[10,5]
SRM 480	Tungsten-20% Molybdenum	Wafer	W-78, Mo-22	[7]
SRM 481	Gold-Silver	Six wires	Au-100; 80; 60; 40; 20; 0 Ag-0; 20; 40; 60; 80; 100	[8]
SRM 482	Gold-Copper	Six wires	Au-100; 80; 60; 40; 20; 0 Cu-0; 20; 40; 60; 80; 100	[8]
SRM 483	Iron-3% Silicon	Small sheet	Fe-97, Si-3	[9]
RM 31	Glass Fibers for Microanalysis	Fibers	10 compositions of various oxides	

[a] The standards are presently labelled NBS SRM's but will be labelled NIST SRM's when current supplies are depleted and new material is recertified.

detection limits of the WDS detector make it more useful for testing elements at low concentration.

The resulting map, displayed on an image analysis system, would show no variations for a homogeneous specimen. The map can also be evaluated statistically to determine the extent of homogeneity of the region studied. When several digital maps are prepared from each representative specimen, a relatively accurate estimate of the homogeneity (or inhomogeneity) of the material can be obtained. This procedure was used recently to re-evaluate some SRM 483 specimens (Fe-3Si). The relative standard deviation calculated from the maps was 1.5 percent, which agreed fairly well with the theoretical relative standard deviation of 1.0 percent predicted by Poisson statistics.[1]

III. Metal Alloy SRM's

A list of the currently available SRM's for microanalysis is given in table 1, where the concentrations listed are nominal values in weight percent. All concentrations subsequently mentioned will also be in weight percent.

The earliest SRM's were mostly binary metal alloys with compositions that were known to exhibit reasonable macro- and micro-homogeneity, and that could be accurately quantitated in bulk. In addition, these early standards were materials that had industrial applications making them immediately useful in quantitative analyses and quality control.

The first standard for microanalysis, SRM 480, was issued in 1968. It was a filament material consisting of an 80 percent tungsten-20 percent molybdenum alloy [7] prepared by a powder metallurgy process from high-purity metal powders. The resulting alloy was formed into a wire 1 mm in diameter and embedded in a pure molybdenum rod 5 mm in

---

[1] A Report of Analysis of SRM 483 by the Electron Probe Microanalyzer which describes these results is available from the author.

diameter that was subsequently covered by electroplating with a 1-mm-thick tungsten layer. Wafers 1-mm thick were cut from the rod for standard specimens. Each wafer then included the pure elements as well as the alloy.

The next two standards, SRM's 481 and 482 [8], were issued in 1969. Each includes six wires 0.5 mm in diameter. The nominal weight percent compositions in SRM 481 are pure gold, 80Au-20Ag, 60Au-40Ag, 40Au-60Ag, 20Au-80Ag, and pure silver. In SRM 482 the nominal compositions are pure gold, 80Au-20Cu, 60Au-40Cu, 40Au-60Cu, 20Au-80Cu, and pure copper. These SRM's have proved to be extremely useful not only as quantitative standards, but also in evaluating correction procedures for electron probe microanalysis (EPMA). The K, L, and M x-ray lines can all be observed with these alloys.

In 1971, SRM 483, iron-3 percent silicon [9], a transformer steel, was issued. This SRM is supplied in small sheets about 3 mm $\times$ 3 mm and less than 0.5 mm thick. The low silicon concentration makes this a useful standard for the analysis of silicon in an intermediate atomic number matrix. As long as there is a linear response between this standard and the unknowns, it is especially useful in the analysis of silicon in steels where the silicon concentration is often very low—much less than 3 percent.

The original iron-chromium-nickel alloy, SRM 479 [10], was issued in 1972. This is a common alloy used in the steel industry. Because of its popularity, this SRM was reissued in 1980 as SRM 479a [5]. Specimens are about 4.5 mm in diameter and about 0.8-mm thick. This is the only ternary metal alloy SRM. The nominal concentrations are 71 percent iron, 18 percent chromium, and 11 percent nickel; it is useful in the analysis of minor concentrations of chromium and nickel.

Cartridge brass, an industrial brass alloy which contains 73 percent copper and 27 percent zinc, was issued in two forms as SRM 478 in 1974. One form is chill cast and is supplied as a cube 6 mm on an edge. The other is wrought and is supplied as a cylinder 6 mm in diameter and 6 mm high. The copper and zinc have been certified for microhomogeneity [11].

IV. Glass SRM's

The more recent SRM's issued by NIST have been glasses. There are several reasons why glasses make useful standards. Many different oxides or other glass-forming compounds can be combined into a variety of different concentration ranges to form homogeneous vitreous solids. Therefore, glasses enable the design and preparation of standards with desired characteristics without the limitations imposed by metal alloys. More than just two or three elements can easily be incorporated into a single glass; thus the coexistence of several oxides or other glass-forming compounds in the same matrix is possible. Also, small additions (up to about 2 wt%) of many oxides can be made to most glasses without disturbing their homogeneity. This results in interesting standards containing several low-concentration elements. For some elements, such as the halogens, strontium, cesium, and several rare earth elements, this appears to be the only reliable mechanism for fabricating a standard.

Glasses can also be made into several useful shapes, including fibers having diameters of 1–50 μm (fig. 3), spheres from less than 1 μm to as much as 200 μm in diameter (fig. 4), and thin films [12,13].

The first glass standard, Mineral Glasses for Microanalysis—SRM 470, was issued in 1981 [14]. This SRM consists of two glasses in the form of bars about 2 mm $\times$ 2 mm $\times$ 20 mm. The two glasses in SRM 470, K-411 and K-412, contain the same oxides frequently found in naturally-occurring minerals. Both contain the oxides of magnesium, silicon, calcium, and iron while K-412 also includes aluminum oxide. The nominal concentrations are

Figure 3. Secondary electron image of glass fibers.

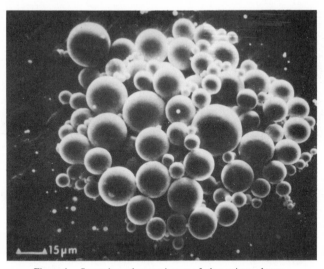

Figure 4. Secondary electron image of glass microspheres.

Table 2. SRM 470, mineral glasses for microanalysis (nominal compositions in weight percent)

Glass	SiO$_2$	FeO	MgO	CaO	Al$_2$O$_3$
K-411	55	15	15	15	--
K-412	45	10	20	15	10

listed in table 2. They are very useful in mineral and ceramics analyses as well as in the analysis of other glasses, oxide-containing materials, and corrosion products.

Glass K-411 has been used to prepare sputtered glass films [13] which were issued in 1987 (SRM 2063) as standards for the analytical electron microscope (AEM). These films were formed by bombarding the bulk glass, K-411, with argon ions and collecting the sputtered glass on carbon support films on 200-mesh copper grids. To minimize matrix effects, the final film thickness was limited to 0.1 μm. Since the integrity of the original bulk glass is not necessarily maintained in the sputtering process, an additional film, with sufficient mass for later quantitation by chemical analysis, was formed simultaneously during the sputtering process.

Glass K-411 is presently being used for the preparation of glass spheres which are being characterized by EPMA for possible certification as an SRM. These spheres are prepared by passing glass shards through a furnace in a gas stream [12,16]. The diameters of the resulting spheres range from 1–12 μm.

A more complex group of glass standards was issued in 1984 [15]. These five standards, SRM's 1871–1875, include a total of 15 different glasses. Each SRM contains three glasses with the same matrix, but two of the glasses contain several additional oxides at concentration levels below 2 weight percent. Because of the complexity of these materials, not all major elements have been certified. In some cases, where the results from only one method of analysis were available, information values are provided. Also, the minor elements are not certified. A list of the elements and their concentrations in these SRM's is given in table 3.

One Research Material, RM 31, is presently available for microanalysis. RM 31 consists of a group of 10 glass fibers with compositions similar to two glasses (the two K-400 glasses) from each of the SRM's 1871–1875.

V. Application of SRM's to Data Reduction Calculations

As previously mentioned, the usefulness of SRM's goes beyond quantitative analysis, quality control applications, and instrument calibration. Because the concentrations of their components are accurately known, these SRM's are useful in the evaluation of several matrix corrections such as those for atomic number, absorption, and fluorescence (ZAF), used in data reduction programs such as FRAME [17] and COR [18].

The tungsten-20 molybdenum alloy, SRM 480, can be used to evaluate the atomic number correction for tungsten for the L and M x-ray lines. Because of the predominantly iron matrix in the iron-3 silicon alloy (SRM 483), a large atomic number correction for silicon is required in addition to a correction for the absorption of silicon x rays by iron. Both nickel and iron in the iron-chromium-nickel alloy (SRM 479a) excite fluorescent emission from chromium. At the same time nickel is strongly absorbed by iron and iron is strongly absorbed by chromium. The cartridge brass, SRM 478, can be used to simultaneously observe the high energy K-lines and low energy L-lines of both copper and zinc in wavelength dispersive analyses (WDS) while the overlap of the Cu$K\beta$ x-ray line, which occurs under the Zn$K\alpha$ line in the energy dispersive (EDS) spectrum, makes this a useful material to study the overlap correction in EDS.

Table 3. Glasses for microanalysis (composition in weight percent)

Elem.	SRM 1871 Glass (Wt. %)			SRM 1872 Glass (Wt. %)			SRM 1873 Glass (Wt. %)			SRM 1874 Glass (Wt. %)			SRM 1875 Glass (Wt. %)		
	K-456	K-493	K-523	K-453	K-491	K-968	K-458	K-489	K-963	K-495	K-490	K-546	K-496	K-497	K-1013
Pb	65.67	63.28	63.10	54.21	54.69	54.74	---	(1.32)	---	---	(1.47)	---	---	(0.86)	---
Si	13.37	(13.09)	(12.94)	28.43	(0.11)	---	23.05	(22.23)	(21.96)	---	(0.19)	---	---	(0.13)	---
Ge	---	---	(0.24)	---	26.10	25.93	---	---	(0.39)	---	---	(0.50)	---	---	(0.34)
Ba	---	---	(0.61)	---	---	(0.46)	41.79	39.53	39.21	---	---	(0.99)	---	---	(0.52)
Zn	---	---	---	---	---	---	3.01	2.93	2.95	---	---	---	---	---	---
P	---	---	(0.24)	---	---	(0.21)	---	---	(0.33)	---	---	(0.42)	32.98	31.59	32.26
Mg	---	---	(0.12)	---	---	(0.22)	---	---	(0.34)	---	---	(0.17)	6.65	6.49	5.86
Al	---	(0.13)	---	---	(0.10)	---	---	(0.11)	---	10.89	(10.2)	(10.1)	6.47	5.97	6.08
B	---	[0.04]	---	---	[0.03]	---	---	[0.06]	---	(23.0)	(21.5)	(21.6)	---	[0.05]	---
Zr	---	(0.38)	(0.33)	---	(0.26)	(0.48)	---	(0.40)	(0.61)	---	(0.53)	(0.52)	---	(0.32)	(0.45)
Ti	---	(0.20)	(0.21)	---	(0.14)	(0.16)	---	(0.27)	(0.32)	---	(0.31)	(0.39)	---	(0.22)	(0.21)
Ce	---	(0.53)	---	---	(0.59)	---	---	[0.80]	---	---	(1.46)	---	---	(0.94)	---
Ta	---	(0.64)	---	---	(0.52)	---	---	(0.95)	---	---	(1.02)	---	---	(0.71)	---
Fe	---	(0.25)	---	---	(0.17)	---	---	(0.35)	---	---	(0.38)	---	---	(0.26)	---
Li	---	[0.0005]	---	---	[0.0005]	---	---	[0.0009]	---	(2.3)	(2.2)	(2.2)	---	[0.0005]	---
Ni	---	---	(0.25)	---	---	(0.20)	---	---	(0.33)	---	---	(0.39)	---	---	(0.31)
Eu	---	---	(0.73)	---	---	(0.64)	---	---	(0.95)	---	---	(1.21)	---	---	(0.53)
U	---	---	(0.23)	---	---	(0.05)	---	---	(0.16)	---	---	(0.24)	---	---	(0.15)
Th	---	---	(0.08)	---	---	(0.12)	---	---	(0.06)	---	---	(0.16)	---	---	(0.10)
Cr	---	---	(0.20)	---	---	(0.19)	---	---	(0.31)	---	---	(0.14)	---	---	(0.14)
O	(20.35)	(20.58)	(20.82)	(16.73)	(16.45)	(16.67)	(31.86)	(31.84)	(31.96)	(63.49)	(60.75)	(61.36)	(53.90)*	(52.46)*	(53.05)*
Total	(99.39)	(99.12)	(100.10)	(99.37)	(99.16)	(100.07)	(99.71)	(100.79)	(99.88)	(99.68)	(100.01)	(100.39)	(100.00)	(100.00)	(100.00)

Values in parentheses are for information only, they are *not certified.*
Values in brackets were calculated from the weight of material added to the melt, they are *not certified.*
*Oxygen values in SRM 1875 were calculated by difference, not by the stoichiometry of the oxides as was done for the other glasses.

The gold-silver and gold-copper alloys, SRM's 481 and 482, respectively, were fabricated specifically for use in the evaluation of data reduction procedures. Five types of x-ray lines, AgL, AuL, and AuM, and CuK and CuL, can be observed. Since these standards cover a range of compositions, k-ratio (corrected intensity of the unknown/corrected intensity of the pure standard) plots can be prepared for the alloys of each SRM at different analytical excitation potentials [8]. For all alloys, atomic number corrections are necessary in addition to substantial absorption corrections for the CuL, AgL, and AuM lines. The AuL lines excite fluorescence from the CuK line and continuum fluorescence enhances the CuK and AuL lines. Since the concentrations are well established from independent methods of analysis, deviations from the curve of a plot of k-ratios vs. concentration for a set of alloys at a given excitation potential would signify errors in the ZAF correction procedure.

The Mineral Glasses for Microanalysis, SRM 470, are especially useful in the EDS analysis of materials containing magnesium, aluminum, and silicon where the x-ray lines of these elements overlap in the low-energy region of the spectrum. For data reduction evaluation, these materials are useful where diminished interelement effects are desirable. Since all elements are present as oxides, interelement effects are minimized compared to the binary and ternary metal alloys discussed previously. At a minimum, there is some excitation of a fluorescence emission of the magnesium K line by the aluminum and silicon K x-ray lines as well as aluminum K by silicon K. Some small absorption effects also occur.

SRM's 1871–1873, which are part of the Glasses for Microanalysis series, are useful in the analyses of low atomic number elements in the presence of higher atomic number elements. In SRM 1871, which is a lead silicate matrix (average at. no. 56), the presence of lead introduces large atomic number and absorption effects on the silicon K line. This glass has been extensively studied under several different experimental conditions and different ZAF correction procedures have been used. None of the procedures corrected the enhancement observed in the silicon K line intensity caused by the presence of the lead. The fluorescence enhancement of silicon K by the lead M lines should be minimal. The lead-germanate matrix of SRM 1872 (average at. no. 55) is unusual. There is an atomic number effect on the germanium L lines; absorption of this line also occurs. The germanium K line is excited by the lead L lines. We have had to use COR in the analysis of this glass since FRAME does not make this fluorescence correction. SRM 1873, which is composed of the oxides of barium, silicon, and zinc, exhibits few anomalies except for an atomic number effect on, and a large absorption of, the silicon K line.

The glasses in SRM's 1874 and 1875 have low average atomic numbers. The lithium-aluminum-borate glasses in SRM 1874 (average at. no. 8) require no correction factors, but since the electron beam significantly damages this glass, it is better suited for secondary ion mass spectrometry (SIMS) analysis than for the electron microprobe. The aluminum-magnesium-phosphate glasses in SRM 1875 (average at. no. 11) are more stable under the electron beam. Some corrections are needed for this glass matrix. The potassium K line excites fluorescence emission of the magnesium and aluminum K lines, and there is some absorption of the potassium K x rays.

The glass fibers described previously, RM 31, can be useful in particle analysis studies. If laid flat in a direction parallel to the energy dispersive detector, these fibers are similar to bulk material; but if laid perpendicular to the detector, they behave more like particles. These observations have been confirmed by Monte Carlo [19] calculations.

The glass spheres make useful standards in particle analyses. Because the geometry of the sphere is known, they can be used in comparing experimental data with the results of Monte Carlo calculations.

VI. Conclusion

During the last 20 years, NIST has certified a variety of SRM's for use in microanalysis. Most contain two or more elements that are homogeneous on the micrometer scale. These standards, with matrices not readily available commercially, are useful in quantitative analysis, as well as in instrument calibration and evaluation of matrix correction procedures.

VII. References

[1] R. B. Marinenko and E. B. Steel, NBS standards for microanalysis: certification and applications, J. Trace and Microprobe Techniques **4**(3), 1986, pp. 129–145.

[2] R. B. Marinenko, E. B. Steel, J. A. Small, and D. E. Newbury, Standards for microbeam analysis, Bull. of the Elect. Microsc. Soc. of Am. (USA) **13**(1), 1983, pp. 67–73.

[3] G. A. Uriano and C. C. Gravatt, The role of reference materials and reference methods in chemical analysis, Issue 4 in CRC Critical Reviews in Analytical Chemistry, Vol. 6, Cleveland, OH, Chemical Rubber Co., 1977 October p. 361.

[4] R. W. Seward, ed., NBS Standard Reference Materials Catalog 1988–89 Edition, Natl. Bur. Stand. (U.S.) Spec. Publ. 260, 1988 January p. 38.

[5] R. B. Marinenko, F. Biancaniello, L. DeRobertis, P. A. Boyer, and A. W. Ruff, Preparation and Characterization of an Iron-Chromium-Nickel Alloy for Microanalysis, SRM 479a, Natl. Bur. Stand. (U.S.) Spec. Publ. 260-70; 1981 May, 25 p.

[6] R. B. Marinenko, K. F. J. Heinrich, and F. C. Ruegg, Micro-homogeneity studies of NBS Standard Reference Materials, NBS Research Materials, and Other Related Samples, Natl. Bur. Stand. (U.S.) Spec. Publ. 260-65, 1979 September, 73 p.

[7] H. Yakowitz, R. E. Michaelis, and D. L. Vieth, Homogeneity characterization of NBS spectrometric standards IV, preparation and microprobe characterization of W-20%Mo alloy fabricated by powder metallurgical methods, Advances in x-ray analysis, Vol. 12, New York, NY: Plenum Press; 1969. 418–438 and Natl. Bur. Stand. (U.S.) Spec. Publ. 260-16, 1969 January, 24 p.

[8] K. F. J. Heinrich, R. L. Myklebust, and S. D. Rasberry, Preparation and Evaluation of SRM's 481 and 482 Gold-Silver and Gold-Copper Alloys for Microanalysis, Natl. Bur. Stand. (U.S.) Spec. Publ. 260-28, 1971 August, 89 p.

[9] H. Yakowitz, C. E. Fiori, and R. E. Michaelis, Homogeneity Characterization of Fe-3Si Alloy, Natl. Bur. Stand. (U.S.) Spec. Publ. 260-22, 1971 February, 22 p.

[10] H. Yakowitz and A. W. Ruff, Preparation and Homogeneity Characterization of an Austenitic Iron-Chromium-Nickel Alloy, Natl. Bur. Stand. (U.S.) Spec. Publ. 260-43, 1972 November, 11 p.

[11] H. Yakowitz, D. L. Vieth, K. F. J. Heinrich, and R. E. Michaelis, Homogeneity Characterization of NBS Spectrometric Standards II: Cartridge Brass and Low-Alloy Steel, Natl. Bur. Stand. (U.S.) Misc. Publ. 260-10, 1965 December, 28 p.

[12] J. A. Small, K. F. J. Heinrich, C. E. Fiori, R. L. Myklebust, D. E. Newbury, and M. F. Dilmore, The production and characterization of glass fibers and spheres for microanalysis, in Scanning Electron Microscopy, Vol. I, AMF O'Hare, IL, in Scanning Electron Microscopy, Inc., 1978, pp. 445–454.

[13] E. Steel, D. Newbury, and P. Pella, Preparation of thin-film glass standards for analytical electron microscopy, in Analytical Electron Microscopy—1981, Geiss, R. H., ed., San Francisco Press, 1981, pp. 65–69.

[14] R. B. Marinenko, Preparation and Characterization of K-411 and K-412 Mineral Glasses for Microanalysis: SRM 470. Natl. Bur. Stand. (U.S.) Spec. Publ. 260-74, 1982 April, 16 p.

[15] R. B. Marinenko, D. H. Blackburn, and J. B. Bodkin, Glasses for Microanalysis: SRM's 1871-1875, Natl. Inst. Stand. Technol. Spec. Publ. 260-112; 1990.

[16] J. A. Small, J. J. Ritter, P. J. Sheridan, and T. R. Pereles, Methods for the production of particle standards, J. Trace and Microprobe Techniques **4**(3), 1986, pp. 163–183.

[17] H. Yakowitz, R. L. Myklebust, and K. F. J. Heinrich, FRAME: An On-Line Correction Procedure for Quantitative Electron Probe Microanalysis, Natl. Bur. Stand. (U.S.) Tech. Note 796, 1973 October, 46 p.

[18] J. Henoc, K. F. J. Heinrich, and R. L. Myklebust, A Rigorous Correction Procedure for Quantitative Electron Probe Microanalysis (COR2). Natl. Bur. Stand. (U.S.) Tech. Note 769, 1973 October, 46 p.

[19] D. E. Newbury and R. L. Myklebust, Monte Carlo electron trajectory calculations of electron interactions in samples with special geometries, in Electron Beam Interactions With Solids for Microscopy, Microanalysis, and Microlithography, AMF O'Hare, IL, Scanning Electron Microscopy, Inc., 1982, pp. 153–63.

QUANTITATIVE ELEMENTAL ANALYSIS OF INDIVIDUAL
MICROPARTICLES WITH ELECTRON BEAM INSTRUMENTS

JOHN T. ARMSTRONG

Division of Geological and Planetary Sciences
California Institute of Technology
Pasadena, CA 91125

I. Introduction

Chemical characterization of individual microparticles is of importance to a number of applications including air pollution and occupational health research, pathology, geology and cosmochemistry (e.g., background tropospheric aerosols and interplanetary dust particles), experimental petrology, corrosion and pigments research, forensic chemistry, fallout and explosives studies, and a variety of materials research areas. A number of microbeam analysis techniques have proven useful for such chemical characterization (e.g., optical microscopy, laser Raman spectroscopy, LAMMA, ion microprobe analysis). One of the most commonly used techniques is x-ray emission analysis with electron microbeam instruments (electron microprobe, SEM, analytical electron microscope). Size distribution, morphometric, electron diffraction and *qualitative* elemental analysis of microparticles have become straightforward applications for electron microbeam instruments; however, quantitative elemental analysis of individual microparticles has remained one of the most difficult applications for these instruments and has been seriously pursued by relatively few researchers. This paper will consider analytical techniques and correction procedures to enable quantitative analysis of individual microparticles and the magnitude of analytical error to be expected in using these procedures.

II. Theoretical

In order to be able to quantitatively analyze individual microparticles with electron microbeam instruments, it is necessary to correct for the effects of sample size and shape on emitted x-ray intensity. For conventional thick, polished specimens, correction for such effects is relatively simple. Such samples are infinitely thick with respect to electron penetration and, if the electron beam is normal to the sample surface, the correction for x-ray absorption in the sample is a simple geometric function

$$I'_A(\rho z) = \phi_A(\rho z) e^{-\mu_A \csc\psi \rho z} \tag{1}$$

where

$$\rho = \text{density}$$
$I'_A(\rho z) = $ emitted x-ray intensity from the layer at depth Z in the sample
$\phi_A(\rho z) = $ generated primary x-ray intensity from layer at depth Z in the sample
$\mu_A = $ mass absorption coefficient by the sample for element A's x rays
$\psi = $ take-off angle, angle between the sample-to-detector axis and the plane of the sample surface.

Electron Probe Quantitation, Edited by K.F.J. Heinrich and
D. E. Newbury, Plenum Press, New York, 1991

261

To calculate the total emitted primary x-ray intensity, one simply needs an accurate expression for $\phi(\rho z)$ and integrate the eq (1) from $\rho z = 0$ to ∞ or total depth of electron penetration (although as discussed below, derivation of a universally reliable expression for $\phi(\rho z)$ is by no means simple). In the case of the microparticle, electrons that would have stayed in the bulk specimen generating primary x rays may pass through the bottom or scatter out of the sides of the particle. Moreover, the x-ray absorption path length becomes a complicated function of x, y, and z related to the particle's size and shape. Thus, eq (1) becomes for the particle.

$$I'_A(\rho z) = \int_{y=\alpha_1(\rho z)}^{\alpha_2(\rho z)} \int_{x=\beta_1(\rho y,\rho z)}^{\beta_2(\rho y,\rho z)} \phi_A(\rho x,\rho y,\rho z) e^{-\mu_A g(\rho x,\rho y,\rho z)} d\rho x \, d\rho y \tag{2}$$

where $g(\rho x,\rho y,\rho z)$ is the distance from the point of x-ray generation to the particle surface along the sample-to-x-ray detector axis.

In order to calculate $\phi(\rho x,\rho y,\rho z)$ it is necessary to estimate the spacial and energy distribution of the energetic electrons in the sample. For a multi-element sample containing a weight fraction C_A of element A, the number of x rays, dI, produced per electron path length, ds, is

$$dI = C_A \omega_A p_{iA} \frac{N_A \rho}{A_A} Q_A \, ds \tag{3}$$

where

N_A = Avogadro's number
ρ = sample density
A_A = atomic weight of element A
$C_A \left(\dfrac{N_A \rho}{A_A} \right)$ = number of A atoms per unit volume
Q_A = ionization cross section for A atoms, a function of the electron energy, E, and the energy required to produce the ionization, E_C
ω_A = fluorescence yield, probability of ionization resulting in x-ray emission of A, nominally a constant
p_{iA} = probability of x-ray emission being of the particular line of interest.

If one knows the number of electrons that pass through each point x, y, z, their energy distribution and their angular distribution of travel through the x, y, z volume element, one can then derive an expression for $\phi(\rho x,\rho y,\rho z)$ from eq (3):

$$\phi(\rho x,\rho y,\rho z) = C_A \frac{N_A}{A_A} \omega_A p_{ij} \int_{E=E_0}^{E_c} \int_{\theta=0}^{2\pi} \int_{\gamma=0}^{2\pi} \frac{n(E,\rho x,\rho y,\rho z,\theta,\gamma)}{d\rho s/dV} \times \frac{Q_A(E)}{dE/d\rho s} d\gamma \, d\theta \, dE \tag{4}$$

where

$n(E,\rho x,\rho y,\rho z,\theta,\gamma)$ = the number of electrons of energy E, scattering angle relative to the beam axis θ, and azimuthal angle in the plane normal to the beam axis γ
$dE/d\rho s$ = the mean electron energy loss while traveling $d\rho s$
$d\rho s/dV$ = the distance traveled by the electron going through the volume element dV at point x,y,z

E_0=electron beam accelerating potential

E_C=critical excitation potential for ionization of an electron from the shell of interest.

There currently are no proposed general analytical expressions for $n(E...)$ although this function can be calculated statistically by Monte Carlo methods incorporating probability functions for electron scattering (see below). If all of the necessary functions were known, the total emitted primary (electron-produced) x-ray intensity could be determined by (A) combination of eqs (2) and (4), (B) determination of an analytical expression to describe the particle boundary (e.g., choice of a geometric model, such as a sphere) for calculation of the x-ray path length, g, and the limits of integration of x and y, and (C) integration with respect to ρz from 0 (surface) to the particle thickness, T (e.g., Armstrong and Buseck, [1]).

Calculation of the total emitted secondary x-ray emission, from fluorescence by characteristic x-ray lines and the x-ray continuum, is considerably more complicated than calculation of the primary emitted intensity. For the case of characteristic fluorescence, the j-lines of element B that are sufficiently energetic to fluorescence x rays of the k-line of element A are produced in the specimen in accord with the function, $\phi_{Bj}(\rho x,\rho y,\rho z)$. These x rays can travel anywhere in the particle and be absorbed with the possibility of exciting a k-line x ray of element A. The fraction of secondarily excited x rays produced in the sample that are emitted in the direction of the x-ray detector is then a function of where in the particle the production occurred. Armstrong and Buseck [2] derived the general expression for calculation of this emitted intensity:

$$I'_{f,Ak}=C_A\frac{r_A^k-1}{r_A^k}\omega^k_A\, p_{Ak}\frac{\Delta\Omega}{4\pi}\sum_B\sum_j I_{Bj}\,\mu^j_{B,A}$$

$$\times\int_{\rho z=0}^{T}\frac{1}{a_0}\int_{y=\alpha_1(\rho z)}^{\alpha_2(\rho z)}\int_{x=\beta_1(\rho y,\rho z)}^{\beta_2(\rho y,\rho z)}\int_{\xi=0}^{2\pi}\int_{\theta=0}^{\pi}\int_{s=\gamma_1(\rho z,\rho y,\rho x,\xi,\theta)}^{\gamma_2(\rho z,\rho y,\rho x,\xi,\theta)}$$

$$\times\left[\phi_{Bj}(\rho x,\rho y,\rho z)\tan\theta\, e^{-\mu^j_B\sec\theta\,\rho s}\, e^{-\mu^k_A\, g(\rho z,\rho y,\rho x,\xi,\theta,s)}\right]ds\, d\theta\, d\xi\, d\rho x\, d\rho y\, d\rho z \qquad (5)$$

where

$I'_{f,Ak}$=emitted characteristic fluorescence x-ray intensity of element A's k-line

I_{Bj}=produced primary intensity of j-line of element B which excites element A

r_A^k=absorption edge jump ratio for k-line of element A

$\mu^j_{B,A}$=mass absorption coefficient of j-line of element B by element A

μ^j_B=mass absorption coefficient of j-line of element B by matrix

$\dfrac{\Delta\Omega}{4\pi}$=fraction of fluorescent radiation leaving sample that reaches detector

$g(z,y,x,\xi,\theta,s)$=path length of secondary x ray from point of production to point of emission from sample.

Calculation of this expression involves the same difficulties as described for the primary x-ray emission in terms of expression of the x-ray production function, $\phi_{Bj}(\rho x,\rho y,\rho z)$. In addition, calculation of fluorescence for a particle is extremely tedious because of the six-fold integration. Calculation of fluorescence by the continuum is similar to characteristic fluorescence with the addition of integration with respect to x-ray energy and inclusion of a function of continuum x-ray production as a function of energy and position in the sample.

III. Methods of Particle Analysis

The various methods proposed to perform quantitative microparticle analysis with electron beam instruments have been reviewed by Armstrong [3] and Small [4]. This section will describe how the most commonly used procedures and the most rigorous procedures can be employed and the following section will compare analytical results using several of these procedures.

A. IGNORE PARTICLE EFFECTS

The easiest way to attempt quantitative analysis of particles is to ignore the effects of sample size and shape. One can then analyze particles relative to conventional thick, polished specimens or make up particle standards by grinding up pieces of standards and dispersing them on a substrate. Using thick, polished samples will certainly produce systematic errors in the analyses because of the differences in x-ray production and absorption. Use of particle standards should lessen errors; however, depending on the similarity in size and shape between the sample particles and standard particles, the differences in relative x-ray production and absorption should still produce systematic errors. In order to estimate the magnitude of errors for these types of analyses, it is necessary to analyze suites of standard particles with respect to thick standards or other sets of particle standards and/or go through rigorous calculations of particle corrections (e.g., through particle ZAF or Monte Carlo calculations described below). The magnitude of errors encountered in these types of analyses is discussed in section IV.

B. VERY-SMALL PARTICLE ANALYSIS

For particles considerably smaller than 1 μm in diameter mounted on a thin, low-Z substrate (such as a carbon film on a beryllium TEM grid) there is virtually no absorption or secondary fluorescence. If these particles are analyzed with a high accelerating potential (e.g., 100 to 200 keV in a TEM) the mean free path between inelastic scattering events becomes similar to the particle thickness so that little energy is lost as the electron beam passes through the specimen. In such a case the emitted x-ray intensity can be calculated simply from eq (3), where the ionization cross section is evaluated for $E = E_0$, the accelerating potential of the electron beam. If one has thin film standards in order to calculate the detection efficiency of the different elements, and if one can accurately estimate the sample and standard thicknesses, then accurate analyses can be performed using eq (3). The main limitation of this procedure is in the maximum size of the particles that be analyzed. For particles greater than 100 to a few hundred nm, absorption and inelastic scattering effects start becoming significant and this approach cannot be accurately employed. A review of analytical electron microscopy procedures in analysis of thin films and particles can be found in Goldstein [5].

C. PEAK-TO-BACKGROUND METHOD

An interesting approach to particle analysis is based on correcting for particle size and shape effects by measurement of both peak intensities and background bremsstrahlung or continuum radiation. This approach has been described by Small et al. [6–8], Statham and

Pawley [9], and Small [4]. This procedure is based on the assumption that emitted characteristic x rays and continuum (background) x rays are influenced by particle effects to the same extent, i.e., that the relative depth and lateral distributions of characteristic and continuum x-ray production are the same. If this is the case then the ratio of the emitted peak-to-background radiation, (P/B), for a particle (ptc) is the same as that for a thick, polished specimen (tps) of the same composition, i.e.

$$\left(\frac{P}{B}\right)_{\text{ptc}} = \left(\frac{P}{B}\right)_{\text{tps}}. \tag{6}$$

Rearranging eq (6) results in the basis for the peak-to-background analytical method

$$P'_{\text{tps}} = P_{\text{ptc}}\left(\frac{B'_{\text{tps}}}{B_{\text{ptc}}}\right) \tag{7}$$

and

$$k = \frac{P'_{\text{tps}}}{P_{\text{std}}} = \frac{P'_{\text{tps}}}{B'_{\text{tps}}} \times \frac{B'_{\text{tps}}}{P_{\text{std}}} = \frac{P_{\text{ptc}}}{B_{\text{ptc}}} \times \frac{B'_{\text{tps}}}{P_{\text{std}}} \tag{8}$$

where P'_{tps} and B'_{tps} are the calculated peak and background intensities for a hypothetical thick, polished specimen the same composition as the particle; P_{ptc} and B_{ptc} are the measured peak and background intensities for the particle; P_{std} is the measured peak intensity of a conventional thick, polished standard that is analyzed with the particles; and k is the calculated k-ratio of relative intensity of thick, polished sample to standard that can be input into a conventional ZAF or other microprobe correction program to calculate composition. B'_{tps} is calculated by the equation

$$B'_{\text{tps}} = \sum_i C_i B_{i,\text{E}} \tag{9}$$

where $B_{i,\text{E}}$ is the background intensity measured (or calculated) for a pure-element thick, polished specimen.

There are a number of disadvantages to this procedure. (1) The primary assumption that peak and background generated radiation has the same spatial distribution is suspect; different physical process are involved in production of characteristic and bremsstrahlung radiation and there is no reason why the two should have the same lateral and depth distribution. (The differential depth distribution of continuum radiation is measurable, but there are not sufficient experimental results reported in the literature to determine the similarity between characteristic and continuum production.) (2) The background *must* be at the same energy as the characteristic peak energy in order to have the same absorption by the sample, therefore, the background has to be interpolated from measurements of intensity at real background positions for the sample. (3) Samples cannot be prepared for many pure elements, such as oxygen or sodium, and so the pure-element backgrounds need to be calculated from measurement of compounds. Work by Small et al. [10,11] shows that this is not a straightforward task. (4) Ratioing peak-to-background means ratioing a high intensity to a low intensity, and the analytical results suffer from the poorer precision of the background measurement. (5) Because of the complicated variations in crystal reflection and detector efficiency, it is generally impractical to use this procedure with crystal-WDS detectors, and EDS detectors alone must be used. (6) Care must be taken that the apparent bremsstrahlung background in an EDS spectrum is truly due to continuum x rays alone and not to artifacts produced by pulse pile-up, incomplete charge collection, backscattered electrons entering the detector, etc. These artifacts produce significant components of the apparent "background" on many EDS units. (7) All of the background must come from the particle, *none* must come from the substrate. Therefore, either the particle must be infinitely big in lateral and depth dimensions with respect to the excited volume by the electron beam or the particle must be mounted on a very thin, low-Z substrate (such as a thin carbon film on a beryllium TEM grid).

The advantages of the peak-to-background procedure are that it is easy to calculate, requires little in the way of special calibration, and is the only proposed procedure (other than Monte Carlo calculations) for doing point analyses on *portions* of particle (or rough specimen) surfaces. In comparative studies, the developers of the peak-to-background method report error histograms for analyses of standards that have standard deviations of about 10 percent relative (about 50 to 100% greater than the magnitude of errors reported using the more complicated particle ZAF correction procedures described in the next section (e.g., Small [4])).

D. PARTICLE ZAF CORRECTION METHOD

Armstrong and Buseck [1,2,12,13] and Armstrong [3] proposed an analysis and correction procedure for microparticles involving development of a series of analytical expressions to solve eqs (4) and (5) for particles of given sizes and shapes. Subsequently, Armstrong [14–17] and Storms et al. [18] have proposed a series of modifications to these expressions. In the Armstrong-Buseck method, particle sizes are measured or estimated and particle shapes are approximated as corresponding to idealized geometric shapes such as cubes, spheres, cylinders, square pyramids, etc. For these geometric models, exact analytical expressions are derived for the limits of integration and x-ray path length function for eqs (2) and (5) and approximate expressions are derived for $\phi(\rho x, \rho y, \rho z)$ in eq (4). With these analytical expressions, eqs (2) and (4) can be solved by numerical integration enabling quantitative analysis. The accuracy and ease of use of this method depends upon the analytical procedures chosen as described in the following sections.

Point vs. broad or rastered beam analyses.   If one can express accurately the boundary of a particle with a series of equations and if one can position the beam on a spot of the particle with precisely known orientation, then it is theoretically possible to accurately correct for x-ray absorption and obtain an accurate analysis. In practice, such control of beam positioning and precise determination of particle surface boundary variations is not obtainable. When using a point beam, the effect of particle shape on emitted x-ray intensities, especially for larger particles, can be huge. As a laboratory exercise, I have had my students perform point analyses of a 100 μm spherical particle (or a rough fracture surface) of a Ni-Al alloy at an accelerating potential 20 keV. The relative emitted intensities of Al $K\alpha$ to Ni $K\alpha$ vary by as much as a factor of 10 (1000% variation!) depending upon where on the particle (or fracture surface) the beam is placed. Defocusing the beam or rastering it over the whole particle surface has a tendency to average out the effects of particle irregularities (as well as some of the effects of electron transmission and scattering from the particle) and makes it conform better to the idealized models of the particle shape. This is our preferred approach for particle analysis.

When using a rastered beam, care must be taken to raster a larger area than the particle, since the dwell times at some of the raster limits are considerably different than for the middle portions of the raster. Similarly, when using a defocused beam, the beam should be larger than the particle so that the Gaussian falloff in electron density at the edges of the beam diameter does not result in a nonuniform beam density hitting the particle surface.

Ideally, all elements should be measured with the same spectrometer so that the effective particle geometry for irregularly-shaped particles is the same for all elements. This is the normal situation for measurements with an energy dispersive spectrometer or a single, multiple-crystal spectrometer on an SEM. However, electron microprobes commonly have several crystal spectrometers, each of which covers only a portion of the total range of wavelengths used in analysis. In this latter case, either (1) all of the elements should be measured with spectrometers that are close to each other, (2) each element should be measured on as many spectrometers as possible and the results averaged, or (3) the particle should be rotated to several positions and all of the elements should be analyzed with their respective spectrometers and the results averaged.

There are some drawbacks to broad beam or rastered beam analysis. Some of the electrons from the beam will not hit the particle so care must be taken regarding the choice of substrate materials. Additionally, the absolute emitted intensities will be a function of the fraction of the beam which strikes the particle. Since this is not known, the results must be normalized and one loses an absolute measure of the quality of the analysis and correction procedures. These drawbacks are more than compensated for, however, by the averaging effect of x-ray pathlength and electron scattering and resulting improvement in the scatter of analytical results (see below).

Choice of substrate and coating materials. Since rastered or defocused beam analysis involves a portion of the bombarding electrons penetrating into the substrate in which the particles are situated, the choice of substrate materials is clearly important. When a choice of substrates is possible, care should be taken to insure that the substrate is conducting, does not degrade under electron bombardment, and does not contain any of the elements that are to be analyzed in the particle. For larger particles, ideal substrates are smooth, low-Z materials such as pyrolytic graphite or polished beryllium planchets. For particles smaller than the excited volume of the electron beam, due to the difficulty in correcting for differences in electron backscattering between particle and substrate, it is ideal if the substrate is of similar average atomic number as the particles, but does not contain any elements that are to be analyzed in the particle or that could cause fluorescence of x-ray lines to be analyzed in the particle.

For particles smaller than about 40 μm, the adhesion to the planchet due to surface charge assisted by adhesion provided by the carbon coat is adequate to keep the particles in place. Larger particles need to be held to the planchet with a binding substance such as carbon paint or a concentrated sucrose-citric acid solution. The binder needs to completely dry before the particles are carbon-coated. Bulk fine particle samples can be dispersed on a carbon planchet by (a) placing them in suspension in a fast-evaporating solvent like Freon TF, (b) ultrasonicating the solution to keep the particles in suspension, and (c) pipetting an aliquot of the particle suspension onto a carbon planchet and allowing the solution to evaporate. Even if the particles are semiconducting the contact between particle and substrate may cover a very small area and the particle could charge during analysis. Thus, coating the particles with a thin layer of a conducting, low-Z material such as carbon generally improves the quality of the analyses.

Choice of standard type. Since the Armstrong-Buseck particle correction procedure calculates electron transmission and scattering and x-ray path length as a function of sample size and shape, either conventional thick, polished specimens, thin films or particles can be used as standards. Typically, the most convenient type standard to use and the one which normally involves the least amount of error is the thick, polished specimen. Use of thin films as standards requires a very accurate estimate of their thickness and correction for differential electron scattering if they are not mounted on the same type of substrate as the particles. Use of particle standards involves a proliferation of error in the estimates of standard as well as sample size, shape and density. The biggest source of error in using thick specimens as standards is in the extension of analytical $\phi(\rho z)$ expressions from thin particle samples to thick standards. As will be shown below, this can be a problem in attempting to analyze very small particles on a substrate of considerably different average atomic number.

   Method of analysis.   Particle analyses are performed in a very similar manner as employed for conventional bulk specimens, and can utilize either wavelength dispersive or energy dispersive detectors. For particles bigger than about a μm in diameter, wavelength dispersive analysis provides greater precision and sensitivity, while for submicron particles, energy dispersive analysis can provide better sensitivity (Armstrong [3]). As noted above, the beam is either defocused or rastered so that the entire upper surface of the particle is bombarded by electrons and measurements are made, if possible, on a single spectrometer. Peak and background intensities are determined, as for conventional specimens, and k-ratios are calculated, where

$$k_A = \frac{(\text{Peak} - \text{Background})\text{for element A in particle}}{(\text{Peak} - \text{Background}) \text{ for element A in standard}} \cdot \qquad (10)$$

These k-ratios are then corrected for the effects of electron scattering and deceleration and x-ray absorption and fluorescence, as described in the following sections, and concentrations are calculated.

   The parameters needed for the Armstrong-Buseck corrections that are not required for thick specimens are estimates of the particle size, shape and density. The particle diameter, with respect to the specimen-spectrometer axis, can be directly measured by SEM or optical microscope imaging. The ratio of the particle thickness to diameter can be estimated by the general particle shape or directly measured by imaging the particle after tilting the sample stage. The model geometric shape(s) that most closely correspond to the observed particle shape are chosen using a fairly simple procedure described below.

   Particle densities must be calculated because x-ray absorption is a function of the mass per unit area traversed by the x ray (density×distance). The particle density can usually only be crudely estimated based on a knowledge of the type of material of which the particle is composed or by a weight concentration average of the pure element densities for those elements found in the particle. Thus, typically, estimation of the particle density has a significant uncertainty; the effect of imprecision in density estimates on the corrected composition can be determined by performing calculations for a set of diameters that bracket the magnitude of uncertainty in the density.

   Particle corrections.   Particle size and shape require changes in all of the corrections normally employed for thick, polished specimens as described above. Correction for a particle analysis by the Armstrong-Buseck method effectively involves (1) calculating, from the emitted particle x-ray intensities, the x-ray intensities that would have been generated from primary electron excitation, had the particle been a thick, polished specimen; (2) ratioing these intensities to the calculated primary electron-excitation generated intensities for the standards; and (3) then using conventional "atomic number" corrections to calculate the relative concentrations of the elements from the estimated primary x-ray intensities. These corrections are done in the conventional iterative fashion, since all of the above steps require input of the sample or standard compositions. For the first iteration, concentrations are assumed to be equal to the ratios of the emitted x-ray intensities from the particle to the pure element, thick, polished specimen, the pure element x-ray intensities being calculated from the emitted x-ray intensities of the standards, normalized to sum 100 percent. In subsequent iterations, these k-ratios are multiplied by the correction factors to give new estimates of concentration until the results converge.

Electron transmission and side scatter correction. Correction for the loss of primary generated x-ray intensity in a particle from energetic electrons transmitting themselves through the particle or scattering out of its sides requires an analytical expression for $\phi(\rho x, \rho y, \rho z)$ in eq (4). The Armstrong-Buseck method involves modifying a $\phi(\rho z)$ expression designed for thin films and bulk specimens to account for particle affects as well.

The $\phi(\rho z)$ expression originally chosen for particle corrections (Armstrong and Buseck [1]) was that of Reuter [19]. This expression was designed for thin films and had the advantage of incorporating a term which corrected for differences in the amount of electron backscattering in the thin film and substrate layers. The expression could be modified for particles and worked well in the analysis of particles relative to thick, polished specimens of the same or similar composition (e.g., Armstrong [3,20]). However, while the expression seems to account well for differences in average atomic number for metals, it did not extrapolate well to oxides (see fig. 1) and did not work well when particle sample and thick standard were greatly dissimilar in composition.

Subsequently, the $\phi(\rho z)$ expression of Reuter was replaced by a $\phi(\rho z)$ expression of Armstrong [15–17]. This expression is based on a modification of Gaussian equations for $\phi(\rho z)$ proposed by Packwood and Brown [21] and Brown et al. [22] that were based on experimentally determined $\phi(\rho z)$ curves and at least loosely parameterized in terms of physical expressions used in Monte Carlo calculations of $\phi(\rho z)$. The Armstrong $\phi(\rho z)$ expression is of the form:

$$\phi(\rho z) = \gamma_0 \, e^{-a^2(\rho z)^2} \, (1 - q \, e^{-\beta \rho z}) \tag{11}$$

where

$$a = 2.97 \times 10^5 \, \frac{\bar{Z}^{1.05}}{\bar{A} \, E_0^{1.25}} \left[\frac{\ln(1.166 \, E/J)}{(E_0 - E_c)} \right]^{0.5} \tag{12}$$

$$\beta = \frac{8.5 \times 10^5 \, \bar{Z}^2}{\bar{A} \, E_0^2 (\gamma_0 - 1)} \tag{13}$$

$$\gamma_0 = \frac{5\pi \, U_0}{(U_0 - 1)\ln(U_0)} \left[\ln(U_0) - 5 + 5 \, U_0^{-0.2} \right] \tag{14}$$

$$q = \frac{[\gamma_0 - \phi(0)]}{\gamma_0} \tag{15}$$

$$\bar{Z} = \frac{\sum_i (C_i Z_i / A_i)}{\sum_i (C_i / A_i)} \tag{16}$$

$$= \text{atom concentration weighted atomic number}$$

$$\bar{A} = \frac{\sum_i (C_i)}{\sum_i (C_i / A_i)} \tag{17}$$

$$= \text{atom concentration weighted atomic weight}$$

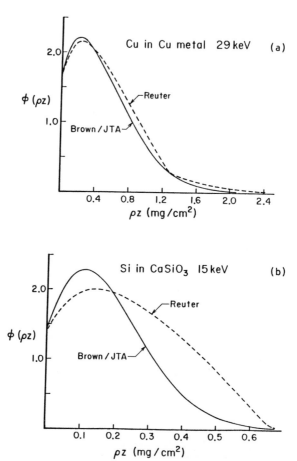

Figure 1. Plot of $\phi(\rho z)$ vs. ρz for (a) Cu $K\alpha$ in Cu metal at $E_0=29$ keV and (b) Si $K\alpha$ in CaSiO$_3$ at $E_0=15$ keV using the Reuter [19] and Armstrong [15,16] [Brown/JTA] equations. While both $\phi(\rho z)$ models produce similar results for Cu metal, which also agrees well with the experimentally-measured $\phi(\rho z)$, the calculated $\phi(\rho z)$ curves are quite different for Si in CaSiO$_3$. For silicates, the Armstrong $\phi(\rho z)$ expression better matches Monte Carlo calculations of $\phi(\rho z)$ and better fits experimentally determined emitted x-ray intensities than does the Reuter model.

and E_0 is the accelerating potential, E_c is the critical excitation potential and J is the mean ionization potential, all in keV. The surface ionization function, $\phi(0)$, can be calculated using either the expression of Reuter [19]

$$\phi(0)=1+2.8\left(1-\frac{0.9}{U_0}\right)\bar{\eta} \tag{18}$$

or the expression of Love et al. [23]

$$\phi(0)=1+\frac{\bar{\eta}}{1+\bar{\eta}}\,[a+b\,\ln(1+\bar{\eta})] \tag{19}$$

where

$$\bar{\eta}=\sum_i C_i\,\eta_i \quad \text{(weight concentration averaged)} \tag{20}$$

$$a=3.43378-\frac{10.7872}{U_0}+\frac{10.97628}{U_0^2}-\frac{3.62286}{U_0^3} \tag{21}$$

$$b=-0.59299+\frac{21.55329}{U_0}-\frac{30.55248}{U_0^2}+\frac{9.59218}{U_0^3} \tag{22}$$

and $U_0=E_0/E_c$. η is the electron backscatter coefficient for an element which can be calculated from a polynomial fit to the data of Heinrich [24] or from the equations of Love and Scott [25]

$$\eta=H1\left[1+H2\,\ln\left(\frac{E_0}{20}\right)\right] \tag{23}$$

where

$$H1=\frac{(-52.3791+150.48371\,Z-1.67373\,Z^2+0.00716\,Z^3)}{10,000} \tag{24}$$

$$H2=\frac{(-1112.8+30.289\,Z-0.15498\,Z^2)}{10,000} \tag{25}$$

and E_0 is in keV. Either $\phi(0)$ expression and either electron backscatter coefficient expression gives similar results. I typically employ the $\phi(0)$ and η expressions of Love and Scott.

The $\phi(\rho z)$ expression of Armstrong has been shown to give good results in correcting for absorption effects in the analysis of silicates, oxides and sulfides (Armstrong [15–17,26,27] and adapts well to the analysis of particles (Armstrong [15]). Figure 1 shows a comparison of $\phi(\rho z)$ curves using the expressions of Reuter and Armstrong for copper metal and for silicon in CaSiO$_3$. Whereas the two expressions agree well for copper metal and also agree well with the experimentally measured $\phi(\rho z)$ distribution, there are significant differences between the curves for the silicates. No $\phi(\rho z)$ distributions have been experimentally measured for multi-element oxides; however the Armstrong $\phi(\rho z)$ much better matches the Monte Carlo calculated $\phi(\rho z)$ and does a much better job in calculating the absorption correction for CaSiO$_3$ than does the Reuter expression (Armstrong [17] and see below).

By setting the integration limits of x, y, and z, the $\phi(\rho z)$ expression can be easily modified to account for those electrons that transmit through the particle or don't strike the particle, but $\phi(\rho z)$ needs to be further modified to account for the effect of electrons scattering from the sides of the particle. This can be done for larger particles by utilizing a fairly simple model. Consider a cubic particle which is considerably larger in diameter than the electron range in it (the distance traveled by the electron until its energy falls below the critical excitation potential of the element of interest). For such a particle, none of the electrons that strike the center of the upper surface of the particle will scatter from the sides of the particles and the distribution of x-ray production for that column of electrons would be equal that for bulk specimen, thus $\phi(0, 0, \rho z)=\phi(\rho z)$. Making the reasonable assumption of an equal scattering possibility at any azimuthal scattering angle, half of the electrons that strike the particle at the middle of any of its upper edges will stay in the particle and half will scatter out. Thus, $\phi(\pm D/2, 0, \rho z)$ and $\phi(0, \pm D/2, \rho z)$ equal 0.5 $\phi(\rho z)$. Similarly 1/4 of the electrons striking the corner of the particle will stay in it and 3/4 will be scattered out. Thus $\phi(\pm D/2, \pm D/2, \rho z)$ equals 0.25 $\phi(\rho z)$. The amount of sidescattering between these points and the center (or a distance equivalent to the electron range toward the center) can be calculated by simple linear or exponential expressions. Armstrong and Buseck [1] incorporates such an expression for cubic particles and shows that the effect using this expression tends to cancel out when the results are normalized and there is not a significant difference from simply assuming $\phi(\rho z)$ is constant over ρx and ρy about the particle surface. In any event, the sidescattered modified (linear variation) cubic particle $\phi(\rho z)$ expression is easy to evaluate and is included as one of the standard particle corrections. The $\phi(\rho x, \rho y, \rho z)$ expression for this case is:

$$\phi(\rho x, \rho y, \rho z)=\left[1-\left|\frac{x}{D}\right|-\left|\frac{y}{D}\right|+\left|\frac{xy}{D^2}\right|\right]\phi(\rho z). \tag{26}$$

For smaller particles, the expression to account for electron sidescatter becomes more complicated. Also, the effect of backscattered electrons will not be properly accounted for with particles of this size, if they are not mounted on a substrate of similar $\underset{\sim}{Z}$. In such cases, Monte Carlo calculations may be required to give best results (see below).

X-ray absorption correction.   Given an analytical expression for $\phi(\rho x, \rho y, \rho z)$ and equations defining the particle surface boundaries, calculation of the amount of x-ray absorption in the particle becomes straightforward if tedious. It simply involves numerical integration of eq (2) with respect to x, y, and z. With current high-speed computers, such calculations can be performed in a matter of seconds; with conventional personal computers (e.g., with a 80287 math co-processor) they can be performed in a matter of a few minutes. Armstrong and Buseck [1] derived the boundary and absorption path length expressions for a variety of geometric models including the (1) cube or rectangular prism (top normal to electron beam, sides parallel and normal to beam-spectrometer axis), (2) tetragonal prism (top normal to electron beam, diagonals parallel and normal to beam-spectrometer axis), (3) triangular prism (rectangular base normal to electron beam, triangular ends parallel to beam-spectrometer axis), (4) square pyramid (square base normal to electron beam, diagonals of square base parallel and normal to beam-spectrometer axis), (5) cylinder (circular bases normal to electron beam), (6) cylinder (circular bases parallel to beam-spectrometer axis), (7) hemisphere (circular base normal to electron beam), and (8) sphere. Armstrong and Buseck showed that the angular models (1–4) could be directly integrated with respect to x and y, leaving numerical integration only with respect to z. Model 5 requires numerical integration with respect to y and z, model 6 with respect to x and z, and models 7 and 8 with respect to x, y, z. The equations of Armstrong and Buseck can be expressed in terms of a particle absorption function $P(\rho,z)$, where

$$P(\rho,z)=\frac{1}{a_0}\int_{y=\alpha_1(\rho z)}^{\alpha_2(\rho z)}\int_{x=\beta_1(\rho y,\rho z)}^{\beta_2(\rho y,\rho z)} h(\rho x,\rho y,\rho z)e^{-\mu_A\, g(\rho x,\rho y,\rho z)}\mathrm{d}(\rho x)\mathrm{d}(\rho y) \tag{27}$$

$$\phi(\rho x,\rho y,\rho z)=h(\rho x,\rho y,\rho z)\phi(\rho z) \tag{28}$$

and

$$f(\chi)_{\mathrm{ptc}}=\frac{\displaystyle\int_{\rho z=0}^{\rho T}\phi(\rho z)P(\rho z)\mathrm{d}(\rho z)}{\displaystyle\int_{\rho z=0}^{\infty}\phi(\rho z)\mathrm{d}(\rho z)} \tag{29}$$

where a_0 is the projected area of the particle with respect to the electron beam, T is the particle thickness, $f(\chi)_{\mathrm{ptc}}$ is the particle absorption correction, and other terms are defined in eqs (1) and (2).

From the work of Armstrong and Buseck [1] with some modifications and corrections proposed by Storms et al. [18], the $P(\rho z)$ equations for the various angular particle models can be expressed as functions of the following terms:

$$\alpha=\mu^j_A \sec\psi \tag{30}$$

$$\beta=\cot\psi\cdot\rho z \tag{31}$$

$$\gamma=\frac{\mu^j_A \sec\psi}{(1+\tan\psi)} \tag{32}$$

$$\zeta=\rho D-2\rho z \tag{33}$$

$$\lambda=\rho D-\beta \tag{34}$$

$$\xi=(\rho D-\rho z)\gamma \tag{35}$$

$$\chi=\mu^j_A \csc\psi\cdot\rho z \tag{36}$$

where ρ=density, D=particle diameter, T=particle thickness, ψ=spectrometer take-off angle, and μ^j_A=mass absorption coefficient of element A's j-line radiation by the matrix.

For the rectangular prism model (#1) with $h(\rho x,\rho y,\rho z)=1$,

$$P(\rho z)=\frac{1}{\rho D}\left[(\rho D-\beta)e^{-\chi}+\frac{1-e^{-\chi}}{\alpha}\right] \tag{37}$$

for $0\leqslant z\leqslant D\tan\psi$, and

$$P(\rho z)=\frac{1}{\rho D}\left[\frac{1-e^{-\alpha\rho D}}{\alpha}\right] \tag{38}$$

for $z>D\tan\psi$.

For the rectangular prism model (#1) with $h(\rho x,\rho y,\rho z)$ as expressed by eq (26),

$$P(\rho z)=\frac{1}{\rho D}\left[\left(\frac{3\rho D}{4}-\frac{\beta}{2}-\frac{\beta^2}{2\rho D}-\frac{\beta}{\alpha\rho D}\right)e^{-\chi}+\frac{1-e^{-\chi}}{2\alpha}+\frac{1-e^{-\chi}}{\alpha^2\rho D}\right] \tag{39}$$

for $0\leqslant z\leqslant\dfrac{(D\tan\psi)}{2}$,

$$P(\rho z)=\frac{1}{\rho D}\left[\left(\rho D-\frac{3\beta}{2}+\frac{\beta^2}{2\rho D}+\frac{\beta}{\alpha\rho D}-\frac{1}{\alpha}\right)e^{-\chi}+\frac{1-e^{-\chi}}{2\alpha}+\frac{1-e^{-\chi}}{\alpha^2\rho D}-\frac{2e\left(\frac{-\alpha\rho D}{2}\right)}{\alpha^2\rho D}\right] \tag{40}$$

for $\dfrac{(D\tan\psi)}{2}\leqslant z\leqslant D\tan\psi$, and

$$P(\rho z)=\frac{1}{\rho D}\left[\frac{1-e^{-\alpha\rho D}}{2\alpha}+\frac{1+e^{-\alpha\rho D}}{\alpha^2\rho D}-\frac{2e\left(\frac{-\alpha\rho D}{2}\right)}{\alpha^2\rho D}\right] \tag{41}$$

for $z>D\tan\psi$.

For the tetragonal prism model (#2) with $h(\rho x,\rho y,\rho z)=1$,

$$P(\rho z)=\frac{4}{(\rho D)^2}\left[e^{-\chi}\left(\frac{\lambda}{2}\right)^2+\left(\frac{1-e^{-\chi}}{\alpha}\right)\left(\frac{\lambda}{2}\right)+\frac{\beta}{\alpha}\right] \tag{42}$$

for $0\leqslant z\leqslant D\tan\psi$ (note: there are errors in the derivations of this equation both in Armstrong and Buseck [1] and in Storms et al. [18]), and

$$P(\rho z)=\frac{2}{(\rho D)^2}\left[\frac{\rho D}{\alpha}-\frac{(1-e^{-\alpha\rho D})}{\alpha^2}\right] \tag{43}$$

for $z>D\tan\psi$ (note: different derivations in Armstrong and Buseck [1] and Storms et al. [18] yield the same expression).

For the triangular prism model (#2) with $h(\rho x,\rho y,\rho z)=1$ and $T=\dfrac{D}{2}$,

$$P(\rho z)=\frac{1}{\rho D}\left(\frac{e^{-\gamma\rho z}-e^{-\xi}}{2\gamma}+\frac{\zeta}{2}e^{-\gamma\rho z}\right). \tag{44}$$

For the square pyramid model (#4) with $h(\rho x,\rho y,\rho z)=1$ and $T=\dfrac{D}{2}$,

$$P(\rho z)=\frac{1}{(\rho D)^2}\left[\left(\frac{\zeta}{\gamma}+\frac{\zeta^2}{2}\right)e^{-\gamma\rho z}+\frac{e^{-\xi}-e^{-\gamma\rho z}}{\gamma^2}\right]. \tag{45}$$

For the cylinder resting on its circular base (model #5) with $h(\rho x, \rho y, \rho z) = 1$,

$$P(\rho z) = \frac{4}{\pi(\rho D)^2} \left\{ \frac{2}{\alpha} (1 - e^{-x}) \sqrt{\beta \lambda} + \frac{2}{\alpha} \left(\frac{\rho D}{2} - \sqrt{\beta \lambda} \right) \right.$$

$$+ 2e^{-x} \left[\sqrt{\beta \lambda} \left(\frac{(\rho D)^2}{4} - \beta \lambda \right)^{1/2} + \frac{(\rho D)^2}{4} \sin^{-1} \left(\frac{2\sqrt{\beta \lambda}}{\rho D} - \beta \sqrt{\beta \lambda} \right) \right]$$

$$\left. - 2 \int_{\rho y = \sqrt{\beta \lambda}}^{\frac{\rho D}{2}} \frac{e^{-\alpha \sqrt{(\rho D)^2 - 4(\rho y)^2}}}{\alpha} \, d(\rho y) \right\} \tag{46}$$

for $0 \leqslant z \leqslant D \tan \psi$ (note: incorporates corrections of Storms et al. [18]), and

$$P(\rho z) = \frac{8}{\pi(\rho D)^2} \left[\frac{\rho D}{2\alpha} - \int_{\rho y = 0}^{\frac{\rho D}{2}} \frac{e^{-\alpha \sqrt{(\rho D)^2 - 4(\rho y)^2}}}{\alpha} \, d(\rho y) \right] \tag{47}$$

for $z > D \tan$.

For the cylinder resting on its side (fiber, model #6) with $h(\rho x, \rho y, \rho z) = 1$ and $T = D$,

$$P(\rho z) = \frac{1}{\rho D} \int_{\rho x = -\sqrt{\rho D_{\rho z} - (\rho z)^2}}^{\sqrt{\rho D_{\rho z} - (\rho z)^2}} e^{-\mu_A \sqrt{(\rho x_2 - \rho x)^2 + (\rho \omega_2 - \rho \omega)^2}} \, d(\rho x) \tag{48}$$

where

$$\rho \omega = \left(\frac{\rho D}{2} \right) - \rho z \tag{49}$$

$$\rho x_2 = \frac{-bm + \left[\frac{m^2 (\rho D)^2}{4} - b^2 + \frac{(\rho D)^2}{4} \right]^{1/2}}{m^2 + 1} \tag{50}$$

$$\rho \omega_2 = m \rho x_2 + b \tag{51}$$

$$m = \tan \psi \tag{52}$$

and

$$b = \rho \omega - \rho x \tan \psi. \tag{53}$$

For the hemisphere and sphere models (#7 and #8) with $h(\rho x, \rho y, \rho z) = 1$, $T = \frac{D}{2}$ for the hemisphere and $T = D$ for the sphere,

$$P(\rho z) = \frac{8}{(\rho D)^2} \int_{\rho y = \alpha_1}^{\alpha_2} \int_{\rho x = \beta_1}^{\beta_2} e^{-\mu_A \sqrt{(\rho x_2 - \rho x)^2 + (\rho \omega_2 - \rho \omega)^2}} \, d(\rho x) \, d(\rho y) \tag{54}$$

where

$$\rho\omega_2 = \left[\frac{(\rho D)^2}{4} - (\rho x)^2 - (\rho y)^2\right]^{1/2} - \rho z - \rho x \tan \psi \tag{55}$$

$$\rho x_2 = \frac{-b + \sqrt{b^2 - 4ac}}{2a} \tag{56}$$

$$a = 1 + \tan^2 \psi \tag{57}$$

$$b = 2\rho\omega_2 \tan \psi \tag{58}$$

$$c = (\rho y)^2 + (\rho\omega_2)^2 - \frac{(\rho D)^2}{4} \tag{59}$$

$$\alpha_1 = 0 \tag{60}$$

$$\beta_1 = -\beta_2 \tag{61}$$

and for the hemisphere

$$\alpha_2 = \left[\frac{(\rho D)^2}{4} - (\rho z)^2\right]^{1/2} \tag{62}$$

$$\beta_2 = \left[\frac{(\rho D)^2}{4} - (\rho y)^2 - (\rho z)^2\right]^{1/2} \tag{63}$$

while for the sphere

$$\alpha_2 = \left[\frac{(\rho D)^2}{4} - \frac{(\rho z)^2}{4}\right]^{1/2} \tag{64}$$

$$\beta_2 = \left[\frac{(\rho D)^2}{4} - (\rho y)^2 - \frac{(\rho z)^2}{4}\right]^{1/2}. \tag{65}$$

The angular models (#1–4) are the quickest to evaluate and represent reasonable extreme limits of particle shape (e.g., the square pyramid is a more extreme model for the effect of a sloping particle upper surface than is the hemisphere). For routine work, these models are all that need be calculated. Particles with flat tops with respect to the electron beam and flat sides with respect to the spectrometer axis are well represented by the rectangular prism model. Particles with flat tops with respect to the beam and sloping or curved sides with respect to the spectrometer axis are well represented by the tetragonal prism model. Particles with sloping or curved tops with respect to the electron beam and flat sides with respect to the spectrometer axis are well represented by the triangular prism model. Particles with sloping or curved tops with respect to the electron beam and sloping or curved sides with respect to the spectrometer axis are well represented by the square pyramid model. In practice, since usually no simple geometric model perfectly fits a real particle shape, the wisest policy is to process the data with more than one particle model and use the variation in results as a measure of the uncertainty in the analysis due to particle geometric effects.

Correction of absorption is obviously dependent upon the values used for mass absorption coefficients. In performing particle corrections, I typically use the tabulated mass absorption coefficients of Heinrich [28] for K-lines for $Z \geqslant 11$, L-lines for $Z \geqslant 32$, and M-lines for $Z \geqslant 74$. I use the tabulated mass absorption coefficients of Henke and Ebisu [29] for K-lines for $Z < 11$ and L-lines for $Z < 32$.

Characteristic fluorescence correction.   Given an expression for $\phi(\rho x, \rho y, \rho z)$ and equations defining the particle boundary, the amount of characteristic fluorescence in the particle can be calculated in the same way as was done for x-ray absorption. Evaluation of the general characteristic fluorescence correction [eq (5)] can be quite tedious, involving evaluation of a six-fold integral. Armstrong and Buseck [2] performed such evaluations for spherical particles of sizes ranging from 1 to 100 μm and having a variety of compositions (e.g., FeNi$_3$, Cu$_5$FeS$_4$, FeCr$_2$O$_4$, KCl, K$_3$PO$_4$), at accelerating potentials of 15 and 20 keV, and take-off angles of 18°, 40° and 52.5°. They showed that it was not possible to directly solve any of the integrals and that six-fold numerical integration was required. Calculations were performed for both $h(\rho x, \rho y, \rho z) = 1$ and for a linear variation in $\phi(\rho x, \rho y, \rho z)$ with tangential distance from the surface, i.e.,

$$h(\rho x, \rho y, \rho z) = 0.5 + \frac{\left[\frac{D^2}{4} - (x^2 + y^2 + w^2) \right]^{1/2}}{2 \left(\frac{D^2}{4} \right)^{1/2}}. \tag{66}$$

They showed that results were insensitive to the model of $h(\rho x, \rho y, \rho z)$ chosen.

Armstrong and Buseck [2] showed that the relative amount of emitted characteristic fluorescence to emitted primary radiation in a 10 μm particle was about half that in a thick, polished specimen of the same composition, that a 5 μm particle has about 1/3 the amount of fluorescence as a thick, polished specimen, and that a 1 μm particle has about 10 percent of the amount of fluorescence of a thick, polished specimen. The relative amount of fluorescence in particles to that in thick, polished specimens was found to be insensitive to accelerating potentials and to be primarily a function of the amounts of absorption of the exciting and excited x rays.

Armstrong and Buseck [2], using polynomial fits of the calculated particle fluorescence data, proposed the following expressions to calculate the relative amounts of emitted characteristic fluorescence to emitted primary radiation in particles to those in thick, polished specimens:

$$\frac{I'_{f,\text{ptc}}/I'_{p,\text{ptc}}}{I'_{f,\text{tps}}/I'_{p,\text{tps}}} = A + Bx + Cx^2 \tag{67}$$

where

$$x = 1 - e^{-\mu_B \rho r} \tag{68}$$

$$A = 0.0260 \tag{69}$$

$$B = 1.1409 + 0.2012\, y \tag{70}$$

$$C = 0.2471 + 0.2741\, y - 0.01315\, y^2 \tag{71}$$

$$y = \frac{\chi_{Ak}}{\mu_{Bj}} \tag{72}$$

and where r is the particle radius, μ_{Bj} is the mass absorption coefficient of the exciting element's radiation by the matrix, $\chi_{Ak} = \mu_{Ak} \csc \psi$, and μ_{Ak} is the mass absorption coefficient of the excited element's radiation by the matrix.

Given the generally small amounts of characteristic fluorescence in particles and the magnitude of other uncertainties in particle analysis, calculation of characteristic fluorescence in particles of other shapes is not warranted and the simplified spherical particle correction given above can be employed. For thick, polished specimens, I use the characteristic fluorescence correction of Reed [30] as modified by Armstrong [17]. Specifically, the assumed constant absorption edge jump ratio values of Reed are replaced by the least squares fits tabulated jump ratio data by Armstrong:

$$r_K = 53.46\, Z - 18.01 \tag{73}$$

for K-lines,

$$\frac{(r_L - 1)}{r_L} = 0.9548 - 0.0026\, Z \tag{74}$$

for L-lines.

The function for (I_{Bj}/I_{Ak}) by Green and Cosslett [31] is replaced by a fit of new data by Armstrong:

$$\frac{I_B}{I_A} = \frac{(U_B - 1)^{1.59}}{(U_A - 1)^{1.59}} \tag{75}$$

for $(U_B - 1)/(U_A - 1) \leqslant 2/3$, and

$$\frac{I_B}{I_A} = \frac{1.87(U_B - 1)^{3.19}}{(U_A - 1)^{3.19}} \tag{76}$$

for $(U_B - 1)/(U_A - 1) > 2/3$ with $U = E_0/E_c$ (for $E_0 = 15$ or 20 keV and 0.7 keV $\leqslant E_c \leqslant 12$ keV).

Errors in the original expressions of Reed resulted in underestimation of the amount of characteristic fluorescence for elements such as magnesium, aluminum, and silicon at 15 or 20 keV by a factor of over 50 percent. The amount of fluorescence in a thick, polished specimen is calculated by the above modifications of the Reed expressions and then multiplied by the relative particle fluorescence factor calculated by eqs (67–72).

Correction for fluorescence by the continuum is generally not specifically considered for thick, polished specimens and algorithms for the absorption correction such as the Gaussian $\phi(\rho z)$ equations probably inadvertently incorporate some correction for continuum fluorescence. To the extent that the peak-to-background assumption is correct, namely that primary and continuum fluorescence radiation have the same spatial distribution, then as long as particle samples and thick, polished standards are similar in composition, there should be no significant correction for continuum fluorescence. The correction is not evaluated in the Armstrong-Buseck method.

Atomic number correction. Since the absorption coefficient in the Armstrong-Buseck method has been formulated in such a way as to calculate the ratio of the emitted primary x-ray intensity to that generated in a thick, polished specimen of the same composition as the particle, the atomic number correction needs to calculate the difference in the primary generated x-ray intensities between the standard and a thick, polished specimen of the same composition as the particle. Thus, a conventional atomic number correction designed for thick, polished specimens can be used for this procedure, i.e.,

$$\frac{I'_{A,tps}}{I'_{A,std}} = \frac{C_{A,ptc}}{C_{A,std}} \times \frac{[R/S]_{A,tps}}{[R/S]_{A,std}} \tag{77}$$

where $R_{A,tps}/R_{A,std}$ is the backscatter correction for element A, accounting for differences in the reduction of generated primary x-ray intensity due to differences in the number and energy distribution of backscattered electrons from thick, polished specimens, of the compositions of the particle and the standard; $(1/S)_{A,tps}/(1/S)_{A,std}$ is the stopping power correction for element A, accounting for differences in the number of inner shell ionizations per incident electron between thick, polished specimens of the compositions of the particle and the standard; and $I'_{A,tps}/I'_{A,std}$ is the ratio of calculated primary generated intensities of thick, polished specimens of the composition of the particle and the standard:

$$\frac{I'_{A,tps}}{I'_{A,std}} = \frac{I_{A,ptc}/f(\chi)_{A,ptc}}{I_{A,std}/f(\chi)_{A,std}} \tag{78}$$

where $I_{A,ptc}$ and $I_{A,std}$ are the measured emitted intensities of element A from particles and standard, respectively, and $f(\chi)_{A,ptc}$ and $f(\chi)_{A,std}$ are the calculated absorption corrections for particle and standard [eq (29)].

Commonly employed corrections for atomic number effects in thick, polished specimens include those of Duncumb and Reed [32] and Love et al. [33]. Both give similar results for normally analyzed elements in most common silicate minerals. I commonly use the atomic number, corrections of Love et al. [33], i.e.,

$$\frac{1}{S_A} = \frac{\left[1 + 16.05\left(\dfrac{\bar{J}}{E_{c,A}}\right)^{1/2}\left(\dfrac{\sqrt{U_{0,A}}}{U_{0,A}-1}\right)^{1.07}\right]}{\sum\limits_i C_i \dfrac{Z_i}{A_i}} \tag{79}$$

where

$$\ln \bar{J} = \frac{\sum\limits_i C_i \dfrac{Z_i}{A_i} \ln J_i}{\sum\limits_i C_i \dfrac{Z_i}{A_i}} \tag{80}$$

and J_i is the mean ionization potential for element i, and

$$R_A = 1 - \bar{\eta}[I(U_0)_A + \bar{\eta}\, G(U_0)_A]^{1.67} \tag{81}$$

where

$$I(U_0) = 0.33148 \ln U_0 + 0.5596(\ln U_0)^2 - 0.06339(\ln U_0)^3 + 0.00947(\ln U_0)^4 \qquad (82)$$

$$G(U_0) = \frac{1}{U_0} [2.87898 \ln U_0 - 1.51307(\ln U_0)^2 + 0.81312(\ln U_0)^3 - 0.08241(\ln U_0)^4] \qquad (83)$$

$$U_0 = \frac{E_0}{E_{C,A}} \qquad (84)$$

where $\bar{\eta}$ is the concentration-weighted electron backscatter coefficient for the matrix [eq (20)]. When using the Love et al. atomic number correction, I employ the Love and Scott [25] expression for the backscatter coefficient, η [eqs (23–25)], and the Berger and Seltzer [34] expression for the mean ionization potential (with J in keV), i.e.,

$$\frac{J}{Z} = (9.76 + 58.8 \, Z^{-1.19})/1000. \qquad (85)$$

One can question why the absorption and atomic number corrections are separated. It would seem more logical to calculate the ratio of emitted x-ray intensities of sample and standard simply by ratioing the sums of calculated emitted primary x-ray intensity, $I_{p,A}$, and emitted secondary fluorescence x-ray intensity, $I_{f,Ak}$, where $I_{f,Ak}$ is expressed by eq (5) and

$$I_{p,A} = \int_{\rho z=0}^{\rho T} \phi(\rho z) \, P(\rho z) \, d(\rho z) \qquad (86)$$

[see eqs (27–29)]. Indeed, such combinations of absorption and atomic number corrections have been proposed in some models of the Gaussian $\phi(\rho z)$ expression (e.g., Packwood and Brown [21]; Brown and Packwood [35]; Bastin et al. [36,37]). This author feels that there is not enough of an experimental base to justify using current $\phi(\rho z)$ expressions to calculate the dependence of primary x-ray production on matrix composition. The tracer experiments on which the initial derivation of the Gaussian $\phi(\rho z)$ model were based (e.g., Brown and Parobek [38]; Parobek and Brown [39]; also see tabulation in Scott and Love [40]) well determine the *shape* of the $\phi(\rho z)$ curve, which is what is critical for the absorption correction, but not the *absolute integral* under the $\phi(\rho z)$ curve, which is what is critical for an atomic number correction or a combined atomic number-absorption correction. The measured $\phi(\rho z)$ curves are commonly normalized to x-ray emission from a reference thin film. In order to use these curves for absolute intensities, the thickness of the reference thin films must be precisely known and there must not be any differences in the absolute accuracy of thickness measurements among the various investigators whose $\phi(\rho z)$ determinations are used to parameterize the $\phi(\rho z)$ expression. There are no published data that demonstrate that the absolute intensities are known with sufficient accuracy.

Even if the absolute values of the integrals under measured $\phi(\rho z)$ curves were known with sufficient accuracy, there still would be problems in using them to construct an accurate atomic number correction. Due to the nature of the tracer experiments used to measure $\phi(\rho z)$ curves, virtually all measurements have been made for a pure tracer element in a pure element matrix. Virtually no determinations have been made for compounds. Therefore, there is considerable uncertainty regarding how to parameterize the results in terms of matrix "average atomic number." There is even the possibility that the atomic number correction depends upon factors not normally considered, such as the sample conductivity. Armstrong [16,17,26,27] has shown that for analysis of thick, polished specimens of common silicates and oxides, use of integration of any of the currently proposed Gaussian $\phi(\rho z)$ models, including the author's, as an atomic number correction produces poorer results than use of either the Duncumb-Reed or Love et al. conventional atomic number corrections. These are the reasons why I have chosen to employ a conventional atomic number correction for particle analyses.

Data processing. The correction procedures described in this section are incorporated into a correction program, CITPTC™, which is available from the author. (Versions of this program are written in Microsoft QuickBASIC™ and PASCAL and run on computers compatible with the IBM PC™ or Apple MacIntosh™.) Required input data include the elements and lines analyzed, the analytical conditions (E_0, ψ), the compositions of the standards, the compositions of the samples or the intensities for the samples relative to the standards, the standard and sample shapes (including thick, polished specimen, thin film, and particle geometric models described above), the estimated particle size or thin film thickness, and the estimated particle or thin film density. The program calculates intensities of pure element, thick, polished specimens relative to the standards by applying the appropriate ZAF corrections (e.g., Henoc et al. [41]) and calculates correction factors for samples of given composition, or compositions of samples of given intensities relative to the standards. In calculating compositions, the program makes use of the hyperbolic iteration method (Criss and Birks [42]) to reach convergence. Typically, five iterations or fewer are required. The absorption correction is calculated by numerical integration of the $\phi(\rho z)$ and $\phi(\rho z)P(\rho z)$ expressions using the trapezoidal method. The step length for numerical integration is an input variable and should be chosen to be no greater than 1/50 of the particle thickness; a typical value of 0.00001 g/cm^2 is stored as a default (under these conditions numerical integration for thick, polished specimens yields results within 0.1 percent of those calculated using exact integration). (For a typical six-element particle sample relative to compound standards, an individual analysis would take a few seconds to process on an IBM AT-type computer with an 80286 math co-processor.)

For more rapid analysis, Armstrong [14,15,27] has shown that the a-factor approach commonly used for thick specimen mineral analysis can be adapted to particle analysis as well. Matrices of a-factors are calculated for element pairs for each combination of particle diameter and shape model desired by means of the Armstrong-Buseck equations. Armstrong [14,15,27] showed that the results using particle a-factors are comparable to those using the full particle ZAF calculations. For details of the procedure, see these papers.

E.   MONTE CARLO METHOD

Several investigators have shown that Monte Carlo calculations of electron trajectories can be adapted to show the effects of particle size and shape on emitted x-ray intensities (e.g., Newbury et al. [43]). However, systematic studies using Monte Carlo calculations in an attempt to parameterize the effects of particle size and shape have not been performed to date. In fact, only a few results have been published of Monte Carlo calculations to calculate $\phi(\rho z)$ curves for multi-element samples of any type (e.g., Kyser and Murata [44]; Myklebust et al. [45]; Newbury and Yakowitz [46] and references cited therein; Ho et al. [47]; Armstrong [17,26,27]).

Monte Carlo procedures for electron beam microanalysis calculate primary generated x-ray distributions in samples in the general manner described above for eq (4). The procedures use random numbers, repetitively applied to probability functions (cross sections) of electron scattering angle and distance, to calculate individual electron trajectories in materials. These calculations are coupled with an expression for the electron deceleration in the material resulting in compilation of the number of electrons and their energies traveling through each point in the excited volume of the sample. With this information, an expression for the probability of inner shell ionization (ionization cross section) as a function of electron energy can be applied to calculate the spatial distribution of primary x rays, $\phi(\rho x, \rho y, \rho z)$. Monte Carlo calculations can be adapted to samples of varying shapes simply by defining the boundary equations and, at each trajectory step, determining whether the electron has left the sample or not.

Monte Carlo calculations can provide very valuable information regarding electron-specimen interactions, but are not panaceas for quantitative analysis because of the uncertainties in the various equations employed by the procedure. There are two commonly used models for electron elastic scattering, Rutherford scattering and Mott-Massey scattering; both have their limitations and employ semi-empirical terms to provide better matches to experimental data (e.g., Reimer and Krefting [48]). Monte Carlo calculations generally employ a continuous expressions for the energy loss of electrons as they travel through the sample (Bethe, [49]) instead of calculating discrete energy losses at each step based on probability functions of all of the inelastic scattering processes. The effects of one or more of the inelastic scattering processes on electron trajectories are commonly ignored. A variety of expressions exist for factors employed in the Monte Carlo procedures, such as mean ionization potential (e.g., Heinrich [50], pp. 229–232) and ionization cross section (e.g., Powell [51,52]). It has been shown that depending upon which of these expressions is employed, the resulting $\phi(\rho z)$ distribution and calculated relative x-ray emissions can vary by as much as is seen using the various conventional ZAF procedures (e.g., Myklebust et al. [45]; Armstrong [17,26]). Monte Carlo calculations may ultimately yield the most accurate quantitative results, but not until more accurate expressions for the various cross sections and electron retardation are developed. The main value in applying Monte Carlo calculations to particle analyses is in evaluating the various assumptions made by the other correction procedures. In particular, Monte Carlo calculations can be used to evaluate the validity of the relationships between $\phi(\rho x, \rho y, \rho z)$ and $\phi(\rho z)$ made in the Armstrong-Buseck method.

Monte Carlo method for multi-element particles. In order to calculate the effects of particle shape and size on the $\phi(\rho x, \rho y, \rho z)$ distribution in multi-element samples, a series of modifications were made to Monte Carlo programs for electron-specimen interaction developed by David Joy [53]. (These programs were written in PASCAL for use in IBM-PC type computers.) The original programs employed single scattering or multiple scattering calculations using the Rutherford scattering model to calculate electron trajectories in pure element bulk specimens. The single scattering model employs a screened Rutherford cross section for elastic scattering:

$$\sigma_E = 5.21 \times 10^{-21} \frac{Z^2}{E^2} \frac{4\pi}{\alpha(1+\alpha)} \left(\frac{E + m_0 c^2}{E + 2m_0 c^2} \right)^2 \tag{87}$$

where E is the electron energy in keV, $m_0c^2 \approx 511$ keV, and α is a screening factor,

$$\alpha = 3.4 \times 10^{-3} \frac{Z^{0.67}}{E} \qquad (88)$$

(e.g., Newbury et al. [43]). The mean free path between scattering events, λ, is

$$\lambda = \frac{A}{N_A \rho \sigma_E} \qquad (89)$$

where N_A is Avogadro's number. The electron distance traveled between scattering events, s, is calculated by applying a random number to eq (89),

$$s = -\lambda \ln(R) \qquad (90)$$

where R is a random number between 0 and 1. The scattering angle, ϕ, is calculated by applying a random number to the differential form of eq (87) which simplifies to

$$\cos \phi = 1 - \left[\frac{2\alpha R}{(1+\alpha-R)} \right] \qquad (91)$$

where α is defined in eq (88). The azimuthal scattering angle, ψ, has a uniform probability at all values; it is calculated by generating another random number

$$\psi = 2\pi R. \qquad (92)$$

The electron energy loss is calculated by the Bethe [49] equation

$$\frac{dE}{d\rho S} = -\frac{78,500\, Z}{EA} \ln \frac{1.166E}{J} \qquad (93)$$

where ρS is the path length in g/cm^2, E is the electron energy in keV, Z is the atomic number, A is the atomic weight, and J is the mean ionization potential in keV. For low electron energies this equation breaks down and it is replaced for $E < 6.34J$ by the expression of Rao Sahib and Wittry [54]

$$\frac{dE}{d\rho S} = -\frac{62,400Z}{A\sqrt{EJ}}. \qquad (94)$$

In this single scattering program, random numbers are generated and electron trajectories are calculated until the electron leaves the sample or falls below a minimum energy (typically 500 eV).

For the multiple scattering program, the approach of Curgenven and Duncumb [55] as modified by Myklebust et al. [45] is employed. The electron range (full distance traveled by the electron until it loses its energy) is calculated by numerical integration of the Bethe equation [eq (93)]. The electron range is divided into 50 steps and the average energy at each step is evaluated through eqs (93) and (94). For each electron, the average elastic scattering angle is calculated at each (ρ_S, E) step from the screened Rutherford model by

$$\tan\left(\frac{\phi}{2}\right) = \frac{0.0072\, Z}{1.5(0.263Z^{0.4})E} \left(\frac{1}{\sqrt{R}} - 1\right) \qquad (95)$$

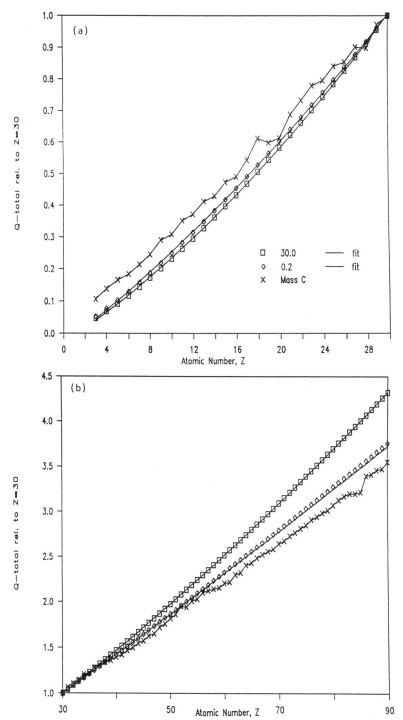

Figure 2. (A and B): Plot of the relative elastic scattering cross sections at 30 and 0.2 keV as a function of Z for (a) $Z < 30$ and (b) $Z > 30$. The cross sections are relative to that for $Z = 30$. Also plotted for comparison is the mass, relative to mass 30, as a function of Z. Note the parallel relation between the relative mass and the relative cross

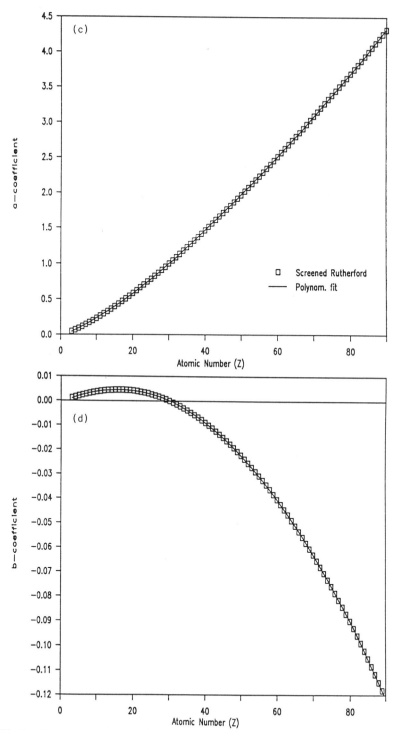

section. The lines through the points plot the polynomial fit to the data given in eq (98). (C and D): Plot of the calculated a and b values for determination of the relative elastic scattering cross section [eq (90)] as functions of Z. The lines through the points plot the polynomial fits for the values given in eqs (99–102).

(Joy [53]), or

$$\tan\left(\frac{\phi}{2}\right) = \frac{0.0072\,Z}{(0.263Z^{0.4})E_0E}\left(\frac{1}{\sqrt{R}}-1\right) \tag{96}$$

(Myklebust et al. [45]) where E_0 and E are in keV and R is a random number between 0 and 1. A second random number is generated to obtain the azimuthal angle from eq (92) and the trajectory for the ρs step is then calculated. The multiple scattering model typically takes about 20 to 25 percent of the time to evaluate compared to the single scattering model.

To adapt these programs to evaluate multi-element specimens, it is necessary to decide at each electron step which element's atom is influencing the scattering. The relative probability that the electron will be scattered by element A, P_A, can be assumed to be equal to the fractional atom concentration-weighted elastic scattering cross section,

$$P_A = \frac{C_A\,\sigma_{E,A}/A_A}{\sum_i C_i\,\sigma_{E,i}/A_i} \tag{97}$$

(Kyser and Murata [44], where σ_E is expressed by eq (87).

At each electron step, the elastic scattering cross sections of each of the elements at the appropriate electron energy can be calculated through eq (87), the relative scattering probabilities calculated through eq (97), and a random number generated to choose the element. Since these evaluations are being performed in the innermost loop of the Monte Carlo program, they result in a substantial increase in the required computer time. Evaluation of the dependence of the elastic scattering cross section on Z and E suggests that this process can be considerably simplified.

Figure 2 shows plots of the elastic scattering cross sections relative to that for $Z=30$, $\sigma_{E,Z=i}/\sigma_{E,Z=30}$, as a function of Z at $E=30$ keV and 200 eV. Also shown on figure 2 is simply a plot of the masses of the different elements relative to mass 30, A_i/A_{30}. Several things are obvious in the figure: (1) there is a smooth relation between $\sigma_{E,Z=i}/\sigma_{E,Z=30}$ and Z that should be expressible as a simple polynomial; (2) the dependence of $\sigma_{E,Z=i}/\sigma_{E,Z=30}$ upon electron energy is slowly varying and for $\Delta Z<30$ could be assumed constant between 30 keV and 200 eV (between 30 keV and 1 keV, there is insignificant variation for $\Delta Z<60$); (3) for $\Delta Z<20$, the relative elastic scattering cross section could be at least crudely estimated as being equal to the relative mass, meaning that the relative scattering probabilities in a multi-element sample could be considered at least roughly equal to the mass concentrations.

Figures 2c and 2d show how the relative elastic scattering cross section data fits to an expression of the sort

$$\frac{\sigma_{E,Z=i}}{\sigma_{E,Z=30}} = a + \frac{b}{E}. \tag{98}$$

The least squares fit of this expression, which can be seen in the figure to be quite good, yields

$$a = -0.03232 + 0.02273\,Z + 3.944\times10^{-4}\,Z^2 \tag{99}$$

for $Z=3$ to 30,

$$a = -0.2485 + 0.03664\,Z + 1.58\times10^{-4}\,Z^2 \tag{100}$$

for $Z=30$ to 90,

$$b = -8.189 \times 10^{-4} + 6.739 \times 10^{-4} Z - 2.146 \times 10^{-5} Z^2 \tag{101}$$

for $Z=3$ to 30, and

$$b = 9.965 \times 10^{-5} + 6.785 \times 10^{-4} Z - 2.256 \times 10^{-5} Z^2 \tag{102}$$

for $Z=30$ to 90, with E in keV. The fit of the $\sigma_{E,Z=i}/\sigma_{E,Z=30}$ values to these polynomials is seen as the solid lines in figures 2a and 2b.

In practice, if the atomic numbers of the various elements in a sample are similar, the relative atom concentration-weighted elastic scattering cross sections can be assumed to be independent of energy and equal to the relative mass concentrations. If the range of atomic numbers is moderate, the relative cross sections can be assumed to be independent of energy, at least down to 1 or 2 keV, and the relative probabilities can be calculated from eq (98) for some typical high electron energy and for a low energy. Only if the range of atomic numbers is extreme is it necessary to evaluate eq (98) at all or a number of electron energies. Use of any of these procedures, however, will considerably shorten the computer time required to process a multi-element specimen.

The other change that needs to be made to adapt the Monte Carlo program for multi-element specimens is modification of the electron retardation expression [eqs (93) and (94)] to

$$\frac{dE}{d\rho s} = \sum_i C_i \frac{-78,500 \, Z_i}{EA_i} \ln\left(\frac{1.166 \, E}{\bar{J}}\right) \tag{103}$$

for $E \geqslant 6.34 \, \bar{J}$ and

$$\frac{dE}{d\rho s} = \sum_i C_i \frac{-62,400 \, Z_i}{A_i(E \bar{J})^{1/2}} \tag{104}$$

for $E < 6.34 \, \bar{J}$,

where $\bar{J}$ is defined in eq (80). The choice of mean ionization potential has a considerable impact on the Monte Carlo results. Several expressions were evaluated; in the data shown below, the expression of Berger and Seltzer [see eq (85)] was used, consistent with the particle ZAF calculations.

Simulation of particle boundaries is straightforward. Particle models including the cube, cylinder resting on its circular base, and sphere have been performed. Assuming the top center of the particle is the origin, then for the cube, if $|x| > D/2$, $|y| > D/2$, $Z < 0$, or $Z > D$, the electron has left the particle. For the cylinder, if $x^2 + y^2 > D^2/4$, $Z < 0$, or $Z > T$ (thickness), then the electron has left the particle. For the sphere, if $x^2 + y^2 + (z - D/2)^2 > D^2/4$, then the electron has left the particle. In order to simulate the requirement for the Armstrong-Buseck ZAF method that the electron beam be defocused or rastered so that the entire particle surface is uniformly bombarded by electrons, random numbers are generated for each electron to determine the initial x and y positions, i.e.,

$$x = R_1 D - \frac{D}{2} \tag{105}$$

$$y = R_2 D - \frac{D}{2}.$$ (106)

For the sphere and the cylinder models, these coordinates are tested to see if they fit the requirement $x^2 + y^2 \leqslant (D^2/4)$. If not, a new pair of random numbers are generated.

In order to convert the electron and energy distribution to primary x-ray distribution, an expression for the ionization cross section, $Q(E)$, must be used [see eqs (3–4)]. A variety of expressions for $Q(E)$ have been proposed in the literature. Powell [51,52] reviews a number of these. The shape of the $\phi(\rho z)$ curve is significantly affected by the model chosen for $Q(E)$ (e.g., Armstrong [17,26]. For the particle data discussed below, five different ionization cross-section models were applied to the Monte Carlo calculated electron energy distributions:

$$Q(E)_A = 7.92 \times 10^{-20} \frac{\ln(U_A)}{U_A E_{C,A}^2}$$ (107)

(Green and Cosslett [56] for K-lines, $E_{C,A}$ in keV,)

$$Q(E)_A = 7.92 \times 10^{-20} \frac{\ln(U_A)}{U_A^{0.7} E_{C,A}^2}$$ (108)

(Hutchins [57] for K-lines,)

$$Q(E_A) = \frac{6.51 \times 10^{-20} Z_{nl} b_{nl}}{U_A E_{C,A}^2} \ln\left(\frac{U}{0.41 + 0.59 \, e^{(1-U)}}\right)$$ (109)

(Worthington and Tomlin [58] with Z_{nl}=number of electrons in the shell, b_{nl}=0.35 for K-shell and 0.25 for L-shell,)

$$Q(E)_A = \frac{6.51 \times 10^{-20} Z_{nl} \ln(U_A)}{1.18(U_A + 1.32) E_{C,A}^2}$$ (110)

(Fabre de la Ripelle as reported in Powell [52] for K-lines,) and

$$Q(E)_A = 6.51 \times 10^{-20} Z_{nl} g(U_A)$$ (111)

where

$$g(U_A) = \frac{1}{U}\left(\frac{U-1}{U+1}\right)^{3/2}\left\{1 + \frac{2}{3}\left(1 - \frac{1}{2U}\right)\ln\left[2.7 + (U-1)^{1/2}\right]\right\}$$ (112)

(Gryzinski [59].)

The $\phi(\rho z)$ distribution for each element is determined by evaluating eq (3) for each electron passing through each $\rho z + \Delta \rho z$ layer. The $\phi(\rho z)e^{-\mu g}$ distribution is determined for each element by calculating the $\phi(\rho z)$ and using the equations of Armstrong and Buseck [1] to determine g for each electron at each x,y position passing through the layer. For the data discussed below, the particle thickness or electron range, whichever is less, was divided into 50 ρz layers in the Monte Carlo calculations. Smooth $\phi(\rho z)$ distributions were typically obtained for 30,000 calculated electron trajectories.

(A faster alternative to that described in the previous paragraph, but one that requires much more computer memory, is to set up a matrix of (x,y,z,E) for the particle and keep track of the accumulated path lengths traversed by electrons in each $x+\Delta x$, $y+\Delta y$, $z+\Delta z$, $E+\Delta E$, cell. Then after all the trajectories have been calculated, the $\phi(\rho x,\rho y,\rho z)$, $\phi(\rho z)$, and $\phi(\rho z)e^{-\mu_A g}$ distributions could be calculated for each element using the ionization cross section and x-ray path length functions. For a $50\times50\times50\times50$ matrix, the minimum required for a reasonably accurate integral calculation, 6.25×10^6 matrix elements are required.)

IV. Results

A. CALCULATION OF PARTICLE ZAF CORRECTIONS

Using the Armstrong-Buseck correction procedures described above, it is possible to simulate the effects of particle size and shape on emitted x-ray intensities for samples and standards of given composition. Performing such calculations is valuable for determining (1) how bad are the results that would be obtained for particle analyses if no corrections or only conventional thick, polished specimen ZAF corrections were applied, (2) what is the variation in results obtained for different particle models and sizes encompassing the uncertainty in the density, size and shape determination, (3) how appropriate are the standards employed for the particle analysis, and (4) what are the best set of analytical conditions (accelerating potential and take-off angle) that can be employed to minimize the effect of particle size and shape on the relative emitted intensities. Armstrong and Buseck [1] showed that in presenting results of calculations of emitted x-ray intensities from particles, it is useful to plot a relative intensity factor, $R_{A/B}$, as a function of diameter for the different particle models, where

$$R_{A/B}=\frac{k_A}{k_B} \tag{113}$$

and k_A [see eq (10)] is the intensity of element A in a particle relative to that in a thick, polished specimen of the same composition. By using relative k-ratios, factors like the fraction of electrons in the beam striking the sample cancel out. If the normalized intensities of different elements in a particle were the same as those in a thick, polished specimen of the same composition (the requirement for simple normalization or use of conventional ZAF corrections for particles) then $R_{A/B}$ should equal 1. The greater the deviation of $R_{A/B}$ from 1, the more important it is to correct for the particle effects.

Figures 3 to 5 show plots of $R_{Mg/Si}$, $R_{Ca/Si}$ and $R_{Fe/Si}$ as a function of diameter for rectangular prism, tetragonal prism, triangular prism and square pyramid model particles of the composition of NBS (NIST) K-411 standard glass (best estimate of composition in oxide weight percent based on analyses relative to Caltech primary standards: MgO=15.21%, SiO$_2$=54.59%, CaO=15.42% and FeO=14.55%). The calculations were made for an accelerating potential of 15 keV and a spectrometer take-off angle of 40°. As can be seen in the figures, there are large variations in the $R_{A/B}$ values depending on the size and shape. Relative intensity ratios differ from those in a thick, polished specimen of the same composition by as much as 35 percent. This is by no means the largest variation encountered under normal analytical conditions; differences by a factor as large as 10 have been reported between particle and thick, polished specimen relative intensity ratios (e.g., Armstrong [3,20]; Armstrong and Buseck [12,13,60]). Figures 3–5 show that for small particles (5 μm diameter), the effect of particle size is much more significant than the particle shape on

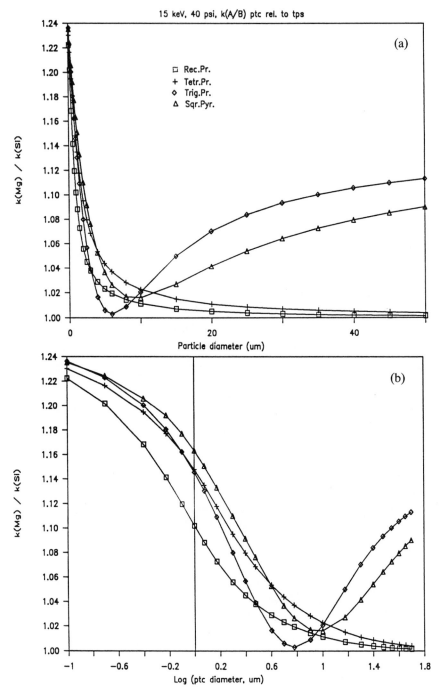

Figure 3. Comparison of calculated $R_{\mathrm{Mg/Si}}$ values for various particle models for NBS K411 glass as a function of (a) particle diameter, D, and (b) $\log_{10}(D)$. Calculations were performed for $E_0 = 15$ keV and $\psi = 40°$. The models used in the calculations are (1) rectangular prism [Rec.Pr], (2) tetragonal prism [Tetr.Pr], (3) triangular prism [Trig.Pr], and (4) square pyramid [Sqr.Pyr].

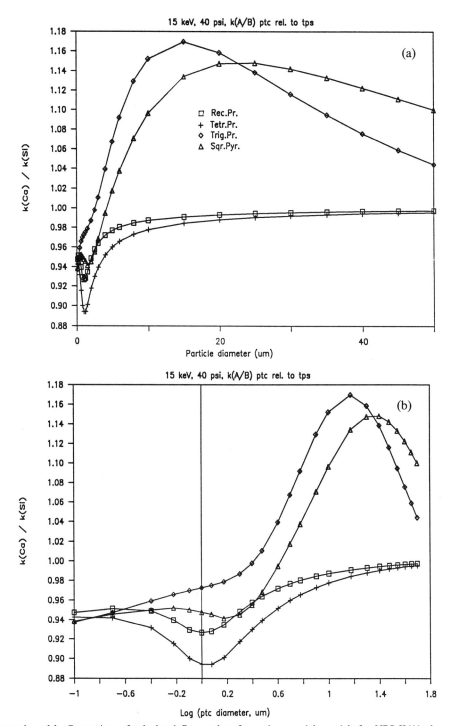

Figure 4a and b. Comparison of calculated $R_{Ca/Si}$ values for various particle models for NBS K411 glass as a function of particle diameter, D, and $\log_{10}(D)$. Conditions same as in figure 3.

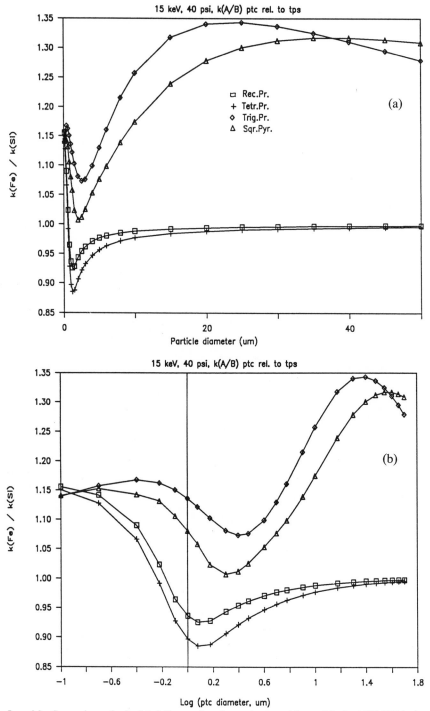

Figure 5a and b. Comparison of calculated $R_{Fe/Si}$ values for various particle models for NBS K411 glass as a function of particle diameter, D, and $\log_{10}(D)$. Conditions same as in figure 3.

relative emitted intensities. For larger particles, the effect of particle shape becomes much more significant than size. The flat topped models, such as the rectangular prism and tetragonal prism, asymptotically approach $R_{A/B}$ values of 1 for large particle diameters, while the nonflat topped models can deviate greatly from 1 for large diameters. Rounded topped or sided models typically lie between the extreme values of $R_{A/B}$ defined by the angular models. For systematic studies of the range of $R_{A/B}$ as functions of size, shape, composition and analytical conditions see Armstrong [3,20] and Armstrong and Buseck [1,12,13,60].

Armstrong and Buseck [1] showed that the effect of varying $\phi(\rho z)$ to account for electron sidescatter resulted in a smaller variation in $R_{A/B}$ then shifting from one particle model to another. Figure 6 shows the effect of changing the model for $\phi(\rho z)$ on $R_{A/B}$ for the various elements relative to Si in rectangular prism particles of the composition of NBS K-411 glass. The models shown are: (1) the Armstrong Gaussian $\phi(\rho z)$ expression described above (a) using the $\phi(0)$ and backscatter coefficients of Love and Scott [25], and (b) using the $\phi(0)$ of Reuter [19] and the backscatter coefficients of Heinrich [24]; (2) the Gaussian $\phi(\rho z)$ expression of Packwood and Brown [21]; (3) the Gaussian $\phi(\rho z)$ expression of Bastin et al. [37]. Also shown are results using the Armstrong/Love-Scott Gaussian $\phi(\rho z)$ [#1a] with an absorption path length expression for a thin film (of thickness D) instead of a rectangular prism particle (of thickness and diameter D). As can be seen in the figure, there is little difference in the $R_{A/B}$ curves depending on $\phi(\rho z)$ model except for very small particles when A and B have significantly different atomic numbers. (The curves shown in fig. 6 continue to converge for $D > 3$ μm.) There is a much greater difference when the absorption path length is changed to that in a thin film (the difference would be even greater if the particle top of sides were angular or curved, instead of flat).

Armstrong and Buseck [1,12,13] and Armstrong [3,20] showed that the four geometric models used in figures 3–5 for most particle compositions and analytical conditions bracket the extreme range of $R_{A/B}$ calculated by all of the geometric models described in the previous section and also bracket most experimentally measured $R_{A/B}$ values for particles. In fact, in their review of the Armstrong-Buseck particle ZAF method, Storms et al. [18] conclude that most of their experimentally measured $R_{A/B}$ values fall toward the middle of the range enclosed by these models and can be reasonably estimated by one of the models that produce intermediate $R_{A/B}$ values, such as the sphere or tetragonal prism. (This author's results tend to show a greater spread in $R_{A/B}$ for a given diameter than do Storms et al.) In any event, the results of Armstrong and Buseck and Storms et al. indicate that the Armstrong-Buseck ZAF procedures predict well the range of effects that particle size and shape have on emitted x-ray intensities and that these effects can be quite large. The next section will show the degree to which individual particle analyses can be corrected for these effects.

B. ANALYTICAL RESULTS WITH ARMSTRONG-BUSECK CORRECTIONS

In order to test the Armstrong-Buseck particle ZAF procedures, Armstrong [3,26] analyzed over 1500 particles of crushed mineral standards obtaining over 4500 $R_{A/B}$ measurements. Samples and standards were obtained by splitting single, gem-quality crystals of mineral standards known to be inclusion-free and homogeneous at the level of microprobe analysis. One portion of the crystal was mounted in epoxy and polished as a standard. The other portion was crushed into a μm-sized powder under alcohol or freon in an agate or a boron-carbide mortar and pestle. Particles were dispersed on graphite planchets as described above, carbon coated, and analyzed relative to the thick, polished specimen of the same composition. Analyses were performed at accelerating potentials of 15 and 20 keV on a Cameca MS-46 electron microprobe using wavelength dispersive spectrometers having a

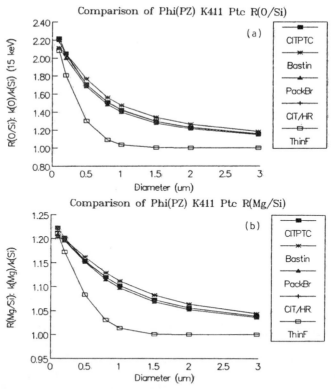

Figure 6. Comparison of calculated (a) $R_{O/Si}$, (b) $R_{Mg/Si}$, (c) $R_{Ca/Si}$, and (d) $R_{Fe/Si}$ values for rectangular prism particles of NBS K411 glass as a function of particle diameter in μm for different models of $\phi(\rho z)$. The $\phi(\rho z)$ models compared are those of Armstrong [15,16] [CITPTC], Bastin et al. [37] [Bastin], Packwood and Brown [21]

take-off angle of 18° and on a MAC electron microprobe using wavelength and energy dispersive spectrometers having an effective take-off angle of 38.5°. (Subsequent confirmatory analyses were performed by the author on a JEOL JSM-35 scanning electron microscope using wavelength and energy dispersive spectrometers having a take-off angle of 35° and a JEOL 733 electron microprobe using wavelength and energy dispersive spectrometers having a take-off angle of 40°. The results with these instruments were comparable with the results described below.)

The primary minerals analyzed in these experiments included anorthite (nominally $CaAl_2Si_2O_8$), titanite (nominally $CaTiSiO_5$), olivine (nominally $[Mg_{1.65}Fe_{0.35}]SiO_4$), rhodonite (nominally $[Mn_{1.3}Fe_{0.5}Ca_{0.2}]Si_2O_6$), orthoclase (nominally $KAlSi_3O_8$), and pyrite (nominally FeS_2). (For the exact compositions of these standards see Armstrong [3,20].) Particles from about 1 μm to 30 μm were analyzed. Particle sizes and shapes were estimated by optical microscopy and/or electron imaging at the time of analysis. The analytical procedures employed were those described above. Since samples and standards were the same composition, the atomic number correction canceled out. In most cases, the magnitude of the characteristic fluorescence correction was small for both particle and thick, polished standard. Therefore, the main correction tested was the absorption correction as described above.

The accumulated results of the particle measurements are given in figures 7 and 8. (Detailed results for individual standards can be found in Armstrong [3,20].) Results are

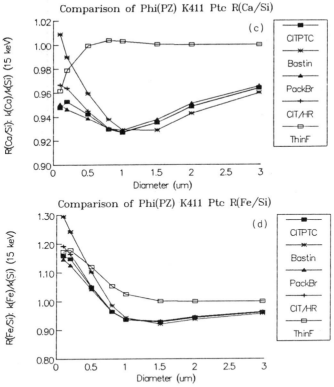

[PackBr], and Armstrong [15,16] using the $\phi(0)$ of Reuter [19] and of Heinrich [24] [CIT/HR]. Also shown in the plots for comparison are calculations using the Armstrong [15,16] $\phi(\rho z)$ with a thin film absorption expression [ThinF]. Calculations were made for $E_0 = 15$ keV and $\psi = 40°$.

shown for applying conventional, thick specimen ZAF corrections to the data (simple normalization) and for applying the Armstrong-Buseck particle corrections. The differences in the results are significant. When conventional corrections are applied, the relative errors in the measured particle compositions range from -70 percent to $+55$ percent. When the Armstrong-Buseck particle corrections are applied, the relative errors range only from -15 percent to $+15$ percent and sharply peak about 0 percent error. Two standard deviations of the relative errors are 33 percent when conventional corrections are applied and 5.2 percent when the Armstrong-Buseck corrections are applied.

Figure 8 shows the mean percentage relative errors and two standard deviations of the percentage relative errors for each of the major elements analyzed in each of the standards at the two accelerating potentials with ("ASUPTC") and without ("Conven. ZAF") applying the particle corrections. The mean errors obtained when the Armstrong-Buseck corrections were applied ranged from only -1.5 percent to $+1.5$ percent showing that the corrections did not result in any systematic bias. On the other hand, the mean errors obtained by applying conventional ZAF corrections ranged from -30 percent to $+30$ percent, showing that some elements had very significant systematic errors. Similarly, the range of standard deviations was much smaller when the Armstrong-Buseck corrections were applied than when they weren't (two standard deviations ranged from 1 to 8 percent, except for one case, when Armstrong-Buseck corrections were applied, and from 1 to 25 percent, except for one case, when they weren't applied). Clearly, the Armstrong-Buseck

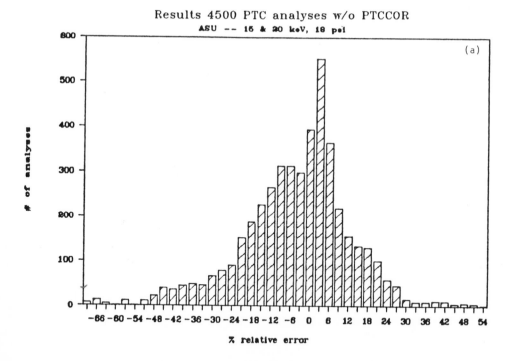

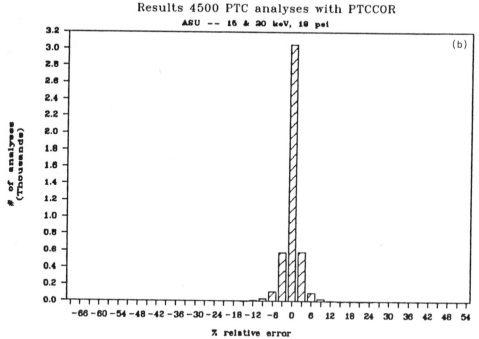

Figure 7. Histograms of percentage relative errors of 4500 oxide weight concentrations calculated from the analyses of 1500 particles using (a) only conventional thick specimen ZAF corrections and (b) using the Armstrong-Buseck particle ZAF corrections. (See text for analytical conditions employed.)

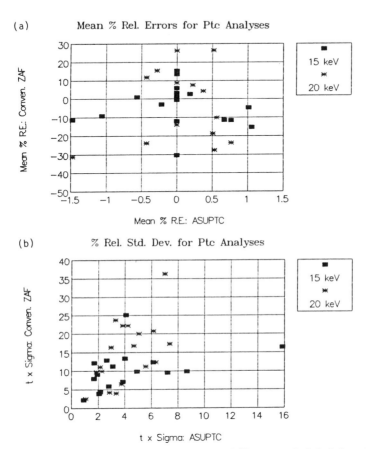

Figure 8. Comparison of (a) mean percentage relative errors and (b) two standard deviations of the range of percentage relative errors for each of the measured elements in each of the standards reported in figure 7 (see text for standard compositions). The comparison is between results obtained using conventional thick specimen ZAF corrections (*y*-axis labeled "Conven. ZAF") vs. those using the Armstrong-Buseck particle corrections (*x*-axis labeled "ASUPTC" for particle correction program). Note the differences in scale between the *x* and *y* axes.

Table 1. Results of analyses of two olivine particles of the same composition obtained by applying conventional and particle corrections

	Actual Oxide wt. %	Conventional Correction		Particle Correction	
		1	2	1	2
MgO	43.94	56.20	47.67	43.77	43.87
FeO	17.11	5.63	12.43	17.30	17.01
MnO	0.30	[0.30]*	[0.30]	[0.30]	[0.30]
SiO$_2$	38.66	37.88	39.60	38.63	38.82
Mg	1.665	2.037	1.766	1.660	1.661
Fe	0.364	0.114	0.258	0.368	0.361
Mn	0.006	[0.006]	[0.006]	[0.006]	[0.006]
Si	0.983	0.921	0.984	0.983	0.986
O	4	4	4	4	4
% Fayalite	17.9	5.3	12.7	18.1	17.9

* Mn not analyzed; analyses normalized to 99.7% with assumed MnO = 0.3%.

corrections appear to do a good job in correcting for x-ray absorption in particles. Moreover, by any reasonable criterion, the results of analyzing particles relative to thick standards and applying either no correction or conventional thick specimen corrections are unacceptably bad.

Table 1 emphasizes the problems observed when attempting to estimate particle compositions without applying proper particle corrections. This table shows the results of analyses of two particles (one small, the other large) of the olivine standard listed above, both with and without applying particle corrections. The errors obtained when particle corrections are not applied are large. The FeO concentration is 67 percent too low for particle 1 and 27 percent too low for particle 2. When not applying particle corrections there is no way of determining that the two particles are the same composition. Particle 1 does not correspond well to the stoichiometry of an olivine, it would be somewhat difficult to determine that the particle was an olivine from that analysis. Perhaps even more significantly, because of compensating errors, the atom proportions for particle 2 well match a olivine composition which could make an analyst think that the results were good. The calculated fayalite (Fe$_2$SiO$_4$) content obtained for the two particles, when particle corrections are not applied, are 5.3 percent and 12.7 percent instead of the correct value of 17.9 percent. On the other hand the errors in the two analyses, when corrected for particle effects using the Armstrong-Buseck corrections are less than 1.5 percent; the results clearly show that the two particles have the same composition with calculated fayalite compositions of 18.1 percent and 17.9 percent compared with the correct value of 17.9 percent.

C. MONTE CARLO CALCULATIONS OF X-RAY EMISSION FROM PARTICLES

Figures 9 and 10 show typical results of using Monte Carlo calculations to generate $\phi(\rho z)$ and $\phi(\rho z)e^{-\chi\rho z}$ curves for thick, polished specimens. A multiple scattering model using the Green-Cosslett ionization cross section [eq (107)] was employed. Simulations were performed for $E_0 = 15$ keV and $\psi = 40°$ using 10,000 electron trajectories in each case. Figure 9 shows the variation in $\phi(\rho z)$ and $\phi(\rho z)e^{-\chi\rho z}$ for various elements in the mineral anorthite, nominally CaAl$_2$Si$_2$O$_8$. Figure 10 shows the variation of $\phi(\rho z)$ and $\phi(\rho z)e^{-\chi\rho z}$ curves for Si $K\alpha$ in different matrices: quartz (SiO$_2$), kyanite (Al$_2$SiO$_5$), wollastonite (Ca-

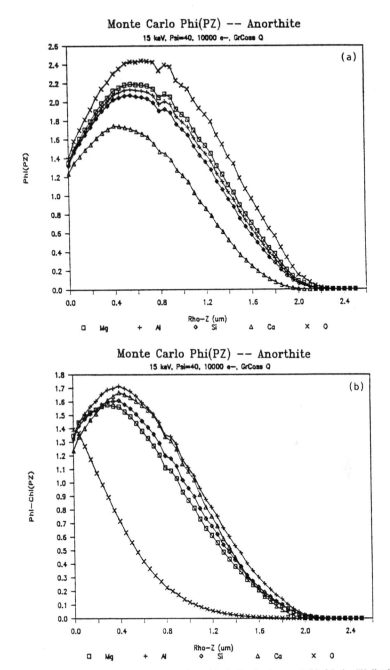

Figure 9. Comparison of Monte Carlo calculations of (a) $\phi(\rho z)$ distributions and (b) $\phi(\rho z)e^{-\chi\rho z}$ distributions for Mg, Al, Si, Ca, and O $K\alpha$ x rays in a thick specimen of anorthite ($CaAl_2Si_2O_8$). Calculations were made using a multiple-scattering model with the Green-Cosslett ionization cross section at 15 keV and a 40° take-off angle. The simulation employed 10,000 electron trajectories.

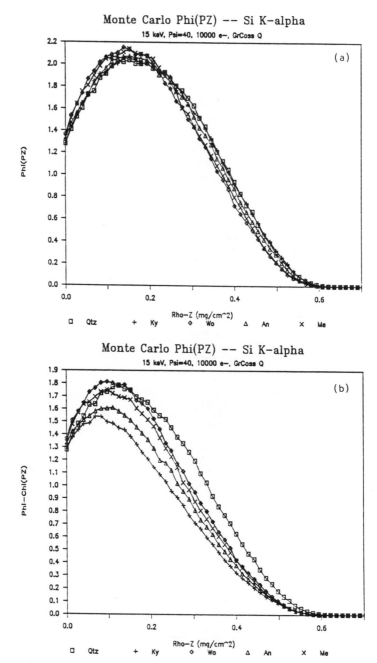

Figure 10. Comparison of Monte Carlo calculations of (a) $\phi(\rho z)$ distributions and (b) $\phi(\rho z)e^{-\chi\rho z}$ distributions for Si $K\alpha$ in quartz (Qtz), kyanite (Ky), wollastonite (Wo), anorthite (An), and melilite (Me) (see text for compositions of the minerals). Calculations were performed at the same conditions as in figure 9.

SiO$_3$), anorthite (CaAl$_2$Si$_2$O$_8$) and melilite (Ca$_2$MgSi$_2$O$_7$). The differences in the shapes of the various curves result in significant differences in the absorption corrections in these cases.

Figure 11 shows the range of $\phi(\rho z)$ curves for thick, polished specimens calculated using (a) various ionization cross section expressions with multiple scattering Monte Carlo simulations (GC=Cosslett, PO=Fabre/Powell, WT=Worthington Tomlin) and (b) Gaussian $\phi(\rho z)$ equations (JTA=Armstrong, BAS=Bastin [37], PB=Packwood and Brown). The calculations were performed for magnesium, aluminum and silicon in melilite (Ca$_2$MgSi$_2$O$_7$) at E_0=15 keV and ψ=40°; 10,000 electron trajectories were calculated for the Monte Carlo simulations. Single scattering Monte Carlo calculations using the same ionization cross sections under the same conditions produce $\phi(\rho z)$ curves that peak at a shallower depth than the Gaussian model, as opposed to the multiple scattering models that peak at a greater depth. As can be seen in the figures, the Monte Carlo calculations result in $\phi(\rho z)$ curves of considerably differing shapes for the same element in the same sample, as do the Gaussian models. Similar variations are observed in the Monte Carlo simulations when varying the mean ionization potential J, $\phi(\rho z)$ curves with larger values of J for an element peak and tail off to zero at lower depths than do those with smaller values of J (e.g., Myklebust et al. [45]). These results underscore the current uncertainties in using Monte Carlo calculations for quantitative analysis (see Armstrong [17]).

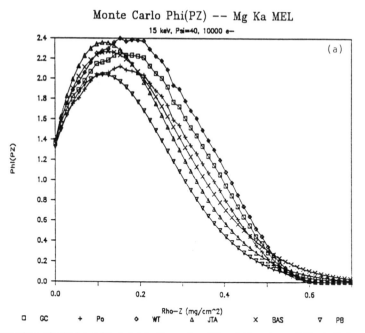

Figure 11. Comparison of calculated $\phi(\rho z)$ distributions using different Gaussian and Monte Carlo models for (a) Mg $K\alpha$, (b) Si $K\alpha$ and (c) Ca $K\alpha$ in a thick specimen of melilite (Ca$_2$MgSi$_2$O$_7$). The $\phi(\rho z)$ models compared are (1) Monte Carlo with Green-Cosslett ionization cross section [GC], (2) Monte Carlo with Fabre/Powell ionization cross section [Po], (3) Monte Carlo with Worthington-Tomlin ionization cross section [WT], (4) Armstrong Gaussian expression [JTA], (5) Bastin Gaussian expression [BAS], and (6) Packwood-Brown Gaussian expression [PB]. The Monte Carlo calculations employed multiple scattering and 10,000 electron trajectories. The analytical conditions used in the calculations are the same as those for figures 9–10.

Monte Carlo Phi(PZ) -- Si Ka MEL

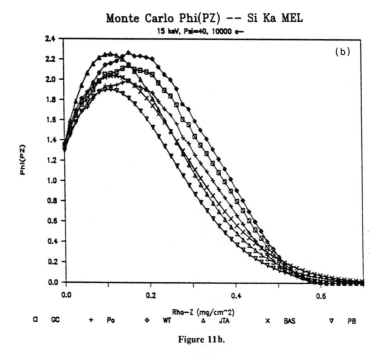

Figure 11b.

Monte Carlo Phi(PZ) -- Ca Ka MEL

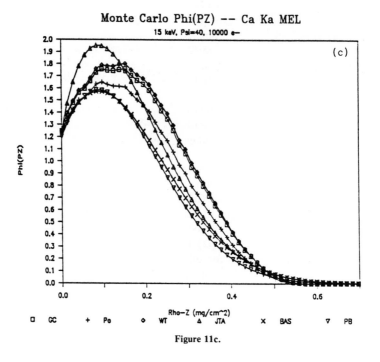

Figure 11c.

Figure 12 shows calculated Si $K\alpha$ $\phi(\rho z)$ curves in cubic particles of melilite with diameters varying from 0.1 to 20 μm. Figure 13 shows the same curves for $CaK\alpha$ in melilite. The Monte Carlo calculations employed single scattering with the Green-Cosslett ionization cross section at $E_0 = 15$ keV and $\psi = 40°$. As can be seen in the figures there is a significant difference in the shape of the $\phi(\rho z)$ curves as a function of particle shape. This difference is due to the absence of backscattered electrons from lower depths producing x rays near the surface of small particles. The simulation model employed in the Monte Carlo simulation was a particle over a substrate that did not scatter electrons (an extreme case, but one reasonably appropriate for particles mounted on beryllium or carbon planchets and very appropriate for particles mounted on thin carbon films). Gaussian model $\phi(\rho z)$ curves for these cases look like the bulk specimen $\phi(\rho z)$ curves for these elements truncated to $\phi(\rho z) = 0$ at the particle thickness. A Monte Carlo simulation with a substrate of the same average atomic number as the particles would produce curves similar to the Gaussian model, but such a case is seldom appropriate for real particle samples mounted on substrates. These results suggest that the Gaussian $\phi(\rho z)$ model, as currently constructed, may fail for very small particle diameters. This is one of the topics of the following section.

D. COMPARISON OF ARMSTRONG-BUSECK AND MONTE CARLO METHODS

Figure 14 underscores the difference in calculated $\phi(\rho z)$ curves for small particles using the Gaussian and Monte Carlo (over low-Z matrix) models. This figure compares the $\phi(\rho z)$ curves for oxygen, silicon, and calcium in cubic particles of melilite having diameters of 0.1, 1, and 20 μm. The Monte Carlo simulation conditions are the same as those for figures 12–13. As can be seen, the $\phi(\rho z)$ curves for the three elements are superimposed for 0.1 μm particles (i.e., oxygen, silicon, and calcium in melilite have identical $\phi(\rho z)$ distributions relative to infinitely thin specimens of the same composition). The $\phi(\rho z)$ curves for oxygen, silicon, and calcium in 20 μm particles of melilite have considerably different values and look similar to those for a thick, polished specimen of melilite (e.g., figs. 9, 11). These data suggest that the relative generated x-ray intensities for small particles calculated in the absorption correction of the Armstrong-Buseck procedure using a Gaussian $\phi(\rho z)$ model that does not incorporate a correction for substrate backscatter can result in a significant error.

On the other hand, figure 15 shows that there is good agreement between the Gaussian and Monte Carlo $\phi(\rho z)$ expressions in calculating the fraction of absorbed x rays in particles. This figure shows the relative percentage difference between the fraction of absorbed to generated x rays as a function of depth for oxygen, magnesium, silicon, and calcium in cubic particles of melilite with diameters ranging from 1 to 20 μm. The Armstrong Gaussian $\phi(\rho z)$ and the multiple scattering Monte Carlo $\phi(\rho z)$ with Green-Cosslett ionization cross section were the models compared. Calculations were performed for $E_0 = 15$ keV and $\psi = 40°$. The results show agreement in all cases for absorption in oxygen in melilite (a case of an extremely high matrix absorption coefficient) of better than 4 percent relative between the Gaussian and Monte Carlo $\phi(\rho z)$ models. The relative agreement was better than 2.5 percent for magnesium, 2 percent for silicon and 0.7 percent for calcium. This level of agreement is much better than the uncertainty due to errors in the estimate of particle size, shape and density, and suggests that the particle absorption correction in the Armstrong-Buseck method is reasonable.

Finally, the differences in the combined corrections using the Gaussian and Monte Carlo $\phi(\rho z)$ models can be seen in figures 16 and 17. These figures show plots of $R_{A/B}$ vs. diameter for cubic particles of NBS K-411 glass for O/Si, Mg/Si, Ca/Si, and Fe/Si. Calcu-

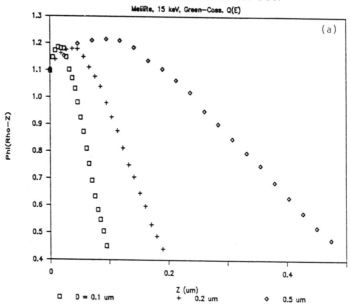

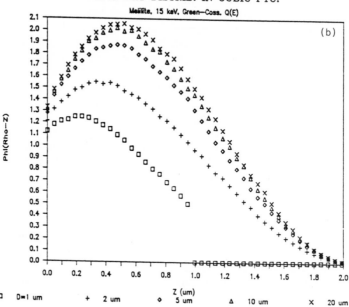

Figure 12. Calculated $\phi(\rho z)$ distributions for Si $K\alpha$ from (a) 0.1, 0.2, 0.5 μm and (b) 1, 2, 5, 10, and 20 μm cubic particles of melilite. The distributions were determined by Monte Carlo calculations for an accelerating potential of 15 keV using single scattering, Green-Cosslett ionization cross section, and 30,000 electron trajectories.

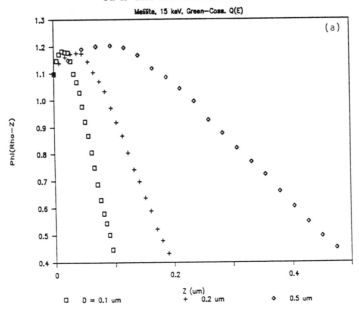

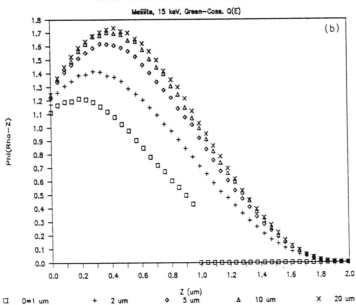

Figure 13. Calculated $\phi(\rho z)$ distributions for Ca $K\alpha$ from cubic particles of melilite. Conditions are the same as those in figure 12.

lations were made for $E_0 = 15$ keV and $\psi = 40°$ for the Armstrong (Phi CIT) and Bastin et al. (Phi BAS) $\phi(\rho z)$ expressions and the single scattering Monte Carlo calculations using the Green-Cosslett (MC-GC), Hutchins (MC-Hu), Fabre/Powell (MC-Fa), and Gryzinski (MC-GZ) ionization cross-section expressions. Very good agreement was seen for $R_{O/Si}$ and $R_{Mg/Si}$ at all diameters between the Gaussian and Monte Carlo models. The outlier curves were those of the Monte Carlo-Gryzinski and the Gaussian-Bastin models, all other curves superimposed with agreement of better than 2 percent relative for all diameters. (The $R_{O/Si}$ and $R_{Mg/Si}$ curves continue to converge for $D > 3$ μm.) For these cases the effects of x-ray absorption dominate.

In the cases where the difference in Z between the two elements is larger and one of the elements is not heavily absorbed, $R_{Ca/Si}$ and $R_{Fe/Si}$, the effects of electron scattering become more important and differences are seen between the Gaussian and Monte Carlo models for small particles. For particles of 1 μm diameter and smaller, there are significant differences among all of the models shown and significant differences in the shapes of the $R_{A/B}$ vs. D curves between the Gaussian and Monte Carlo models. The mean difference in $R_{Ca/Si}$ between the Gaussian and Monte Carlo model is about 10 percent for particles smaller than 1 μm, while that for $R_{Fe/Si}$ is about 20 percent. For particles larger than 1.5 μm the results using the two types of models begin to converge, so that the difference in $R_{Ca/Si}$

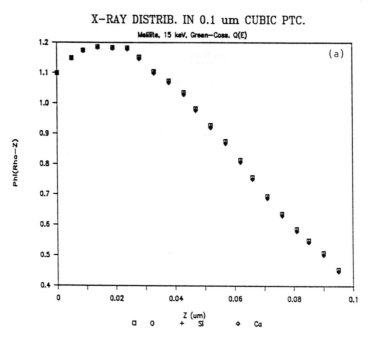

Figure 14. Comparison of $\phi(\rho z)$ distributions of O, Si, and Ca $K\alpha$ in (a) 0.1 μm diameter, (b) 1 μm diameter and (c) 20 μm diameter cubic particles of melilite. The distributions were determined using Monte Carlo calculations under the conditions given in figure 12. Note that for small particle diameters, all of the elements have the same $\phi(\rho z)$ distribution relative to their individual emissions from an infinitely thin film.

X-RAY DISTRIB. IN 1 um CUBIC PTC.

Melilite, 15 keV, Green–Coss. Q(E)

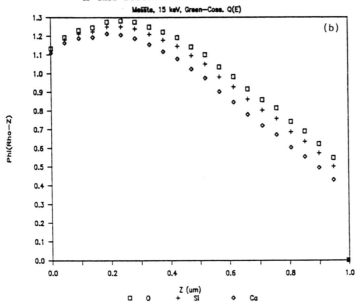

(b)

X-RAY DISTRIB. IN 20 um CUBIC PTC.

Melilite, 15 keV, Green–Coss. Q(E)

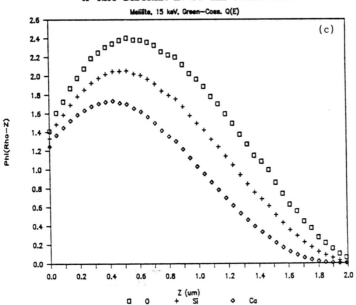

(c)

O X–RAY ABSORB. IN CUBIC PTC.

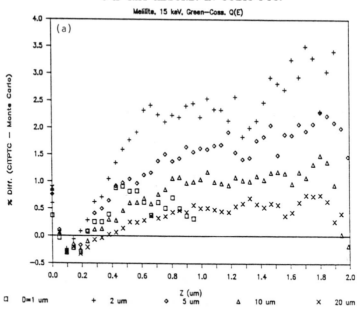

Mg X–RAY ABSORB. IN CUBIC PTC.

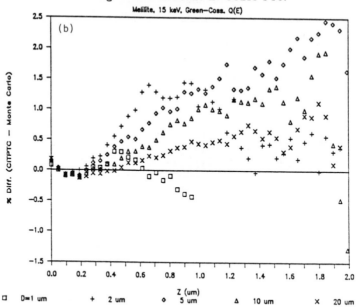

Figure 15. Comparison of the percentage difference in values calculated for $P(\rho z)$ (amount of absorption at layer ρz, using the Armstrong Gaussian $\phi(\rho z)$ expression as opposed to using Monte Carlo calculations, as a function of ρz. The comparisons are made for (a) O $K\alpha$, (b) Mg $K\alpha$, (c) Si $K\alpha$, and (d) Ca $K\alpha$ in cubic particles of melilite having diameters of 1, 2, 5, 10, and 20 μm. Calculations were performed for an accelerating potential of 15 keV and a take-off angle of 40°. The conditions for the Monte Carlo calculations are the same as those for figure 12.

Si X-RAY ABSORB. IN CUBIC PTC.

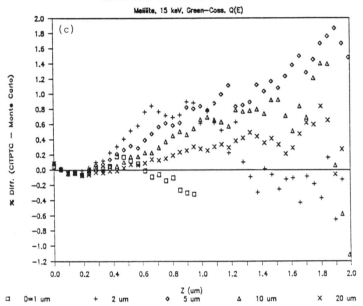

Melilite, 15 keV, Green–Coss. Q(E)

Ca X-RAY ABSORB. IN CUBIC PTC.

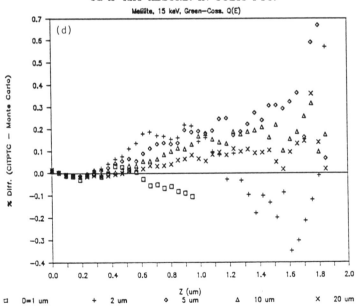

Melilite, 15 keV, Green–Coss. Q(E)

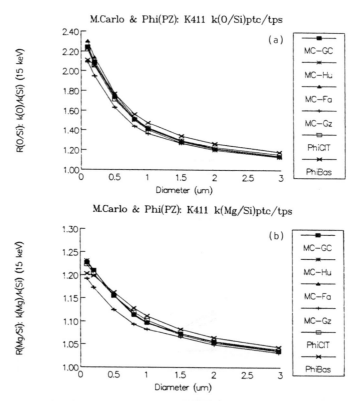

Figure 16. Comparison of calculated (a) $R_{O/Si}$, (b) $R_{Mg/Si}$, (c) $R_{Ca/Si}$, and (d) $R_{Fe/Si}$ for rectangular prism particles of NBS K411 glass as a function of particle diameter (0.1 to 3 μm) using various Gaussian and Monte Carlo $\phi(\rho z)$ models. The models employed are (1) single-scattering Monte Carlo calculations using the Green-Cosslett ionization cross section [MC-GC], (2) Monte Carlo calculations with the Hutchins ionization cross section [MC-Hu], (3)

and $R_{Fe/Si}$ between Gaussian and Monte Carlo models is about 3 to 5 percent for 3 μm particles and 0.5 to 1 percent for 20 μm particles.

V. Conclusions

Comparison of Monte Carlo and Gaussian $\phi(\rho z)$ calculations suggest that the Armstrong-Buseck particle ZAF corrections with the current Armstrong model for $\phi(\rho z)$ can be used to accurately analyze particles larger than 1–2 μm in diameter. For elements of greatly differing Z in particles smaller than 1 μm, significant errors can result in using the current Armstrong-Buseck method unless the particles are mounted on a substrate with a similar atomic number. Monte Carlo calculations show promise in developing an accurate correction procedure for small particles; however, correct expressions for the ionization cross section and mean ionization potential need to be developed to make these corrections exact. Particles smaller than 1 μm can be accurately analyzed at high keV (100–200 keV) in an analytical electron microscope using conventional procedures for that instrument.

For particles larger than 1 μm in diameter, a correction procedure such as the Armstrong and Buseck ZAF procedure or the peak-to-background method should be employed.

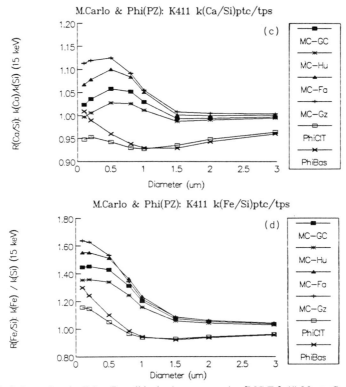

M.Carlo & Phi(PZ): K411 k(Ca/Si)ptc/tps (c)

M.Carlo & Phi(PZ): K411 k(Fe/Si)ptc/tps (d)

Monte Carlo calculations using the Fabre/Powell ionization cross section [MC-Fa], (4) Monte Carlo calculations using the Gryzinski ionization cross section [MC-Gz], (5) the Gaussian expression of Armstrong [PhiCIT], and (6) the Gaussian expression of Bastin et al. [PhiBas]. Calculations were performed for an accelerating potential of 15 keV and a take-off angle of 40°; Monte Carlo simulations employed 30,000 electron trajectories.

The errors resulting when particle corrections are not applied are systematic and can be as high as several hundred percent. In extreme cases, using point analyses, it is possible to make order of magnitude errors in estimating relative concentrations. Utilizing proper correction procedures that are relatively easy to implement and that usually require only seconds of computer processing times, errors in particle analyses can be reduced to the ± 5 to 8 percent relative level, i.e., only a factor of two or three or so worse than when analyzing thick, polished specimens.

The author wishes to thank David Joy for sharing his Monte Carlo programs; Dale Newbury, Bob Myklebust, and Chuck Fiori for helpful conversations about Monte Carlo programming; and Peter Buseck and Art Chodos for their past advice in assisting the author in the development and testing of particle correction procedures. This work was supported in part by the National Aeronautics and Space Administration through grant NAG 9-43. Division Contribution Number 4787 (683).

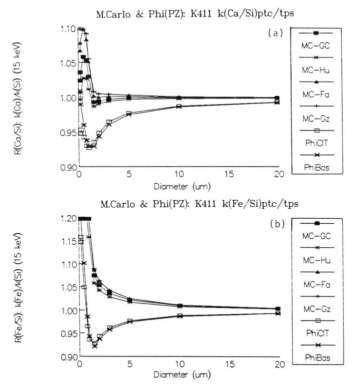

Figure 17. Comparison of (a) $R_{Ca/Si}$ and (b) $R_{Fe/Si}$ calculated by Gaussian and Monte Carlo models for cubic particles of melilite with diameters ranging up to 20 μm. The models and conditions employed are the same as for figure 16 except that Monte Carlo calculations for particles larger than 3 μm use the multiple scattering approximation. Note that the $R_{Ca/Si}$ and $R_{Fe/Si}$ values, which diverged for small particles between the Monte Carlo and Gaussian models, converge for larger particles.

VI. References

[1] J. T. Armstrong and P. R. Buseck (1975), Quantitative chemical analysis of individual microparticles using the electron microprobe: Theoretical, Anal. Chem. **47**, 2178–2192.

[2] J. T. Armstrong and P. R. Buseck (1985), A general characteristic fluorescence correction for the quantitative electron microbeam analysis of thick specimens, thin films and particles, X-Ray Spectrom. **14**, 172–182.

[3] J. T. Armstrong (1978), Methods of quantitative analysis of individual microparticles with electron beam instruments, Scanning Electron Microscopy 1978, **1**, 455–467.

[4] J. A. Small (1981), Quantitative particle analysis in electron-beam instruments, Scanning Electron Microscopy 1981, **1**, 447–461.

[5] J. I. Goldstein (1979), Principles of thin film x-ray microanalysis, in Introduction to Analytical Electron Microscopy, J. J. Hren, J. I. Goldstein, and D. C. Joy, eds., Plenum Press, New York, 83–120.

[6] J. A. Small, K. F. J. Heinrich, C. E. Fiori, R. L. Myklebust, D. E. Newbury, and M. F. Dilmore (1978), The production and characterization of glass fibers and spheres for microanalysis, Scanning Electron Microscopy 1978, **1**, 445–454.

[7] J. A. Small, K. F. J. Heinrich, D. E. Newbury, and R. L. Myklebust (1979), Progress in the development of the peak-to-background method for the quantitative analysis of single particles with the electron probe, Scanning Electron Microscopy 1972, **2**, 807–816.

[8] J. A. Small, K. F. J. Heinrich, D. E. Newbury, R. L. Myklebust, and C. E. Fiori (1980), Procedure for the quantitative analysis of single particles with the electron probe, in Characterization of Particles, K. F. J. Heinrich, ed., NBS Spec. Publ. 460, 29–38.

[9] P. J. Statham and J. B. Pawley (1978), New method for particle x-ray microanalysis based on peak to background measurements, Scanning Electron Microscopy 1978, **1**, 469–478.

[10] J. A. Small, S. D. Leigh, D. E. Newbury, and R. L. Myklebust (1986), Continuum radiation produced in pure-element targets by 10–40 keV electrons: An empirical model, in Microbeam Analysis—1986, A. D. Romig, Jr. and W. F. Chambers, eds., San Francisco Press, 289–291.

[11] J. A. Small, D. E. Newbury, and R. L. Myklebust (1987), Test of a Bremsstrahlung equation for energy-dispersive x-ray spectrometers, in Microbeam Analysis—1987, R. H. Geiss, ed., San Francisco Press, 20–22.

[12] J. T. Armstrong and P. R. Buseck (1975), The minimization of size and geometric effects in the quantitative analysis of microparticles with electron beam instruments, Proc. 10th Annual Conf., Microbeam Analysis Society, 9A-F.

[13] J. T. Armstrong and P. R. Buseck (1977), Quantitative individual particle analysis: A comparison and evaluation of microprobe techniques, in Proc. VIII Int. Conf. on X-Ray Optics and X-Ray Microanal., 41A-H.

[14] J. T. Armstrong (1980), Rapid quantitative analysis of individual microparticles using the a-factor approach, in Microbeam Analysis—1980, D. B. Wittry, ed., San Francisco Press, 193–198.

[15] J. T. Armstrong (1982), New ZAF and a-factor correction procedures for the quantitative analysis of individual microparticles, in Microbeam Analysis—1982, K. F. J. Heinrich, ed., San Francisco Press, 175–180.

[16] J. T. Armstrong (1984), Quantitative analysis of silicate and oxide minerals: A reevaluation of ZAF corrections and proposal for new Bence-Albee coefficients, in Microbeam Analysis—1984, A. D. Romig, Jr. and J. I. Goldstein, eds., San Francisco Press, 208–212.

[17] J. T. Armstrong (1988), Quantitative analysis of silicate and oxide minerals: Comparison of Monte Carlo, ZAF, and $\phi(\rho z)$ procedures, in Microbeam Analysis—1988, D. E. Newbury, ed., San Francisco Press, 239–246.

[18] H. M. Storms, K. H. Janssens, S. B. Torok, and R. E. Van Grieken (1989), Evaluation of the Armstrong-Buseck correction for automated electron probe x-ray microanalysis of particles, X-Ray Spectrom. **18**, 45–52.

[19] W. Reuter (1972), The ionization function and its application to electron probe analysis of thin films, in Proc. 6th Int. Conf. X-Ray Optics and Microanalysis, G. Shinoda, K. Kohra, and T. Ichinokawa, eds., Univ. Tokyo Press, Tokyo, 121–130.

[20] J. T. Armstrong (1978), Quantitative electron microprobe analysis of airborne particulate material, Ph.D. Thesis, Arizona State University.

[21] R. H. Packwood and J. D. Brown (1981), A Gaussian expression to describe $\phi(\rho z)$ curves for quantitative electron probe microanalysis, X-Ray Spectrom. **10**, 138–146.

[22] J. D. Brown, R. H. Packwood, and K. Milliken (1981), Quantitative electron probe microanalysis with Gaussian expression for $\phi(\rho z)$ curves, in Microbeam Analysis—1981, R. H. Geiss, ed., San Francisco Press, 174.

[23] G. Love, M. G. Cox, and V. D. Scott (1978), The surface ionisation function $\phi(0)$ derived using a Monte Carlo method, J. Phys. D **11**, 23–31.

[24] K. F. J. Heinrich (1966), Electron probe microanalysis by specimen current measurement, in Optique des Rayons X et Microanalyse, R. Castaing, P. Deschamps, and J. Philibert, eds., Herman, Paris, 159–167.

[25] G. Love and V. D. Scott (1978), Evaluation of a new correction procedure for quantitative electron probe microanalysis, J. Phys. D **11**, 1369–1376.

[26] J. T. Armstrong (1988), Accurate quantitative analysis of oxygen and with a W/Si multilayer crystal, in Microbeam Analysis—1988, D. E. Newbury, ed., San Francisco Press, 301–304.

[27] J. T. Armstrong (1988), Bence-Albee after 20 years: Review of the accuracy of the a-factor correction procedures for oxide and silicate minerals, in Microbeam Analysis—1988, D. E. Newbury, ed., San Francisco Press, 469–476.

[28] K. F. J. Heinrich (1966), X-Ray absorption uncertainty, in The Electron Microprobe, T. D. McKinley, K. F. J. Heinrich, and D. B. Wittry, eds., John Wiley and Sons, New York, 297–377.

[29] B. L. Henke and E. S. Ebisu (1974), Low energy x-ray and electron absorption within solids (100–1500 eV region), in Advances in X-ray Analysis, Vol. 17, C. L. Grant, C. S. Barrett, J. B. Newkirk, and C. O. Ruud, eds., Plenum Press, New York, 150–213.

[30] S. J. B. Reed (1965), Characteristic fluorescence correction in electron-probe microanalysis, Brit. J. Appl. Phys. **16**, 913–926.

[31] M. Green and V. E. Cosslett (1968), Measurements of K, L and M shell x-ray production efficiencies, J. Phys. D **1**, 425–436.

[32] P. Duncumb and S. J. B. Reed (1968), The calculation of stopping power and backscatter effects in electron probe microanalysis, in Quantitative Electron Probe Microanalysis, K. F. J. Heinrich, ed., NBS Spec. Publ. 298, 133–154.

[33] G. Love, M. G. Cox, and V. D. Scott (1978), A versatile atomic number correction for electron-probe microanalysis, J. Phys. D **11**, 7–21.

[34] M. J. Berger and S. M. Seltzer (1964), Tables of energy losses and ranges of electrons and positrons, in Studies of Penetration of Charged Particles in Matter, National Academy of Sciences, National Research Council Publ. 1133, 205–268.

[35] J. D. Brown and R. H. Packwood (1982), Quantitative electron probe microanalysis using Gaussian $\phi(\rho z)$ curves, X-Ray Spectrom. **11**, 187–193.

[36] G. F. Bastin, F. J. J. van Loo, and H. J. M. Heijligers (1984), An evaluation of the use of Gaussian $\phi(\rho z)$ curves in quantitative electron probe microanalysis, X-Ray Spectrom. **13**, 91–97.

[37] G. F. Bastin, H. J. M. Heijligers, and F. J. J. van Loo (1986), A further improvement in the Gaussian $\phi(\rho z)$ approach for matrix correction in quantitative electron probe microanalysis, Scanning **8**, 45–67.

[38] J. D. Brown and L. Parobek (1976), X-ray production as a function of depth for low electron energies, X-Ray Spectrom. **5**, 36–43.

[39] L. Parobek and J. D. Brown (1974), An experimental evaluation of the atomic number effect, in Advances in X-ray Analysis, Vol. 17, C. L. Grant, C. S. Barrett, J. B. Newkirk, and C. O. Ruud, eds., Plenum Press, New York, 479–486.

[40] V. D. Scott and G. Love (1983), Quantitative Electron-Probe Microanalysis, Halsted Press, New York, see pg. 175–176.

[41] J. Henoc, K. F. J. Heinrich, and R. L. Myklebust (1973), A Rigorous Correction Procedure for Quantitative Electron Probe Microanalysis (COR2), NBS Tech. Note 769.

[42] J. Criss and L. S. Birks (1966), Intensity formulae for computer solution of multicomponent electron probe specimens, in The Electron Microprobe, T. D. McKinley, K. F. J. Heinrich, and D. B. Wittry, eds., John Wiley and Sons, New York, 217–236.

[43] D. E. Newbury, R. L. Myklebust, K. F. J. Heinrich, and J. A. Small (1980), Monte Carlo electron trajectory simulation—an aid for particle analysis, in Characterization of Particles, K. F. J. Heinrich, ed., NBS Spec. Publ. 460, 39–62.

[44] D. F. Kyser and K. Murata (1974), Quantitative electron microprobe analysis of thin films on substrates, IBM J. Res. Dev. **18**, 352–363.

[45] R. L. Myklebust, D. E. Newbury, and H. Yakowitz (1976), NBS Monte Carlo electron trajectory calculation program, in Use of Monte Carlo Calculations in Electron Probe Microanalysis and Scanning Electron Microscopy, K. F. J. Heinrich, D. E. Newbury, and H. Yakowitz, eds., NBS Spec. Publ. 460, 105–128.

[46] D. E. Newbury and H. Yakowitz (1976), Studies of the distribution of signals in the SEM/EMPA by Monte Carlo electron trajectory calculations—An outline, in Use of Monte Carlo Calculations in Electron Probe Microanalysis and Scanning Electron Microscopy, K. F. J. Heinrich, D. E. Newbury, and H. Yakowitz, eds., NBS Spec. Publ. 460, 15–44.

[47] Y. Ho, J. Chen, M. Hu, and X. Wang (1987), A calculation method for quantitative x-ray microanalysis for microparticle specimens by Monte Carlo simulation, Scanning Electron Micros. **1**, 943–950.

[48] L. Reimer and E. R. Krefting (1976), The effect of scattering models on the results of Monte Carlo calculations, in Use of Monte Carlo Calculations in Electron Probe Microanalysis and Scanning Electron Microscopy, K. F. J. Heinrich, D. E. Newbury, and H. Yakowitz, eds., NBS Spec. Publ. 460, 45–60.

[49] H. Bethe (1930), Zur Theorie des Durchgangs schneller Korpuskularstrahlen durch Materie, Ann. Phys. Leipz. **5**, 325–400.

[50] K. F. J. Heinrich (1981), Electron Beam X-Ray Microanalysis, Van Nostrand Reinhold, New York.

[51] C. J. Powell (1976), Cross sections for ionization of inner-shell electrons by electrons, Rev. Mod. Phys. **48**, 33–47.

[52] C. J. Powell (1976), Evaluation of formulas for inner-shell ionization cross sections, in Use of Monte Carlo Calculations in Electron Probe Microanalysis and Scanning Electron Microscopy, K. F. J. Heinrich, D. E. Newbury, and H. Yakowitz, eds., NBS Spec. Publ. 460, 97–104.

[53] D. C. Joy (1984), Beam interactions, contrast, and resolution in the SEM, J. Microsc. **136**, 241–258.

[54] T. S. Rao-Sahib and D. B. Wittry (1974), X-ray continuum from thick elemental targets for 10–50 keV, J. Appl. Phys. **45**, 5060–5068.

[55] L. Curgenven and P. Duncumb (1971), Simulation of electron trajectories in a solid target by a simple Monte Carlo technique, Report No. 303, Tube Investments, Saffron Walden, England.

[56] M. Green and V. E. Cosslett (1961), The efficiency of production of characteristic x-radiation in thick targets of a pure element, Proc. Phys. Soc. London **78**, 1206–1214.

[57] G. A. Hutchins (1974), Electron probe microanalysis, in Characterization of Solid Surfaces, P. F. Kane and G. B. Larrabee, eds., Plenum Press, New York, 441–484.

[58] C. R. Worthington and S. G. Tomlin (1956), The intensity of emission of characteristic x-radiation, Proc. Phys. Soc. A **69**, 401–412.

[59] M. Gryzinski (1965), Classical theory of atomic collisions: I. Theory of inelastic collisions, Phys. Rev. Sect. A **138**, 336–358.

[60] J. T. Armstrong and P. R. Buseck (1978), Applications in air pollution research of quantitative analysis of individual microparticles with electron beam instruments, in Electron Microscopy and X-Ray Applications to Environmental and Occupational Health Analysis, P. A. Russell and A. E. Hutchings, eds., Ann Arbor Science, Ann Arbor, 211–228.

THE $f(\chi)$ MACHINE: AN EXPERIMENTAL BENCH FOR THE MEASUREMENT OF ELECTRON PROBE PARAMETERS

J. A. SMALL, D. E. NEWBURY, R. L. MYKLEBUST, C. E. FIORI,
A. A. BELL, AND K. F. J. HEINRICH

National Institute of Standards and Technology
Gaithersburg, MD 20899

I. Introduction

In routine electron probe analysis involving sample targets that are electron opaque, flat, and conductive, the mechanisms describing the interaction of the beam electrons with the target atoms and the subsequent x-ray generation, absorption, and detection are well known. Various correction procedures are currently available for routine quantitative analysis that offer accuracies of 2 percent relative, as determined from studies of well-characterized, homogeneous standards [1].

There are several important classes of samples and analytical conditions for which accurate quantitative analysis is difficult because the various mechanisms of the electron-specimen interactions and the subsequent x-ray generation and absorption are not well known. These areas include the analysis of particles and thin films, the determination of light elements such as carbon, nitrogen and oxygen, and the determination of elements by the use of low energy electron beams. One of the main limitations to further improvements in these areas is that our knowledge of the various physical interactions is derived principally from experiments and theoretical studies performed in the period 1920–1960 [2]. The topic of medium energy (2–50 keV) electron-target interactions has not been rigorously studied since that time. In a number of cases, data important to quantitative analysis procedures are nonexistent or not measured with sufficient accuracy. Examples include the absorption parameter, $f(\chi)$, and the electron backscattering correction, R, especially at low beam energies. Conventional electron microprobes and scanning electron microscopes equipped with x-ray spectrometers are not suitable for such measurements, because the design constrains the analyst to a single value of the angle between the electron beam and the x-ray spectrometer axis.

The construction of an electron optical bench, the "$f(\chi)$ machine," has recently been completed at the National Institute of Standards and Technology. This instrument, which is conceptually similar to an instrument developed by Green in 1962 [3], offers flexible control of the various beam-specimen interaction parameters such as specimen tilt and detector angle. The system consists of a fixed-position electron gun and column, a target stage with 4 degrees of freedom (x, y, z, and tilt) and a detector mount/goniometer with an angular accuracy and precision of 0.1°. This instrument makes it possible to measure experimentally many of the parameters that are critical to the extension of quantitative electron probe analysis to the analytical specimens and system conditions mentioned above.

II. Instrumental

A photograph and a schematic representation of the instrument are shown in figures 1 and 2. Unlike a conventional electron beam instrument in which the electron column is mounted vertically, the column for the bench is mounted horizontally, as shown in figure 3.

Electron Probe Quantitation, Edited by K.F.J. Heinrich and
D. E. Newbury, Plenum Press, New York, 1991

Figure 1. Photograph of $f(\chi)$ instrument.

The symmetry axis for the system is therefore the vertical axis through the center point of the chamber. This arrangement of the column allows for greater freedom of movement for the detectors.

An ETEC[1] scanning electron microscope column was installed on the instrument. The column consists of a tungsten filament gun and three lenses: two condensers and one objective. The gun voltage and the specimen current are adjustable from 2.5 to 30 keV and 10^{-6} to 10^{-11} amps respectively. The electron column extends 7 cm inside the chamber to permit focusing of the electron beam at the target which is mounted at the center point of the chamber. The working distance of the objective lens is 5 cm.

The detector mount is located on the top plate of the vacuum chamber lid, which allows the detectors to be rotated through about 240° with the remaining sector being occupied by the electron column. The chamber top consists of an outer ring that forms the main vacuum seal to the chamber and an inner plug that serves as the detector mount. The inner plug rotates on a compressed "o" ring sleeve that is contained in a groove in the outer ring. As shown in figure 4, the goniometer is attached to the plug and the angle of the detector is determined relative to an indicator attached to the outer ring.

We currently have two x-ray detectors available for use on the instrument. The first of these is a flow proportional detector that mounts directly onto the bottom of the plug as shown in figure 5. In addition to the rotational motion, the flow proportional detector can also be moved in a radial direction, with respect to the target, to increase or decrease the collection solid angle. For the work presented in this paper the detector was located approximately 9 cm from the center of the chamber. The front of the detector is equipped with an aluminized mylar window, a 200-mm-wide vertical slit, and permanent magnets to

---

[1] Certain commercial equipment, instruments, or materials are identified in this paper to specify adequately the experimental procedure. Such identification does not imply recommendation or endorsement by the National Institute of Standards and Technology, nor does it imply that the materials or equipment identified are necessarily the best available for the purpose.

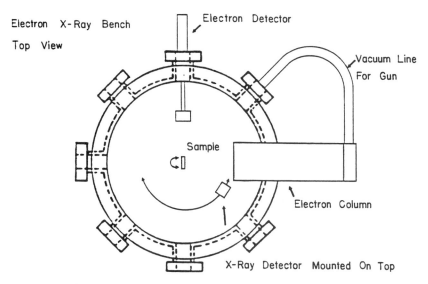

Figure 2. Schematic representation of the instrument (top view).

minimize the signal generated by scattered electrons. The preamplifier for the detector was mounted inside the sample chamber so that the signal line was as short as possible (approximately 12 cm) to minimize the noise level.

The second detector is the Si-Li detector shown in figure 6. It is equipped with a Be window and has a resolution of about 155 eV at the 5.89 keV Mn Kα x-ray peak. The detector is currently being modified to include an electron trap and entrance slit. It is mounted 5 cm from the center axis of the chamber and thus requires an entrance slit width of about 100 μm to maintain the desired angular resolution.

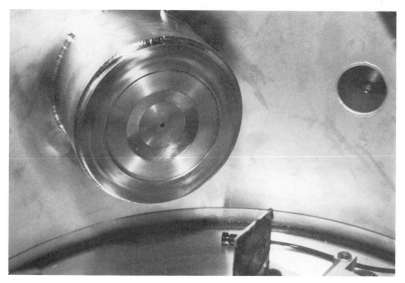

Figure 3. Electron column mount.

Figure 4. Detector goniometer.

Of critical importance to the experimental setup are (1) the positioning of the target at the center axis of the chamber, (2) the determination of the incident angle between the target and the electron beam, and (3) the determination of the take-off angle of the x-ray spectrometer relative to the specimen surface.

(1) Specimen centering:

The electron column is mounted on the chamber so that the beam is directed along the central axis of the chamber. To place the target at the central axis of the chamber, the beam

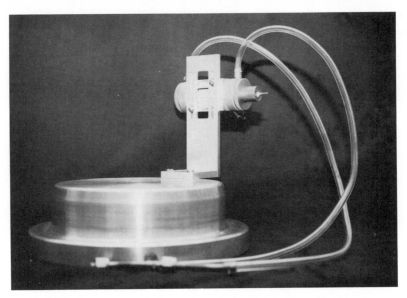

Figure 5. Detector mounting plug with flow-proportional counter in place.

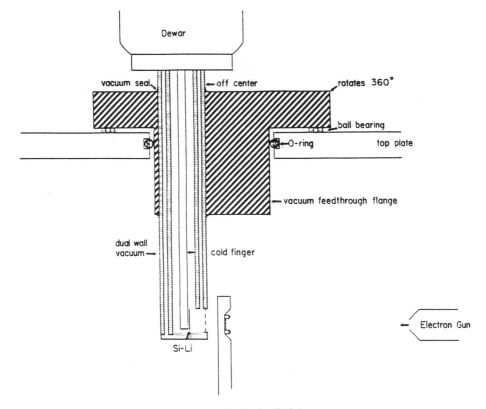

Figure 6. Design for the Si-Li detector.

of a 200 mW HeNe spotting laser was directed from ports 2 and 7 in figure 2, along a diameter perpendicular to the beam axis. The specimen holder was placed so that the laser beam just grazed the surface to define the central axis position. Stage motions in the *x-y* plane then permitted the precise and accurate positioning of the sample on the electron beam axis.

(2) Specimen rotation about vertical axis:

To determine the rotation of the specimen about the vertical axis, which defines the angle of incidence of the electron beam, the alignment laser was placed in a port 45° from the electron gun port (port 2, fig. 2). The specimen was then rotated around the vertical axis until the reflected laser beam entered the 45° port on the other side of the gun (port 8, fig. 2) as shown in figure 7. With this condition of perpendicularity established, the 0° point of the stage goniometer was calibrated. The mechanical rotation permitted a wide range of angles between the specimen and the electron beam to be selected with an estimated uncertainty of 0.1°.

(3) X-ray take-off angle

The x-ray emergence angle or detector take-off angle, *y*, is defined as the angle between the detector and the plane defined by the surface of the target. The 0° position of the

Figure 7. Laser alignment of x-ray target.

detector relative to the specimen and the calibration of the detector goniometer were determined mechanically by positioning the spotting laser in port 7, figure 2. The lid was placed on the instrument and the detector was positioned such that the maximum intensity of the laser beam could be observed through the detector slit at the center of port 2, figure 2. This condition defined 0° and the goniometer was used for the measurement of take-off angles to a precision of about 0.1°.

We verified the mechanical calibration of the detector goniometer to an accuracy of 0.1° by plotting the x-ray intensities for iron vs. the values of $f(\chi)$ determined from the duplex model (see below) for the same angles. The Fe data and the duplex model were chosen for this procedure since:

(1) The Fe Kα x-ray at 6.4 keV has the lowest absorption of the targets studied to date.

(2) The fit of the duplex model to the iron data was the best fit for all targets and all models studied.

The resulting plot is shown in figure 8. If the model is a good description of the absorption in the specimen, then a plot of the measured data vs. predicted $f(\chi)$ should be a straight line passing through the origin. Linear regression applied to the plot in figure 8 does indeed produce an excellent fit with a linear multiple correlation coefficient of 0.997. Any offset of the y intercept from zero was translated into an angular displacement by back

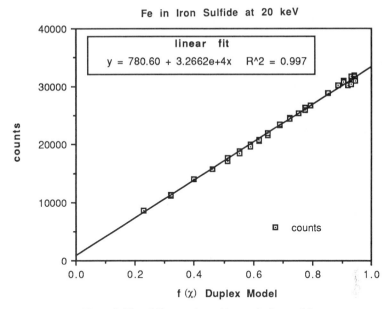

Figure 8. Plot of Fe x-ray intensities vs. duplex model.

calculating, from the duplex model, the angle necessary to shift the linear fit so that the y intercept would be equal to zero. Following this procedure for the iron data resulted in a calculated angular uncertainty of 0.06°.

III. Experimental

The initial experiment carried out in the instrument is the study of the x-ray absorption term $f(\chi)$, where for a given set of conditions $f(\chi)$ is defined in eq (1) as the ratio of the absorbed x-ray intensity, I', to the unabsorbed x-ray intensity generated by the election beam, I.

$$f(\chi) = I'/I \tag{1}$$

In literature expressions, $f(\chi)$ is defined in terms of several parameters that are dependent on the specimen and the experimental set-up

$$f(\chi) = f[\chi, E_0, E_c, Z, A] \tag{2}$$

where χ is a function of the mass absorption coefficient μ/ρ and the x-ray emergence angle Ψ:

$$\chi = (\mu/\rho) \csc \Psi \tag{3}$$

In addition, E_0 is the primary beam energy, E_c is the critical excitation potential for the x-ray line of interest, Z is the mass averaged atomic number of the target, and A is the mass averaged atomic weight of the target [4].

For this study, we measured the emitted x-ray intensities from electron opaque targets at several different x-ray emergence angles. Since it is not possible to measure the total x-ray intensity generated in a target by the electron beam, it is necessary to apply a correc-

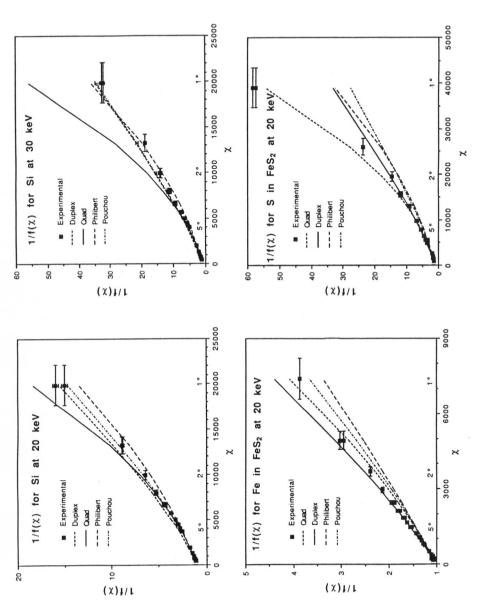

Figure 9. Plots of the experimental data and literature models for $f(\chi)$.

tion factor to the experimental data to obtain the correct $f(\chi)$ values. In our earlier work, we followed Green who used a graphical method to obtain the correct $f(\chi)$ values from the emitted intensity ratios by extrapolating χ to zero [5]. In this work, as shown in eq (4), we use a numerical method for this correction, multiplying the experimental ratios by an $f(\chi)$ term calculated from the quadratic model at the largest emergence angle measured for the sample.

$$f(\chi)_\Psi = \frac{I_\Psi}{I} = \frac{I_\Psi}{I_{max}} f(\chi)_{max}^{quad} \tag{4}$$

where I_Ψ is the x-ray intensity for a given emergence angle Ψ and I_{max} is the intensity at the maximum emergence angle measured.

In general, we prefer to correct the data with an existing model, rather than with a fit of the experimental points, since the value of the ratio at $\chi=0$ from the extrapolation is highly dependent on the minimum angle included in the data set.

The selection of the quadratic model, defined in eq (5) below, is somewhat arbitrary. Any of the various absorption models would be suitable since they all accurately predict experimental results for high x-ray emergence angles. This method of correction was not available to Green at the time of his investigation.

In the initial experiment with this instrument, two different sample compositions were studied: elemental silicon and FeS_2. Measurements were taken on the Si target at beam energies of 20 and 30 keV, and on the FeS_2 target at 20 keV. At least two measurements were taken at each x-ray emergence angle. The uncertainties in $f(\chi)$ are expressed as the combined uncertainties resulting from the propagation of Poisson counting statistics for both the numerator and the denominator. The measurement uncertainty of Ψ is estimated to be 0.1° and is indicated on the plots as horizontal error bars for each angle.

IV. Results

The experimental data and the literature models plotted as $1/f(\chi)$ vs. χ, are shown in figure 9. The curves for the various literature models displayed in figure 9 are not fits to the experimental data. They represent independent calculations of $f(\chi)$ for the respective target and set of experimental parameters. The formulations of the four absorption models studied are listed in eqs (5–8).

Quadratic—Heinrich [6]

$$f(\chi) = (1+1.2\times10^{-6}\gamma\chi)^{-2} \tag{5}$$

where $\gamma=(E_0^{1.65}-E_c^{1.65})$, and the energies are in keV.

- -

Duplex—Heinrich [7]

$$f(\chi) = (1+1.65\times10^{-6}\gamma\chi)^2/[1+\alpha(1.65\times10^{-6}\gamma\chi)] \tag{6}$$

where $\alpha = 0.18+2/\gamma+(8\times10^{-9})\,E_c+0.005\,\sqrt{Z}$ with E_c in eV.

- -

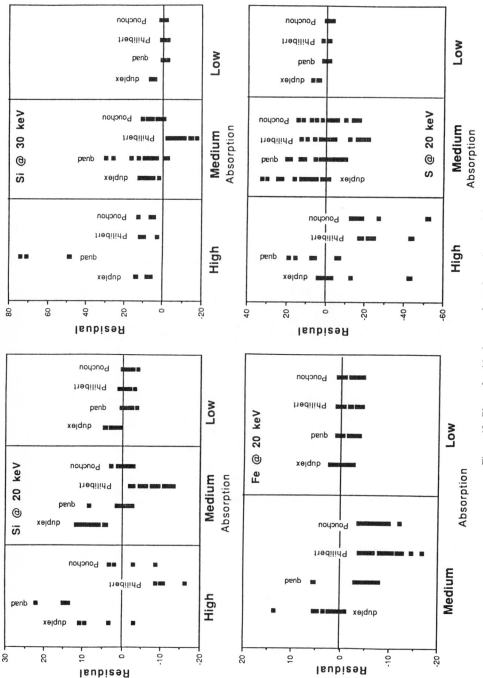

Figure 10. Plots of residual ranges from the various models.

Philibert [8]

$$1/f(\chi) = \left[1+\frac{\chi}{\sigma}\right] \left[1+\frac{h}{1+h}\frac{\chi}{\sigma}\right] \tag{7}$$

where $h = 1.2A/Z^2$, and $\sigma = 4.5\times10^5/\gamma$

- -

Pouchou and Pichoir [9]

$$f(\chi) = [F_1(\chi)+F_2(\chi)]/F \tag{8}$$

where F_1, F_2, and F are complicated integral expressions for which the reader is referred to reference 9.

To evaluate the performance of each model, we constructed residual plots that display the % deviations between the different models and the experimental data defined according to eq (9):

$$\% \text{ res.} = 100\{[f(\chi)_{model}-f(\chi)_{exp}]/f(\chi)_{exp}\}. \tag{9}$$

The evaluation of the performance of a given model was based on two criteria: the magnitude of the residuals and the structure observed in the plot of the residuals vs. the x-ray emergence angle. The better the performance of a given model, the smaller the residuals will be and the less structure there will be in the plots with the residuals distributed randomly around zero [10,11]. For the purpose of this discussion lack of structure is the absence of any discernable pattern in the residuals.

V. Discussion

A. MAGNITUDE

The residuals from the different models are shown in figure 10. These plots show the ranges of the residuals from the models for three distinct regions of χ: high absorption $\chi>10,000$; medium absorption $1,000<\chi<10,000$; and low absorption $\chi<1,000$. There is no high absorption region for iron because the Fe $K\alpha$ x ray at 6.4 keV has a low mass absorption coefficient in FeS_2 and it was not possible to measure accurately angles low enough to provide values of χ in excess of 10,000. The following observations regarding the magnitude of the residuals can be made from these plots.

B. GENERAL OBSERVATION

1. The residuals have the smallest ranges and are the closest to zero for the low χ, high angle, measurements. In general, for the targets studied, the low absorption regions for all models are within the expected experimental uncertainties of ±5 percent.

C. SPECIFIC OBSERVATIONS

For the medium and high absorption regions, it is informative to consider the elemental Si and compound FeS_2 targets separately.

Silicon

Medium Absorption Region

1. At medium absorption for the Si target, the Pouchou and the quadratic models span zero for both beam energies.

2. At 20 keV the Pouchou has narrowest range.

3. At 30 keV the residual range for the duplex, Philibert, and Pouchou models are approximately the same size.

4. The residual range for the quadratic model at 30 keV is substantially broader than the residuals for the other models.

5. The range of residuals for the duplex model is displaced above zero for both beam energies.

6. The range of residuals for the Philibert model is displaced below zero for both beam energies.

High Absorption Region

7. At high absorption for the Si targets the quadratic model severely overestimates the experimental data for both beam energies.

8. All four models overestimated $f(\chi)$ with respect to the experimental data for high absorption region of the 30 keV run.

FeS$_2$

Fe Medium Absorption Region

1. Both the Pouchou and Philibert models have ranges displaced below zero for medium absorption on Fe.

2. The range of the quadratic model spans zero for the medium absorption region of Fe.

3. The range of the duplex model for Fe at medium absorption lies predominantly above zero.

S Medium Absorption Region

1. For medium absorption on S, the range for the duplex model is predominantly above zero.

2. The ranges for the other three models span zero.

S High Absorption Region

3. At high absorption, the Philibert and Pouchou models severely underestimate the experimental data.

4. The duplex range at high absorption spans zero but is quite broad.

5. For the S high absorption region the quadratic model has the narrowest range and spans zero.

D. STRUCTURE

The structure of the residuals for the different models is shown in figures 11–14. These figures are residual plots vs. x-ray emission angle for the various targets. In general the precision between duplicate runs was better than 2 percent. For figures 11–14 the duplicate runs at each emission angle were averaged and displayed as a single bar. The following general observations can be made regarding the structure observed in the residual plots:

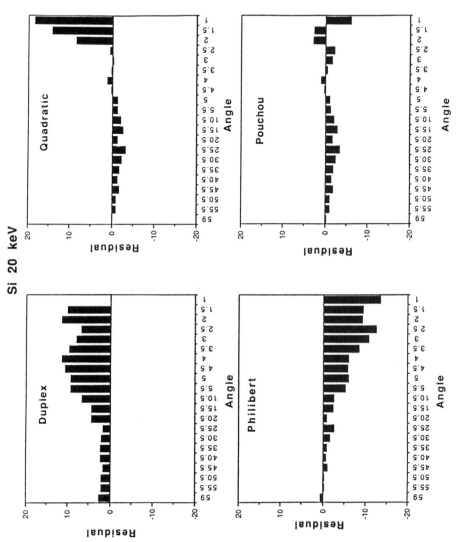

Figure 11. Residual plots vs. angle for Si at 20 keV.

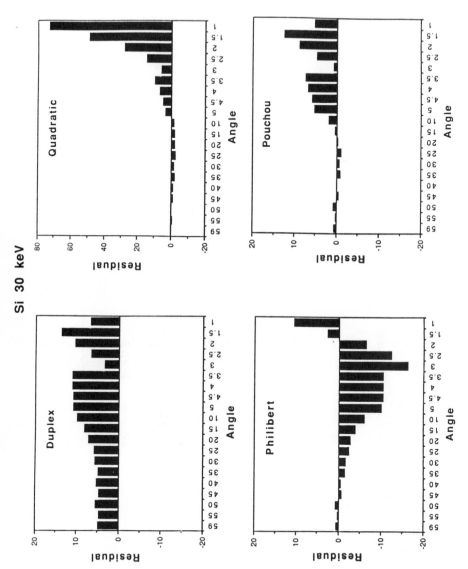

Figure 12. Residual plots vs. angle for Si at 30 keV.

Si at 20 keV, figure 11:

1. The residual plots for both the Pouchou and the quadratic models are almost identical and have minimal structure except for the high absorption region, i.e., angles below 2.5°.

2. Below 2.5°, the residuals for the quadratic model increase dramatically.

3. The Pouchou model, which in general has the best performance for this sample at 20 keV, remains relatively flat for all emergence angles.

4. The residuals for the duplex model are all positive and increasing for angles less than 25.5°.

5. The residual plot for the Philibert model has considerable structure. The residuals increase with increasing absorption.

Si 30 keV, figure 12:

1. Both the quadratic and the Pouchou models have little or no structure for angles $>5°$.

2. Below 5°, both models have increasing positive residuals.

3. The structure of the residual plot for the Philibert model is similar to the 20 keV plot for angles $>2°$. Below 2°, the 30 keV residuals increase dramatically.

4. The structure in the residual plot for the duplex model is the lowest of the models although the plot is displaced above zero.

Fe in FeS₂ at 20 keV, figure 13:

1. Three of the four models, quadratic, Pouchou, and Philibert, all have considerable structure in the residual plots; however, the magnitude of their displacement from zero is different.

2. The shape of the residual plots for the quadratic, Pouchou, and Philibert models is similar with an increase in the negative residuals as a function of decreasing angle.

3. The duplex model has significantly less structure than the other three models with only a slight increase in residuals at low angles.

S in FeS₂ at 20 keV, figure 14:

1. The Pouchou and Philibert models have considerable structure in their residual plots, similar to the structure observed in the Fe plots.

2. The shape of the Philibert and Pouchou residual plots are almost identical with decreasing residuals for angles $<15°$.

3. The residual plot for the duplex model has a positive displacement from zero for angles $>7°$ and negative displacement for angles $<3°$.

4. Of the four models studied the quadratic model has the least structure in the residual plot.

VI. Conclusions

Based on the results of this initial study, we were able to draw the following general conclusions regarding the performance of the different absorption models.

A. For the targets and conditions studied in this experiment, all the models have some structure in their residual plots. The performance of a given model was dependent on the target composition as well as the magnitude of the χ parameter.

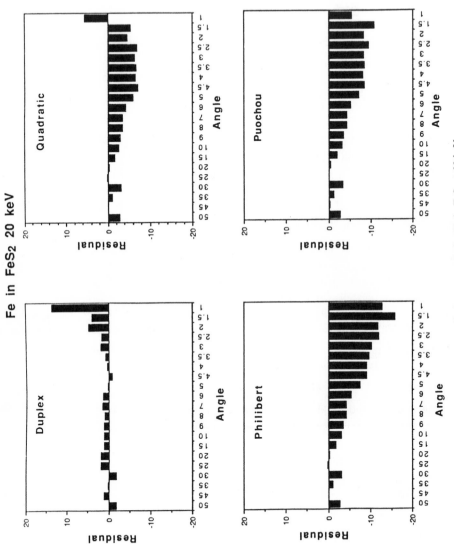

Figure 13. Residual plots vs. angle for Fe in FeS$_2$ at 20 keV.

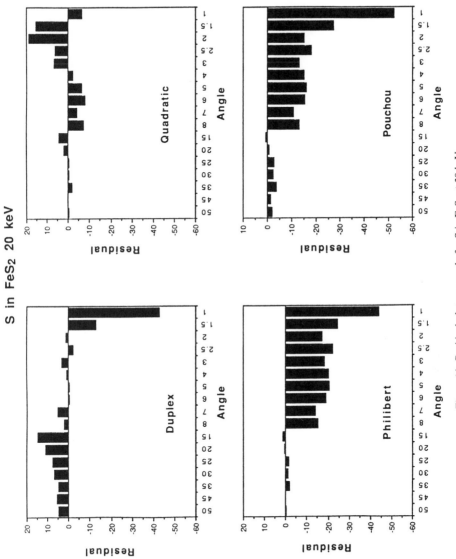

Figure 14. Residual plots vs. angle for S in FeS₂ at 20 keV.

The following is a list of the models that performed best on a given target.

1. Si 20 and 30 keV: Pouchou
2. Fe in FeS_2: duplex
3. S in FeS_2: quadratic

B. Of the four models studied no single model outperformed the others for all targets and conditions.

As a result of these findings, our objectives are to expand the experimental data base and develop an improved model for the absorption term used in microanalysis.

VII. References

[1] J. I. Goldstein, D. E. Newbury, P. Echlin, D. C. Joy, C. Fiori, and E. Lifshin, Scanning Electron Microscopy and X-Ray Microanalysis, New York: Plenum Press, 1981, 330.

[2] K. F. J. Heinrich, Electron Beam Microanalysis, New York: Van Nostrand Reinhold, 1981, pp. 205-254.

[3] M. Green, Ph.D. thesis, University of Cambridge, 1962, p. 117-120.

[4] J. I. Goldstein, 1981, 310.

[5] M. Green, The Target Absorption Correction in X-Ray Microanalysis, in X-Ray Optics and X-Ray Microanalysis, H. H. Pattee, V. E. Cosslett, and A. Engstrom, eds., New York: Academic Press, 1973, pp. 361-377.

[6] K. F. J. Heinrich, 1981, pp. 291-292.

[7] K. F. J. Heinrich, Personal Communication.

[8] J. Philibert, in X-Ray Optics and X-Ray Microanalysis, H. H. Pattee, V. E. Cosslett, and A. Engstrom, eds., New York: Academic Press, 1973, 379.

[9] J. Pouchou and F. Pichoir, Quantitative Analysis of Homogeneous or Stratified Microvolumes Applying the Model PAP, this publication.

[10] F. J. Anscombe and J. W. Tukey, The Examination and Analysis of Residuals, Technometrics 5, 1963, pp. 141-160.

[11] J. J. Filliben, Testing Basic Assumptions in the Measurement Process, ACS Symposium Series, No. 63, Validation of the Measurement Process, J. R. De Voe, ed., 1977, pp. 30-113.

QUANTITATIVE COMPOSITIONAL MAPPING WITH THE ELECTRON PROBE MICROANALYZER

D. E. NEWBURY, R. B. MARINENKO, R. L. MYKLEBUST, AND D. S. BRIGHT

Center for Analytical Chemistry
National Institute of Standards and Technology
Gaithersburg, MD 20899

(Note: Color reproductions of the figures 19–23 here shown in gray tones are available on request from the authors)

I. Introduction

A. DOT MAPPING

Of all the techniques of electron probe microanalysis, the one that has undergone the least change over the history of the field is the technique of producing an image of the distribution of the elemental constituents of a sample, which can be termed compositional mapping. Even today with the dominance of computers for digital data collection and processing in the microprobe laboratory, most compositional mapping is still carried out with an analog procedure that is little changed from the "dot mapping" or "area scanning" technique described by Cosslett and Duncumb in 1956 [1]. The dot mapping procedure can be summarized as follows: (1) As in conventional scanning electron imaging, the beam on the cathode ray tube (CRT) is scanned in synchronism with the beam on the specimen. (2) When the beam is at a particular position on the specimen and an x-ray photon is detected with either a wavelength-dispersive (WDS) or an energy dispersive spectrometer (EDS), the corresponding beam location on the CRT is marked by adjusting the current to excite the phosphor to full brightness. (3) The white dots produced on the CRT display are continuously recorded by photographing the screen to produce the dot map.

While it is simple and effective procedure, dot mapping requires careful attention to the instrumental set-up in order to produce high quality results. As illustrated in figure 1, taken from the work of Heinrich [2], the quality of the recorded image is sensitive to the adjustment of the intensity of the dot on the CRT and on the number of x-ray events accumulated. Because of the relative inefficiency of x-ray generation, the analyst frequently wishes to obtain an elemental map with the minimum acceptable number of x-ray pulses to reduce the amount of time required to accumulate the image. The tendency is therefore to make every x-ray pulse highly visible on the display by adjusting the recorded dot to be as bright as possible, even to the point of "blooming" onto adjacent regions of the phosphor and enlarging the dot. When the dot on the CRT is adjusted for large size and high brightness so that the eye is sensitive to the individual pulses (fig. 1a), a map of the structure can be obtained with relatively few counts (16,000 in this case), but the resulting map has poor spatial resolution and poor contrast sensitivity because the dots tend to overlap; extraneous background counts outside the structure of interest are also visible. When the number of counts is doubled (36,000 in fig. 1b) the overlap of the large dots results in a further loss of information, despite the better statistics of the measurement. Better spatial resolution and contrast sensitivity is achieved by adjusting the dot brightness so that a single x-ray event produces the smallest possible dot on the film, which is just barely visible. More than one event at each location on the film is needed to fully saturate the film. In this condition, the highest spatial resolution and contrast sensitivity is obtained, but in order to visualize a range of concentrations, the number of counts must be greatly increased, e.g., figures 1c (64,000 counts) and 1d (300,000). Experience has shown that between 100,000 and 1,000,000

Electron Probe Quantitation, Edited by K.F.J. Heinrich and
D. E. Newbury, Plenum Press, New York, 1991

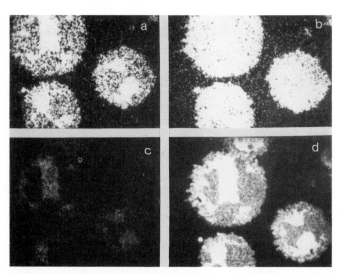

Figure 1. Effects of choice of operating parameters on dot map quality. The effect of the size of the spot written on the CRT is illustrated by the figures arranged vertically; the effect of the total number of x-ray pulses which have accumulated is illustrated horizontally. (a) large CRT dot and 16,000 pulses; (b) large dot and 36,000 pulses; (c) small dot at threshold of visibility and 64,000 pulses; (d) small dot and 300,000 pulses. From Heinrich [2].

x-ray pulses must be accumulated to produce an optimum dot map, depending on the spatial distribution of the constituent [2]. The optimum result clearly involves careful adjustment and a considerable time expenditure to accumulate sufficient x-ray events. For concentration levels below 10 wt. %, the time necessary to obtain the desired number of counts can extend to hours. Moreover, dot mapping is vulnerable to failure, since the full recording time must be expended before the image can be examined; if the brightness of the dot has been set incorrectly, the entire time is wasted since the only record of the experiment is the film recording.

B.   LIMITATIONS OF DOT MAPPING

While it is a powerful technique, the limitations of dot mapping can be summarized as follows:

1)   The final image is qualitative in nature, since the information which is vital to quantitative analysis, namely the count rate at each point in the image, has been lost in the recording process. The area density of dots, as seen in figure 1d, provides only meager quantitative information, and only then in those cases where the structure of interest occupies a significant area fraction of the image.

2)   Because the dot is adjusted to full brightness, no true gray scale information is possible. While ratemeter signals have been used to produce a continuous gray scale x-ray area map, the relative inefficiency of x-ray production restricts the use of ratemeter signals to very high concentration constituents, usually greater than 25 wt. %, for which a low energy x-ray line is available which can be excited efficiently to produce a high count rate.

3)   The recording of the dot map on film media greatly reduces the flexibility of the information for subsequent processing. Registration of multiple images for color superposition on film, for example, is difficult [3].

4) The dot mapping technique has poor detection sensitivity, since no background correction is possible. Characteristic and bremsstrahlung x rays which occur in the energy acceptance window of the WDS or EDS spectrometer are counted with equal weight. With WDS, detection limits in the dot mapping mode are approximately 0.5–1 wt. %, while for EDS systems, which have a much poorer peak-to-background, the limit is approximately 5 wt. % and false results due to apparent changes in concentration caused by the atomic number dependence of the bremsstrahlung can occur.

5) The dot mapping technique has poor contrast sensitivity, depending on the concentration level of a constituent. That is, while it is possible to image a region containing a constituent at 5 wt. % against a background which does not contain that constituent, it is practically impossible to visualize that same 5 wt. % increase above a general level of, for example, 50 wt. %.

6) The time penalty for recording dot maps is significant, and is often a barrier to their use. As noted above, the need for 100,000 to 1,000,000 detected x-ray pulses to form a satisfactory map demands long accumulation time. To map major constituents (10% or higher) typically requires 10 min to 1 h with wavelength dispersive spectrometry, which has the most favorable mapping characteristics, while mapping minor constituents (1% to 10%) requires 1 to 5 h or more. Trace constituents (less than 1%) are usually not accessible to mapping, except in most favorable circumstances. Because most analog scanning systems only permit the display and recording of one signal at a time, even when multiple spectrometers tuned to different elements are available, conventional dot mapping is highly inefficient. Mapping several constituents in the same field of view requires repeating the scanning process for each new map, when in fact the signals are available in parallel.

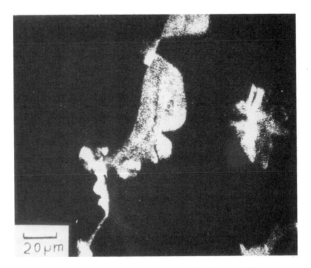

Figure 2. Dot map of the zinc distribution at the grain boundaries of polycrystalline copper produced by diffusion-induced grain-boundary migration. The dot map required 6 h of accumulation time and concentration levels as low as 1 wt. %, as revealed by point analyses, are imaged [4]. (Specimen courtesy of Daniel Butrymowicz, National Institute of Standards and Technology.)

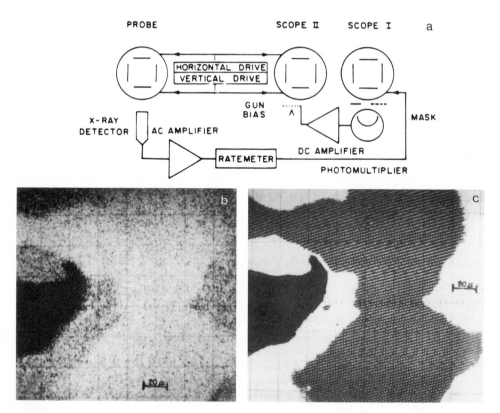

Figure 3. Analog concentration mapping technique of Heinrich [4]. (a) Schematic diagram of system; explanation given in text; (b) conventional dot map and (c) concentration map of iron.

Figure 2 shows an example of the practical limits of performance of conventional dot mapping. The distribution of zinc at the grain boundaries of polycrystalline copper, which results from the process of diffusion-induced grain boundary migration (DIGM), is mapped at concentration levels as low as 1 wt. %, as confirmed by selective point analyses [4]. To map this level of concentration required an accumulation time of 6 h. From the area density of dots, it is clear that the zinc concentration varies in a complex way across the DIGM structure. A complete elucidation of this variation would require careful location of the point analysis locations, a problem made extremely difficult in this case by the complete lack of contrast of this structure in a backscattered electron image.

C. DEVELOPMENT OF COMPOSITIONAL MAPPING

The limitations of dot mapping have been apparent since its inception, and solutions, both analog and digital, have been sought by a number of workers. An example of an analog approach is given in figure 3 [5]. Heinrich devised a "concentration mapping" scheme, making use of an x-ray ratemeter signal to adjust the vertical deflection of a CRT which was viewed by a photomultiplier. A mask was placed across this display to permit only specified ratemeter values to be registered. When signals in this allowed range were received, a second CRT scanned in synchronism with the specimen was modulated to indicate the presence of the element of interest. By calibrating the ratemeter count rate relative to a standard, an image could be produced in which a specific concentration range was mapped. By incorporating circuitry to produce a different display for different photo-multiplier signals, more than one concentration range could be simultaneously mapped. An example of a conventional dot map and a compositional map of the same area produced with this system is shown in figure 3(b) and (c).

Birks and Batt [6] attempted a very early digital solution to the compositional mapping problem. They prepared digital intensity maps by using a multichannel analyzer (MCA) as a storage device for a two-dimensional array of quantitative x-ray counts obtained as the beam was scanned. In this case the data were examined directly as a numerical printout to provide a display of the quantitative information. Marinenko et al. [7] developed a digital control system which permitted the automatic recording of x-ray count data matrices as large as 100×100 points. The data were examined as numerical printouts and not fully quantified and assembled into images.

The rapid commercial development of the computer-based MCA with its increasing computing speed, matrix correction capability, and storage capacity has made digital recording of x-ray counts to create maps a practical possibility [8,9]. Many commercial software packages provide this first stage of formal quantitative compositional mapping, at least for the EDS systems for which most MCA software is designed. In many commercial systems, gray or color scale images can be created from the digital arrays of x-ray count information. These matrix arrays of count rates form the basis for calculating fully quantitative compositional maps.

The recent intensive efforts in establishing quantitative compositional mapping has been accomplished by equipping microbeam analysis laboratories with computer systems of sufficient power to handle both instrument control and the collection and processing of very large matrix data arrays, typically ranging from a minimum of 128×128 arrays for two constituents to 512×512 arrays for as many as 10 constituents [10,11,12]. Most importantly, these systems permit the collection of data from wavelength spectrometers as well as energy dispersive spectrometers. The inclusion of wavelength spectrometry for compositional mapping is critical. While energy spectrometry can be effectively applied to the quantitative mapping of major (>10 wt. %) and many minor (>1 wt. %) constituents, there are often interferences which can only be effectively separated by the higher resolution of WDS. Moreover, for trace elements ($0.01-1$ wt. %), wavelength spectrometers are the only practical solution because of their inherently higher peak-to-background signals and the vastly improved counting situation of WDS compared to EDS. Not only is the pulse-counting deadtime more than a factor of 10 shorter for WDS, permitting limiting count rates in excess of 100,000 counts per second, but because of the narrow energy discrimination of the diffraction process, all of this limiting count rate can be applied to the constituent of interest by WDS. In EDS, the limiting count rate applies to the integrated total x-ray spectrum which reaches the active silicon detector.

II. Quantitative Compositional Mapping

The fundamental concept of quantitative compositional mapping with the electron microprobe can be stated simply: at every point (picture element or pixel) of a matrix scan, a complete quantitative x-ray microanalysis is performed [13]. X-ray count rates are accurately measured at each pixel, and all of the corrections of conventional fixed point analysis are applied. Certain additional corrections, described in the following sections, must be applied to correct for effects which result from scanning. The count rate data from unknowns and standards are corrected for deadtime and backgound, and the resulting charac-

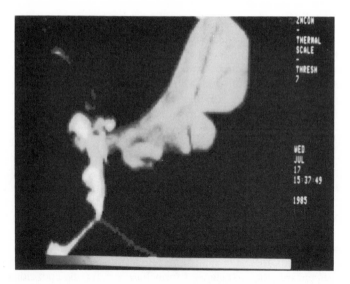

Figure 4. Quantitative compositional map of the same region as shown in figure 2. The concentration data is encoded with a thermal scale, the color sequence of which is presented as a band located along the bottom of the figure. The band represents 0–10 wt. % zinc. (Specimen courtesy of Daniel Butrymowicz, National Institute of Standards and Technology.)

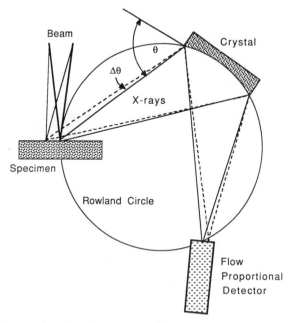

Figure 5. Schematic diagram of wavelength spectrometer, illustrating the detuning of the spectrometer by scanning the beam off the focusing circle of the spectrometer.

teristic (peak) intensities are used to form the k-value, which is the ratio of intensity in the unknown to the intensity in the standard. Martix corrections are applied to the k-values by any of the available procedures: ZAF, empirical sensitivity factors, or $\phi(\rho z)$. The resulting matrices of quantitative concentration values are assembled in a digital image processor into images in which the gray or color scale is related to the true concentration and not merely to the act of detecting an x ray, as in dot mapping, or to raw x-ray counts, as in some MCA systems. As an example of the finished product of quantitative compositional mapping, figure 4 shows the same region as figure 2, but in the form of a true compositional map instead of a dot map [13]. The true compositional microstructure is revealed down to levels of 0.1 wt. %.

While all of the conventional quantitative analysis steps are well established for fixed point analysis [14,15], in order to achieve quantitative compositional mapping, close attention must be paid to two special problems which arise when the beam is scanned rather than maintained at the fixed point defined by the optic axis of the electron column and the focal point of the x-ray spectrometers. These problems involve instrumental effects, wavelength dispersive spectrometer defocusing and energy dispersive spectrometer decollimation, and a specimen-related effect, that of spatially-variable background. Spectrometer defocusing/ decollimation effects are so significant at low magnifications (below 500 diameters magnification) that, if not properly corrected, they will dominate image contrast in a compositional map, even of a major constituent. Background effects are more subtle, but if uncorrected, they will dominate the contrast at the level of minor and trace constituents, resulting in apparent occurrences of elements not actually in the specimen.

A. SPECTROMETER DEFOCUSING/COLLIMATION

1. WDS Case

As illustrated in figure 5, the wavelength spectrometer is a focusing device. The exact condition of focusing is only established when the x-ray emission point on the specimen, the diffracting crystal, and the detector occupy appropriate positions on the focusing circle, knowns as the Rowland circle. Because the crystal has a substantial width (perpendicular to the plane of fig. 5) on the order of 1 cm, the exact focusing condition is actually satisfied along a vector lying in the specimen plane and parallel to the radius of curvature of the cylindrical section of the crystal. When the beam is scanned off the Rowland circle, away from this exact focusing line, the geometric condition necessary to satisfy Bragg's law for diffraction of the desired peak wavelength no longer holds, and the spectrometer is effectively detuned off the peak, so that the measured x-ray intensity decreases. This effect is illustrated in figure 6, which shows the intensity distribution observed when a pure element standard of chromium is scanned at a magnification of 400 diameters, which represents a linear scan excursion of 250 μm. The intensity distribution is observed to be in the form of bands which run parallel to line of focus. The magnitude of the defocusing effect is quite severe, causing the intensity to decrease by 50% at the maximum scan excursion from the focus line at this magnification.

At least four methods exist to correct for wavelength spectrometer defocusing due to displacement of the electron beam impact point: (1) stage scanning, (2) crystal rocking, (3) standard mapping and (4) standard modeling. Methods (1) and (2) represent mechanical methods, while (3) and (4) are mathematical approaches.

a. Stage scanning

The most direct solution to correction of defocusing is to prevent the effect from happening at all by keeping the beam position constant and scanning the stage to bring the analytical point onto the Rowland circle, as described for compositional mapping by Mayr and Angeli [16]. This method has the advantage of measuring all pixels at the maximum possible count rate and with uniform counting statistics.

The stage scanning approach is clearly dependent on the precision of the mechanical stage and is subject to distortions introduced by any hysteresis in positioning. Mechanical stages can typically scan vectors in one direction with negligible positioning errors, but when a matrix of points is scanned, requiring registration of each scan line, there is often significant positioning error. The relative magnitude of the effect increases as the magnification increases and the scan excursion decreases. On most older instruments which have been subjected to extensive use, the cumulative effects of wear generally introduce such a degree of mechanical imprecision as to render stage scanning impractical above an image magnification of a few hundred diameters. Mechanical imprecision in stage positioning can be effectively eliminated through the use of optical encoding on the stage. With the stage position encoded through a readout, the external drive can be applied until the desired stage position is reached in spite of the effects of any hysteresis. Such stages can provide accuracy and reproducibility in a grid scan at the spatial level of 0.5 μm or better, which is satisfactory for mapping at magnifications up to 1,000 diameters or higher.

At low magnifications, below 100 diameters which corresponds to image fields with dimensions greater than 1 mm, the effect of mechanical imprecision becomes negligible. Stage scanning becomes the preferred method of mapping because of count rate considerations. By bringing the analytical point to the spectrometer focus, the maximum count rate is obtained. The other methods discussed below suffer from severely decreased count rates at low magnifications (defocus mapping and defocus modeling) or physical limitations on the crystal size (crystal rocking).

b. Crystal rocking

Figure 5, which shows the defocusing effect, also suggests another mechanical solution involving the crystal. When the beam is scanned off the Rowland circle, the crystal can be

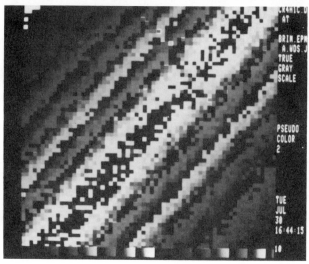

Figure 6. Digital x-ray count map for a chromium standard scanned at a magnification of 400× (250 μm scan deflection). The data are presented in intensity bands (each band represents a narrow range of counts) to reveal the decrease in intensity parallel to the line of focus.

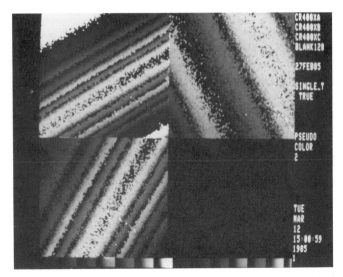

Figure 7. Defocusing maps obtained simultaneously for three different wavelength spectrometers obtained on a chromium standard.

rocked along the Rowland circle in synchronism with the scan on the specimen to bring the effective setting of the spectrometer back to the peak position. In terms of the defocus map shown in figure 6, this amounts to moving the line of focus across the scan field in synchronism with the scan. For more than a decade, such a scheme has been accomplished by analog control for a single wavelength spectrometer [17]. When the situation is considered for more than one wavelength spectrometer, the line of optimum focus is different for each spectrometer because of its relative orientation on the instrument, as shown in figure 7. Swyt and Fiori [18] have discussed techniques for computer control of crystal rocking for multiple spectrometers. Knowledge of the position of the focus line relative to the scan for each spectrometer forms the basis for controlling which crystal(s) is rocked and the extent of the rocking for any scan position. Because the digital scan is discrete rather than continuous, as it would be for analog scanning, the control system is designed to complete the action of crystal rocking prior to the collection of any x-ray counts at a given pixel location.

Like stage scanning, crystal rocking has the advantage of achieving a uniform count rate, and therefore uniform counting statistics, across the image field. At low magnifications (below 100 diameters) the physical size of the crystal limits the extent of rocking, setting a restriction on the applicability of crystal rocking.

c. Standard mapping

An intensity map of a standard, as illustrated in figure 6, inherently contains all of the information needed to correct for the effects of defocusing which are encountered in mapping an unknown that contains an element present in the standard [11,13]. The unknown and standard intensity maps must be obtained under exactly the same conditions of spectrometer/crystal selection, scan orientation relative to the spectrometer, magnification, and beam energy. If the beam current is different between the two maps, it must be accurately measured to permit scaling. By taking the ratio of the unknown intensity map to the standard intensity map on a corresponding pixel-by-pixel basis, a k-value map is created. The decrease in intensity due to defocusing is eliminated in the k-value map. For a correspond-

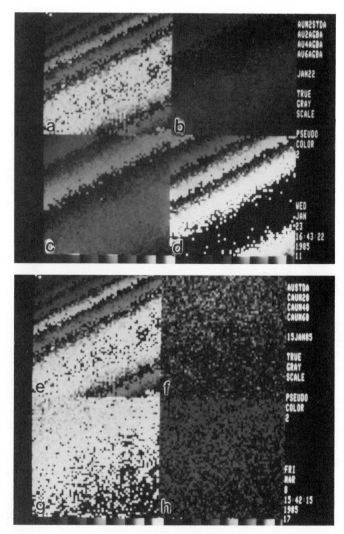

Figure 8. Correction of defocusing by the defocus mapping procedure: a–d are Au $M\alpha$ intensity maps for (a) gold on a pure gold standard, (b) 20Au–80Ag, (c) 40Au–60Ag, and (d) 60Au–40Ag. After correction by the defocus mapping procedure, the corresponding compositional maps are given in (f) 20Au–80Ag, (g) 40Au–60Ag, and (h) 60Au–40Ag.

ing pixel, the decrease is a multiplicative factor which is the same in the numerator and denominator of the k-value; this factor therefore cancels in forming a ratio. An example is shown in figure 8 of the use of defocus mapping to produce a quantitative compositional map in which a significant defocus artifact in the raw intensity map is eliminated.

As seen in figure 8, the defocus mapping procedure provides an effective correction for the defocus artifact. The advantages of the defocus mapping procedure are the direct applicability to any number and configuration of wavelength spectrometers, providing that the conditions noted above are met, and the elimination of any requirement for mechanical scanning. However, the disadvantages of the technique are significant: (1) Because of time constraints, mapping generally uses relatively short dwell times per pixel, compared to

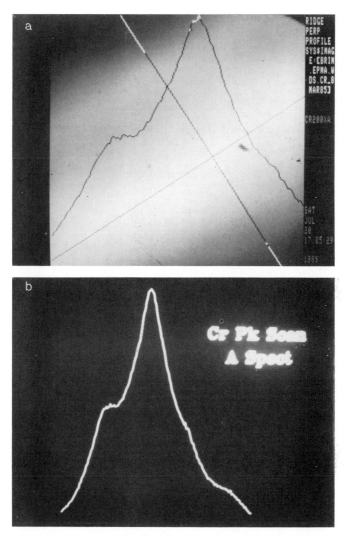

Figure 9. Correspondence between area scanning and wavelength scanning: (a) intensity profile along a vector perpendicular to the line of focus in a standard map of chromium; (b) spectrometer wavelength scan with fixed beam position on chromium.

normal single point analysis integration times, and consequently, the counting statistics measured at a single pixel may be poor when the concentration is low. To improve the statistics of the k-value at that pixel, it is necessary to use similar dwell times for the standard map, even when the concentration of the constituent of interest is much higher in the standard. Typically several standards must be mapped to analyze an unknown, and the time penalty incurred in the standard mapping step is therefore considerable. (2) Although defocus maps for standards can be archived and used for other unknowns, the need arises for a new standard map whenever the scan conditions, particularly the magnification, are changed, making the technique even less time efficient. (3) Defocus mapping requires standards which have a highly polished area as large as the area to be mapped on the unknown.

Any surface defects such as scratches or pits in the unknown or the standard will appear as artifacts in the k-value map, and consequently, in the final compositional map as well.

d.   Defocus mapping

Close examination of a standard map, figure 9(a), reveals that the intensity distribution along a vector taken perpendicular to the line of focus in an area scan closely corresponds to the intensity distribution observed in a conventional wavelength scan of the spectrometer, figure 9(b). In both plots, the combined $K\alpha_1 - K\alpha_2$ peak can be recognized. This geo-

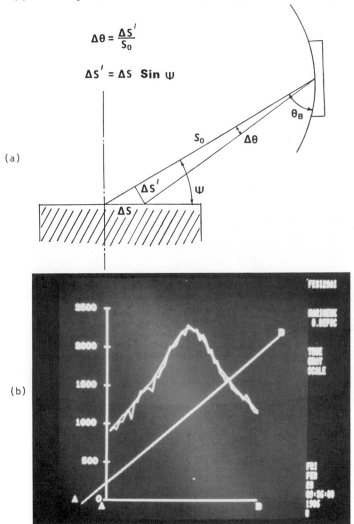

Figure 10. (a) Schematic diagram illustrating the relationship between a deflection Δs from the line of focus in the plane of the specimen and the effective detuning $\Delta\theta$ of the spectrometer from the Bragg angle θ_B. (b) Comparison of a directly measured standard map and a standard map calculated from a wavelength scan. The intensity along the vector AB is plotted for both maps, with the jagged trace for the measured map and the smooth trace for the calculated map. The overlap of the two traces indicates that the peak modeling procedure can accurately calculate a standard map.

metrical equivalence between the situations of (1) area scanning, i.e., beam scanned but with the spectrometer fixed on the peak and (2) wavelength scanning, i.e., beam fixed on the optic axis but with the spectrometer scanned, is illustrated in figure 10(a). This correspondence can be used as the basis of a mathematical method for correcting spectrometer defocusing in which a digital spectrometer wavelength scan can be used to calculate the correction for defocusing at any pixel in the area scan intensity map of an unknown [19,20]. The general problem can be stated thus: for any selection of magnification in the mapping of an unknown, we wish to calculate the standard map which would be appropriate for that situation. With the appropriate standard map in hand, a k-value map can be calculated, and quantification proceeds as in the case for defocus mapping described previously.

The information needed for the defocus modeling procedure consists of the intensity maps for each element of the unknown and wavelength scans of the appropriate pure element or compound standards on the corresponding spectrometers. A standard map for each spectrometer is calculated by the following procedure, as illustrated in figure 10(a).

For a pixel located a perpendicular distance ΔS from the line of best focus, the equivalent spectrometer detuning $\Delta\theta$ can be approximated, since $\Delta\theta$ is an angle on the order of 1°, by the expressions:

$$\Delta\theta \cong \sin \Delta\theta = (\Delta S'/S_0) \tag{1}$$

$$\Delta\theta = (\Delta S/S_0) \sin \Psi \tag{2}$$

where Ψ is the take-off angle of the spectrometer and S_0 is the distance of the beam impact point on the spectrometer to the crystal. S_0 depends on the radius R of the Rowland circle and the value of the Bragg angle to which the spectrometer is tuned:

$$S_0 = 2 R \sin \theta_B \tag{3}$$

The approximation is made that the value of S_0 is a constant for all points in the scan. This is a valid approximation since S_0 has a value on the order of 15 cm while ΔS is approximately 1 mm for a low magnification (100×) map. The construction of the standard map from the wavelength scan data requires that the distance ΔS from the focus line be calculated for each pixel in the scan matrix. The focus line is defined by specific points (X_1, Y_1) and (X_2, Y_2), directly measured for each spectrometer in the intensity maps of the unknown. The distance ΔS of any point (X_3, Y_3) from the focus line is given by the expression:

$$\Delta S = (-D \cdot X_3 + Y_3 - E)/\pm(D^2 + 1)^{1/2} \tag{4a}$$

where

$$D = (Y_2 - Y_1)/(X_2 - Y_1) \tag{4b}$$

$$E = Y_1 - D \cdot X_1. \tag{4c}$$

The location of the pixel relative to the focus line, i.e., above or below the line, is determined by the sign of ΔS, as given by the choice of signs $\pm$ in the denominator of eq (4a). In eq (4a), ΔS is expressed in terms of pixel units. To convert ΔS from pixel units to distance units, the value must be multiplied by a scale factor, SF:

$$SF = L/(M \cdot N) \tag{5}$$

where M is the magnification of the image on the CRT, L is the linear dimension of the scan display on the CRT, and N is the number of points in the scan matrix. The final equation, which relates the amount of defocusing to the pixel location, is given by:

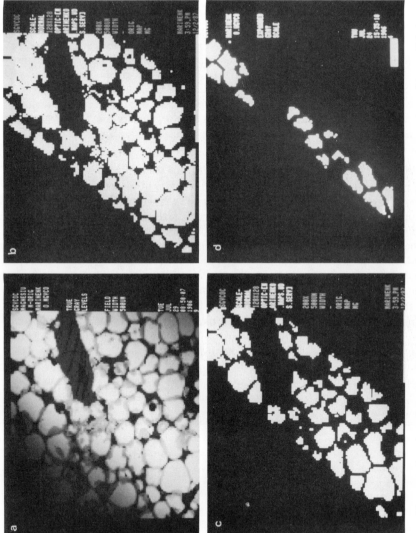

Figure 11. Location of the line-of-optimum-focus in the intensity map of a complex specimen: (a) raw digital intensity map from magnesium in a complex oxide displayed with 255 discrete intensity levels; (b) binary image created with threshold set as indicated in the 0–255 stripe along the bottom of the image; (c) binary image with increased threshold; (d) binary image with further increase in the threshold. (Specimen courtesy of John Blendell, National Institute of Standards and Technology.)

$$\Delta\theta = \Delta S \cdot \sin\phi \cdot SF/S_0 \qquad (6)$$

The standard map is generated on a pixel-by-pixel basis by calculating the value of $\Delta\theta$ appropriate to each pixel from eqs (4) and (6). This value of $\Delta\theta$ is used to select the position relative to the peak in a digital wavelength scan measured on the standard. The value of the intensity measured at this detuning position, relative to the peak intensity, is used to calculate the appropriate standard intensity value for that amount of defocusing. By repeating this procedure for each pixel, a complete standard map is generated for the particular magnification and spectrometer orientation. The accuracy with which a standard map can be calculated from a wavelength scan is illustrated in figure 10(b). In this figure, the intensity distribution along the vector *AB* is plotted for a directly measured standard map (jagged trace) and a calculated standard map (smooth trace). The traces match closely, confirming the accuracy of the defocus modeling procedure. The defocus modeling procedure can be applied down to magnifications of approximately 150 diameters. Below this magnification, the spectrometer is detuned so far off the peak that the intensity falls to an unacceptably low value.

When an unknown is mapped, an important step is the location of the line-of-optimum-focus for each spectrometer. The focus line can be readily located even in a complex image field by the following procedure, which is illustrated in figure 11. The intensity map for the unknown is converted from a 255 gray level ("continuous") image, figure 11(a), into a binary image, figure 11(b), in which pixels with a value equal to or greater than a defined threshold are arbitrarily set to full white (value=255 units) while all pixels below the threshold are set to black (value=0 units). By progressively increasing this threshold, the image collapses to the focus line, as shown in figures 11(c) and 11(d). The position of the focus line can be accurately found even in complex microstructures which introduce significant discontinuities in the binary images. While the position of the focus line in the scanned field is reasonably constant, changes in the electronic scan rotation or in the elevation of the sample can introduce errors which can be eliminated by applying the threshold binary image technique.

The advantages of the defocus modeling procedure are:

1) Great flexibility and efficiency are obtained because a single archived spectrometer wavelength scan can be used to calculate standard maps at any magnification, from approximately 150 diameters to the magnification at which defocusing is negligible, approximately 2000 diameters.

(2) Because the wavelength scan for a particular spectrometer need only be recorded once and then archived, the scan can be recorded with a large number of counts at each angular position. This procedure has the advantage that the counting statistics in the calculated standard map are greatly improved over that in a directly measured map, as can be seen in the comparison of the traces in figure 11. The counting statistics in the calculated standard map are improved to the point that the main contribution to the statistical uncertainty in the k-value map arises only from the intensity map of the unknown.

(3) Artifacts arising from surface irregularities in the standard, such as scratches and pits, are eliminated from the standard map.

e. Counting statistics in WDS mapping

An important problem to consider in WDS mapping is the counting statistics of the intensity measurement which can be obtained. The time spent per pixel in mapping is extremely limited compared to the time taken in conventional fixed point analysis when trace levels are sought. The counting statistics of the pixel intensity measurement will clearly constitute a significant limitation to practical analysis.

A general consideration of practical counting statistics can be developed with the following arguments. If we consider mapping at magnifications below 800 diameters (scan field dimensions of 125 μm or greater) then for a 128×128 digital scan, the pixels will have an effective size on the specimen of 1 μm (edge measurement) or greater. The performance of modern electron microprobe optics is such that a beam of 0.25 μm diameter should be capable of containing 500 nA of electron current at a beam energy of 20 keV [14]. Allowing for the size of the interaction volume added in quadrature with the beam size, the mapping pixels should be independently measured, i.e., the lateral sampling area of the interaction volume projected on the specimen surface should not significantly overlap adjacent pixels, at least for targets of intermediate and high atomic number ($Z > 20$). For low atomic number targets, e.g., carbon and silicon, some overlap will occur, unless the beam energy can be lowered significantly while still retaining sufficient overvoltage for the analysis.

With a 500 nA beam current, it should be possible to obtain a count rate of 100,000 counts per second on a pure element standard. If we allow a pixel dwell time of 2 s (corresponding to a 9.1 h accumulation time for a 128×128 map), then the following counting statistics can be obtained at the various concentration levels, considering only the counts due to the characteristic peak and ignoring any loss of signal due to absorption:

Concentration	Relative Standard Deviation
0.1 (10%)	0.7%
0.01 (1%)	2.2%
0.001 (0.1%)	7.1%

When we consider the four methods of correcting spectrometer defocusing, we note a special advantage with regard to counting statistics that the mechanical methods, stage scanning and crystal rocking, inherently possess as compared to the mathematical methods, standard mapping and defocus modeling. In the mechanical methods, the relative standard deviation statistic of the standard is the same at all pixels, since all pixels are effectively measured at the peak of maximum counting rate. In standard mapping and defocus modeling, the relative standard deviation increases with pixel location away from the line of optimum focus.

An important consideration in observing trace levels of a constituent in an image is the spatial extent of the object in the field of view, i.e., over how many contiguous pixels does it spread. The human eye/brain image processor is excellent at recognizing signals which rise slightly above noise levels, provided that the object is significantly extended in the image [21]. Thus, it is often possible to recognize trace constituents, providing the background is properly corrected, in images at concentration levels for which the single pixel detection limit would be calculated to be much poorer.

2. EDS Case

The energy dispersive spectrometer operates in a line-of-sight mode. Because no focusing is involved, the defocusing artifact encountered with WDS is eliminated. However, at very low magnifications (less than 25 diameters, which corresponds to a scan deflection of 4 mm or greater) a related artifact, collimation effects, may be encountered. EDS detectors are typically collimated to restrict the field of view of the detector to a region on the

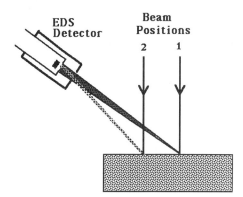

Figure 12. Collimation solid angle effects with energy dispersive x-ray spectrometry. When the beam is scanned to the edge of the scan field, the solid angle of the detector is reduced, lowering the overall count rate.

specimen which is a few millimeters in diameter to prevent the detection of extraneous sources of radiation. When the beam is scanned sufficiently off the optic axis of the microscope, the solid angle of collection of the EDS will be restricted, as illustrated in figure 12. The reduction in collection angle leads to a reduction in the total spectrum count rate relative to the on-axis count rate. Correction for this effect requires an accurate model of the detector acceptance angle as a function of beam position. An experimental approach to developing a correction factor involves measuring the intensity from a pure element standard as a function of scan position to determine a scaling factor as a function of pixel location. By repeating this measurement at several magnifications, an overall mathematical model appropriate to any magnification can be developed. Alternatively, if mechanical stage scanning is available, the collimation solid angle effect can be completely eliminated by always bringing the measurement point to the center of the collimation field.

3. Rough Specimens

The previous discussion of WDS and EDS mapping techniques assumes that the specimen has a planar surface, so that the only effect on spectrometer defocusing/collimation solid angle is due to the lateral displacement of the beam in the x-y plane. However, if the specimen has any topographic features, the influence of the z-dimension can also lead to a decrease in measured x-ray intensity. This is particularly true for the case of the wavelength spectrometer, where the depth of focus in the z-direction is typically on the order of only 10 μm for a vertical spectrometer, and 100 μm for an inclined spectrometer. For such specimens, the best approach to achieve quantitative compositional mapping is to resort to mechanical stage scanning (x-y) to bring the analyzed point to the optic axis of the spectrometer, combined with stage elevation control in the z-direction. The z-position must be automatically controlled during the stage scan. This control can be accomplished by the focus of an optical microscope/TV camera system whose depth of field is less than the depth of field of the wavelength spectrometer. If the specimen has too few features to permit continuous focusing, the analyst can determine the focus position of a series of specific points in the field to be scanned, and a mathematical fit to the specimen surface can be used to determine the elevation during the stage scan. Such an approach can also be used with beam scanning, providing the specimen is not too irregular.

Mapping with energy spectrometry has the advantage that elevation changes of the order of millimeters are needed to encounter significant collimation solid angle effects, thus providing a real advantage when mapping rough samples. However, detector shadowing effects may be encountered with rough specimens if the EDS is placed at a low take-off angle relative to the specimen surface.

Finally, variable surface topography can lead to serious errors in quantitative analysis, and additional corrections to those applied the the case of flat samples are required [14].

B. BACKGROUND CORRECTION

If accurate quantitation is to be achieved for minor and trace constituents, a correction for the background must be made as part of the mapping procedure. The spectral background arises principally from the x-ray bremsstrahlung, which has been described by Kramers as [22]:

$$I_v = kZ(E_0 - E_v) \qquad (7)$$

where I_v is the intensity for a given bremsstrahlung energy E_v, k is a constant, Z is the atomic number and E_0 is the incident beam energy. For targets of mixed composition, the bremsstrahlung intensity varies with the concentration-weighted average atomic number.

In conventional fixed beam analysis, several effective strategies exist for background correction [14,15]. For wavelength spectrometry, the spectrometer can be detuned from the peak position to measure the background on either side of the peak, and the background at the peak position can then be estimated by interpolation. Alternatively for WDS, the background can be measured at the spectrometer peak position on a different, unanalyzed element not present in the unknown, and Kramers' equation can be used to scale the background appropriate to the specimen composition as part of the quantitative analysis iteration procedure. For energy spectrometry, two approaches are popular: background modeling and background filtering. Background modeling makes use of the fact that the EDS measures the complete x-ray spectrum, including the background at all energies, as part of every analysis. By using a modified version of Kramers' equation, the background at the position of any peak of interest can be calculated accurately from the background measured at two preselected energies in the low and high energy regions of the spectrum [23]. Background filtering considers the spectrum as a power spectrum in which the characteristic peaks are contained in the high frequency portion of the distribution, while the slowly changing background is in the low frequency region. A mathematical process such as the digital "top hat" function modifies the original spectrum so as to suppress the low frequency component, the background, while passing the high frequency component, the peaks. In compositional mapping, background correction methods can be based upon these conventional techniques, but additional consideration must be given to the sheer volume of data to be processed.

1. WDS Case

Background correction for WDS could proceed by methods which would mimic conventional correction by detuning. After on-peak, characteristic x-ray intensity maps are collected, the spectrometer(s) could be detuned and the scan could be repeated to collect background intensity maps. However, such a procedure would require an enormous time penalty. If we regard background collection in the most important measurement case of trace and minor elements, typical accumulation times for mapping to this sensitivity are on the order of 10 h or more, as described previously. Clearly, a similar time per pixel would have to be expended counting with the spectrometers detuned to achieve adequate precision. If only one background measurement is made per analytical point, such a procedure doubles the mapping time, and if the spectrometers are detuned above and below the peak, the time is tripled, which is obviously unacceptable.

An alternative procedure when multiple spectrometers are available is to dedicate one spectrometer to the measurement of background, while the others measure characteristic peaks. Background values for the spectrometers tuned to peak positions are then obtained by scaling the value measured on the background spectrometer by efficiency factors determined by separate measurements on pure element targets. While such a procedure avoids the time penalty of the previous technique, it still occupies a valuable wavelength spectrometer purely for background measurement. In many situations, the analyst wishes to use all available spectrometers for measuring elemental constituents of interest. Indeed, losing the use of a spectrometer for background may incur the time penalty of having to repeat the scan in order to measure a remaining constituent.

When we consider the possible reasons for variation in the intensity of characteristic peaks measured during mapping, the intensity at a given point can vary because of the local composition and because of the instrument due to the spectrometer defocusing described in detail above. The background clearly varies because of the composition through Kramers' equation. A question of interest is whether the background also varies in the scan due to spectrometer defocusing. Figure 13 shows the result of an experiment to compare the amount of defocusing between a characteristic peak (Ti $K\alpha$) and a nearby background region obtained by detuning the spectrometer on the low energy side of the Ti $K\alpha$ peak. The intensity map of the characteristic peak shows a significant defocusing artifact, while the intensity map of the background shows a negligible loss of intensity due to defocusing, amounting to less than 2% across the scan field [24]. This result is not surprising, since the background is a very slowly changing function, except near an absorption edge, while the characteristic peak varies rapidly with angle. This experiment suggests that it is not necessary to make a correction for defocusing of the background.

There still remains the obvious problem of correcting for the local variation in the background due to changes in the composition. This correction can be carried out by the following procedure, which is a variant of the conventional fixed point background correction method of using a nonanalyzed element for the background measurement [24,25]:

1) The wavelength spectrometers, A, B, C, etc., are tuned to the characteristic peak positions of all the elements E_i in the unknown. This must include all the major constituents (>1 wt. %), and preferably the minor constituents (>1 wt. %) as well.

2) With the spectrometers at the peak positions, a background measurement is made on another element, E_j with atomic number Z_j which is not present in the unknown.

3) Characteristic intensity maps are recorded for the unknown with each spectrometer.

4) As a first estimate of the background, the single value measured for each spectrometer on E_j is subtracted from the measured characteristic intensity at each pixel. A k-value map is calculated with appropriate standardization and one of the defocus correction methods is applied to eliminate defocusing. The set of k-value maps is used to calculate a first set of compositional maps through application of an appropriate matrix correction procedure: ZAF, $\phi(\rho z)$, or α-factors.

5) A map of the concentration-weighted average atomic number Z is calculated from the compositional maps.

6) A background map is calculated for each spectrometer using the background measured on E_j and scaling for the local composition at each pixel with the multiplicative factor Z/Z_j based on the atomic number dependence of Kramers' equation.

7) The quantitation step is then repeated using the background appropriate to each pixel. This procedure can be applied iteratively, but in practice, the first calculation with the Z/Z_j adjusted background is sufficiently accurate.

A test of this procedure is shown in figure 14, which shows a series of elemental maps for an aligned aluminum-copper eutectic. The compositional maps for Al, figure 14(a), and

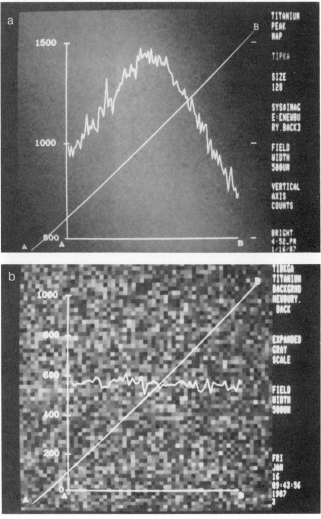

Figure 13. Comparison of wavelength spectrometer defocusing for (a) Ti $K\alpha$ and (b) a background region on the low energy side of the Ti $K\alpha$ peak. The beam current and dwell time per pixel were increased to allow collection of approximately 500 counts per pixel to test for defocusing. Significant defocusing is noted for the intensity map of the characteristic peak, but the defocusing of the background produces less than a 2% change along the vector AB. Field width = 500 μm.

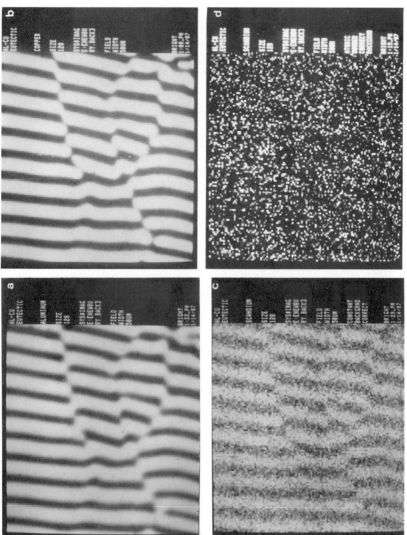

Figure 14. Correction of background in wavelength dispersive mapping. Sample: Al-Cu eutectic, directionally solidified. (a) composition map for aluminum; (b) compositional map for copper; (c) compositional map for scandium, prepared with a constant background correction; the map appears to show scandium segregation to the Cu-rich phase; (d) compositional map for scandium, prepared with variable background correction dependent on local composition; the apparent scandium contrast is eliminated. Contrast expansion has been applied to this map to reveal scandium levels as low as 500 ppm (0.05 wt. %).

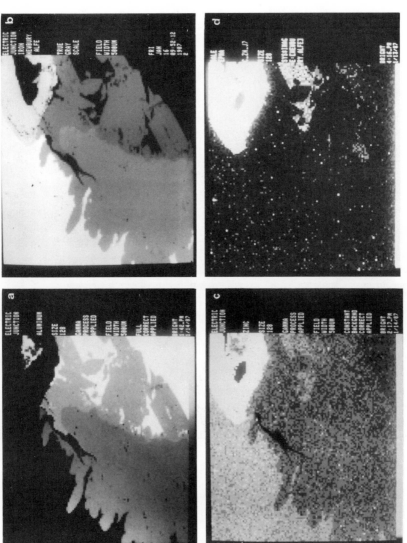

Figure 15. Application of the WDS background correction procedure to reveal the correct distribution of a minor constituent. Sample: reaction zone in an aluminum wire/steel screw from a household electrical outlet. (a) aluminum compositional map; (b) iron compositional map; (c) zinc compositional map prepared with a constant background correction; (d) zinc compositional map prepared with a variable background correction dependent on local composition. The apparent zinc segregation to the Fe-rich zone is seen to be an artifact.

Cu, figure 14(b), reveal the alternating Al-rich and Cu-rich phases in a lath-like structure. A compositional map at the wavelength position of Sc, prepared with a constant correction for background at each pixel, shows an apparent Sc component segregated to the copper-rich phase, figure 14(c). Sc is not present in the sample, and the apparent contrast is purely an artifact of the atomic number dependence of the background. When the above procedure is followed with a compositionally-dependent background correction applied at each pixel, the false Sc contrast is eliminated, as shown in figure 14(d).

An example of a practical application of the WDS background correction procedure is shown in figure 15, which depicts a series of maps from an Al-Fe interaction zone in an aluminum wire/steel screw junction from a failed household electrical outlet [26]. The compositional maps for Al, figure 15(a), and Fe, figure 15(b), reveal distinct zoning. A Zn map, figure 15(c), prepared with a constant background correction, shows a high-Zn region at the surface of the Al-rich zone, and an apparent segregation at a lower concentration level of Zn to the Fe-rich zone as compared to the Al-rich zone. When a compositionally-dependent background correction is applied at each pixel, figure 15(d), the apparent segregation of Zn to the Fe-rich zone is eliminated and seen to be purely an artifact of the atomic number dependence of the background.

2. EDS Case

The background correction problem is more serious for energy spectrometry since the peak-to-background is at least a factor of 10 poorer compared to WDS. Thus, serious artifacts will be encountered even at the level of minor elements if a proper background correction is not made. As an example, figure 16 shows compositional maps of the same Al-Cu eutectic described previously, but obtained in this case with energy spectrometry. An apparent Sc concentration of 1% is observed segregated to the Cu-rich phase.

Background correction in EDS can proceed by either background modeling or background filtering. Either technique will produce satisfactory results. Background filtering is preferred for those specimens where the very nature of the measured x-ray spectrum may change with electron dose [18]. Biological specimens, for instance, will undergo significant mass loss during electron irradiation. Such mass loss affects the absolute intensity of the background and may alter the shape of the background spectrum in a way that is difficult to predict and model. Background filtering can be readily applied since the details for the shape of the spectrum are irrelevant. However, to apply background filtering, either a computationally intensive algorithm must be applied during the collection of data, or else extensive portions of the spectrum must be saved on a channel-by-channel basis for subsequent processing.

For robust specimens, such as those typically encountered in materials science applications, which do not undergo significant damage during electron bombardment, the background modeling procedure can be applied. Background modeling only requires that x-ray counts from low- and high-energy background windows in the spectrum be recorded in addition to the regions for the peaks of interest. The data for each pixel, consisting of the peaks and the two background windows, are processed with the FRAME C algorithm in the main computer to produce the quantitative maps [27]. This approach is illustrated in figure 16 for the Al-Cu eutectic sample. The apparent level of the Sc component in the map prepared without proper background correction is approximately 1 wt. % in the Cu-rich phase. With background modeling via the FRAME C correction, the apparent Sc component is eliminated. A trace Cl constituent was also observed as a contaminant, and after background correction, figure 16(e), was found to be present at a maximum level of 1% in the sample.

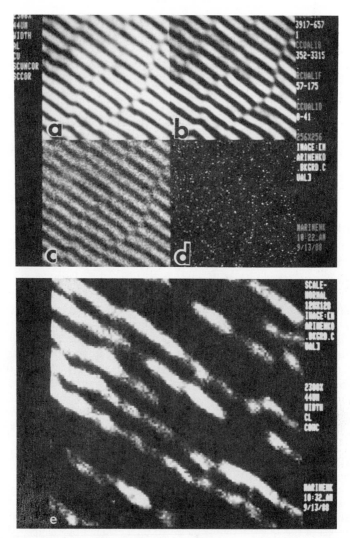

Figure 16. Background correction in energy dispersive compositional mapping: (a) compositional map of aluminum; (b) compositional map of copper; (c) compositional map of scandium with no background correction; (d) compositional map of scandium with FRAME C background correction; (e) compositional map of a trace chlorine constituent.

For beam sensitive specimens such as those encountered in biological and polymer studies, a different strategy must be followed. Fiori et al. [28] have discussed the specialized techniques, based on adaptation of the Hall method, for background correction in mapping of this important class of specimens.

III. Display of Quantitative Images

Once the quantitative compositional maps have been calculated, the matrices of concentration data are used to construct compositional maps with the aid of a digital image processor. Effective techniques for the display of the numerical concentration information

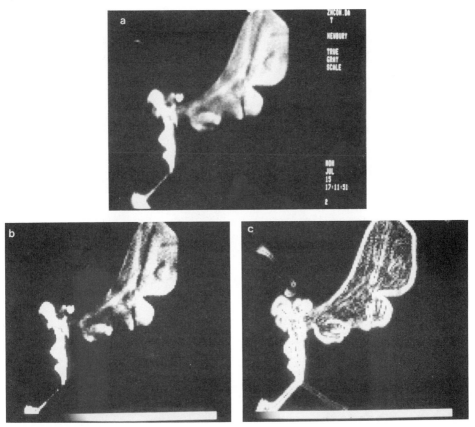

Figure 17. Compositional map of zinc at the grain boundaries of polycrystalline copper (ref). (a) Zinc concentrations in the range from 0–10 wt. % are interpreted as a linear gray scale. (b) Application of a nonlinear ramp function, depicted as the bar along the bottom of the image, representing 0–10 wt. %. (c) Application of a two-dimensional spatial derivative.

as images must be considered. Images may be encoded with a variety of gray or color scales operating on the concentration data as a linear scale. Several transformations are available to enhance the visibility of selected concentration ranges or certain spatial features such as edges. Finally, an intermediate transformation of the concentration data into a multidimensional histogram permits recognition of significant regions in concentration space. A selection of the digital image processing techniques used in our laboratory as applied to specific problems is presented in the following sections.

A. GRAY SCALE IMAGES

1. Linear Scale

Figure 17(a) shows a compositional map of the distribution of zinc at the grain boundaries of polycrystalline copper after the phenomenon of diffusion-induced grain boundary migration (DIGM) has taken place [4]. The concentration data over the scale from 0–10 wt. % are interpreted by a linear assignment to a gray scale, with black corresponding to 0% and white corresponding to 10%. The image is perceived by the viewer as

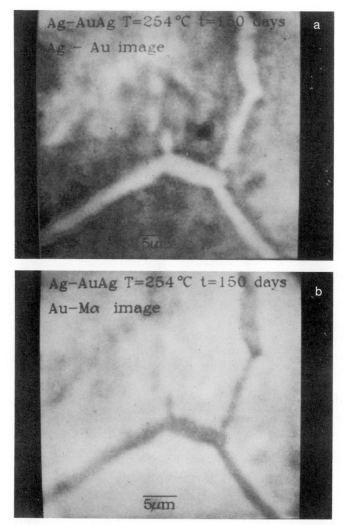

Figure 18. Application of a linear gray scale with offset zero. Sample: 65Ag-35Au alloy which has undergone diffusion induced grain boundary migration. (a) silver compositional map; the enrichment in silver at the boundary is 5 wt.%. (b) Corresponding gold compositional map. (Specimen courtesy of Daniel Butrymowicz, National Institute of Standards and Technology.)

having a continuous intensity scale, although the gray scale is digital and therefore discrete in nature. The natural contrast of the image has been enhanced by matching the gray scale to cover only the concentration range of interest, 0–10%, rather than the maximum range, 0–100%.

2.   Linear Scale with Offset Zero

If the concentration range of interest does not include zero, then the gray scale can be assigned with black corresponding to some nonzero value of the concentration. Offsetting the zero value has the effect of enhancing the contrast in a selected range of the concentration scale. In conventional analog signal processing of scanning electron micrograph images, this is the familiar technique of "black level" enhancement, or differential amplification [14]. An example of the application of this technique is shown in figure 18. The sample is a silver-gold alloy which has undergone the diffusion-induced grain boundary migration phenomenon [29]. In this system, the enrichment of silver at the boundary is approximately a 5% increase above a general background of 65% silver-35% gold. Such a small level of compositional contrast could not be imaged by conventional dot mapping because of the high silver background. However, with the black value of the gray scale offset from a concentration of 0% to a higher value, the enhanced silver at the grain boundary becomes easily visible, as shown in figure 18(a). The corresponding decrease in gold in this region can also be observed by the same method, as shown in figure 18(b).

3. Nonlinear Scale

To further enhance the contrast of the image, a nonlinear scale can be assigned to the concentration data, as shown in figure 17(b). The nonlinear gray scale is depicted as a strip along the bottom of the image. The scale has been chosen to expand the contrast of the limited region of the concentration data approximately 3%–5%, which provides better sensitivity to the compositional structures in the broad portion of the zinc-rich zone. Because of the nature of the nonlinear scale, this expansion of contrast over a limited concentration range is obtained at the sacrifice of contrast sensitivity in the high concentration range, 6%–10%, where all of the range is plotted as white.

4. Spatial Derivative

A two-dimensional digital derivative applied to this concentration data has the effect of highlighting the areas of rapidly changing concentration, as shown in figure 17(c). Such an image processing algorithm has the advantage of revealing features at such low concentrations that they are not easily visible in a linear gray scale presentation, such as the boundary indicated by the arrow in figure 17(c) [24]. The spatial derivative highlights rapid changes in concentration regardless of the absolute value of the concentration, and thus the narrow boundary with a low zinc concentration is readily observed. The action of the spatial derivative on the broad zone of zinc diffusion is less useful. The details which are clearly seen in figures 17(a) and 17(b) are, for the most part, rendered undecipherable by the spatial derivative. The number of highlighted features in the spatial derivative image clearly indicates the complexity of the structure, but the form of the complex structure is obscured by the spatial derivative transformation.

B. COLOR SCALE IMAGES

The sensitivity and gray scale recognition limitations of the human eye/brain image analyzer for gray scale image interpretation can be overcome to a certain extent through the judicious use of color scales. A variety of color scales can be applied to the presentation of single band (i.e., one component) images, including several cartography scales, the visible light color spectrum, the "pseudocolor" scales, and the thermal scale [30]. Of these scales, we have found the thermal scale to be the most useful, with the pseudocolor scale occasionally being utilized for special situations.

1. Thermal Scale

The thermal scale is the sequence of colors observed from a black body heated through a range of temperatures: black, deep red, red, cherry red, orange, yellow, through white.

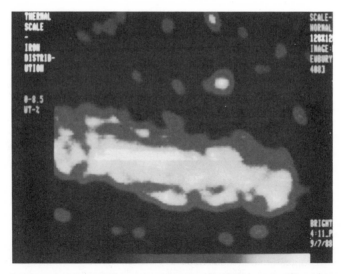

Figure 19. Extended thermal scale applied to a map of a minor (<10%) iron constituent in an ancient precious metal alloy artifact. (Specimen courtesy of Carol Handwerker, National Institute of Standards and Technology.)

This scale provides a good approach to a "logical" color scale, that is, a scale where the "attention value" of a given color varies in a monotonic, roughly linear fashion across the scale. The thermal scale presentation of the zinc/copper data, shown in figure 4, reveals the usefulness of this scale. The concentration data are again assigned in a linear fashion to the thermal scale, shown as a band along the bottom of the image for the range from 0–10 wt. %. This choice of presentation provides a wider dynamic range than a linear gray scale, as seen by comparing figures 4 and 17(a). The low concentration boundary previously observed only in the spatial derivative gray scale image, figure 17(c), is observed directly in the thermal scale image. Moreover, the observer is much better at recognizing a specific color and relating that color to a numerical value in the concentration scale. Thus, not only is the low concentration boundary directly observable in the thermal scale image, but from the deep red color, the concentration can be visually estimated to be near the threshold of the scale, or approximately 0.1 wt. %. The dynamic range of the thermal color scale permits the recognition of features covering approximately 2 decades of concentration, or 0.1–10 wt. % in this case.

The sensitivity for display of the thermal scale can be artificially extended by the addition of "dark" colors such as purple and blue between black and the threshold red. An example is shown in figure 19, in which the extended thermal scale is applied to a compositional map of a minor iron constituent in an ancient precious metal artifact.

2. Pseudocolor Scale

Pseudocolor scales are created by dividing the linear scale into bands with colors assigned to create sharp color contrast between any two adjacent bands. There is no logical sequence to the color assignment, so that "hot" colors, to which the eye is drawn, may appear at any position in the linear scale. The effect of this lack of a logical color sequence is generally to cause confusion in the presentation of a complex structure. This effect can be seen in figure 20, which presents the zinc/copper DIGM image with a particular choice of pseudocolor scale. The low concentration boundary can be detected in this image, but the complex compositional microstructure is difficult to interpret in pseudocolor.

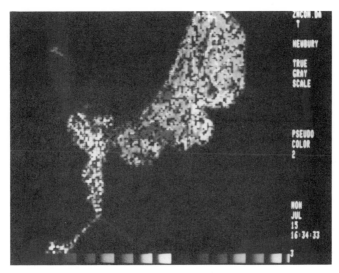

Figure 20. Compositional map of zinc at the grain boundaries of polycrystalline copper. Presentation of the data with a pseudocolor scale. (Specimen courtesy of Daniel Butrymowicz, National Institute of Standards and Technology.)

C.   PRESENTATION OF MULTIBAND DATA

1.   Color Superposition

The simultaneous presentation of several images has traditionally been made by the technique of primary color superposition, in which as many as three compositional maps can be overlaid by assigning each to a different primary color, red, green, or blue [31]. This

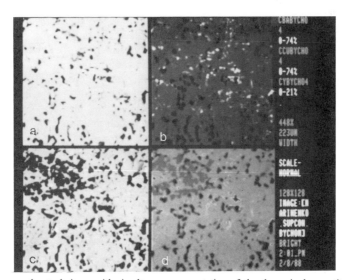

Figure 21. Color overlay technique, with simultaneous presentation of the three single constituent images. (a) barium; (b) copper; (c) yttrium; (d) color overlay with barium = red; copper = green; and yttrium = blue. (Specimen courtesy of John Blendell, National Institute of Standards and Technology.)

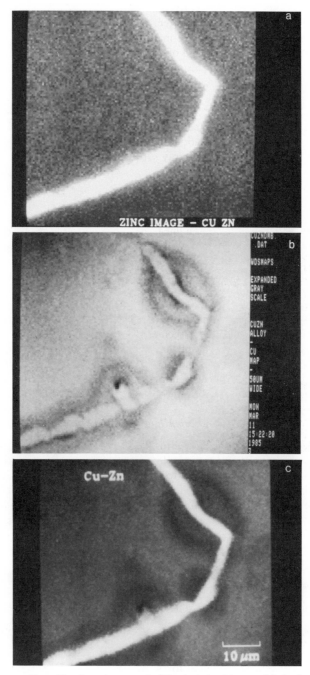

Figure 22. Color superposition with prior enhancement of the single band images. (a) zinc image showing DIGM in a Cu-Zn alloy, with offset-zero linear gray scale enhancement so that 2% Zn=white; (b) corresponding copper image with offset-zero gray scale so that black=94% Cu; (c) color superposition with zinc=red and copper=green. (Specimen courtesy of Daniel Butrymowicz, National Institute of Standards and Technology.)

superposition can be done either indirectly on color film with separate recording of the individual images through color filters, or directly on the separate guns of a color television monitor. The appearance in the composite image of secondary colors (cyan, magenta, or yellow) reveal the positions at which two of the constituents coincide. The appearance of white indicates that all three constituents are present.

The digital mapping procedure can readily incorporate the color superposition technique. As illustrated in figure 21, the presentation of the images can be made even more effective by simultaneously providing, in the same field of view, the individual constituents as gray scale images and the color superposition of the three constituents. This form of presentation permits an easier interpretation of the superimposed color image by providing an immediate reference to the individual gray scale images.

A special capability of the digital mapping procedure further improves the utility of the color superposition technique. The contrast of the individual gray scale images can be enhanced prior to the step of color superposition, as illustrated in figure 22. The sample represents an early stage in the DIGM process in the zinc-copper system. The zinc compositional map, figure 22(a), is presented as a linear gray scale image with black corresponding to zero concentration, and white set at 2 wt. % to provide strong contrast. The contrast of the corresponding copper map, figure 22(b), has also been sharply enhanced by the zero offset technique. This contrast reveals an apparent deficiency in the copper concentration away from the boundary. The logical question to ask is whether this copper deficiency is a result of the zinc diffusion. This question can be readily answered by color superposition in which the enhanced zinc map is presented on the red channel and the enhanced copper map is presented on the green channel. The secondary color yellow reveals the location of zinc-copper regions. It is readily apparent that the area of copper deficiency has resulted from another factor in addition to the zinc. Independent measurements revealed this third element to be oxygen [32]. The contrast enhancement of the single band images prior to color superposition was vital in revealing this unexpected effect.

2.   Concentration Histogram Images

While the thermal scale is reasonably effective in permitting an observer to recognize quantitative concentration relationships in an image, this procedure is necessarily limited to one constituent at a time; it is not possible to overlay a second channel of information for simultaneous viewing. If we examine the individual constituent images which are superimposed in the primary color overlay technique, it is difficult to recognize a quantitative scale in the single constituent images which are intensity modulated in a single primary color, just as it is difficult to recognize specific values in gray scale images. When these single band primary color images are superimposed, it is virtually impossible to assign a quantitative color scale, because of the complex interrelationship of the hue and intensity of the secondary colors.

To overcome this limitation, we have developed multidimensional histogram images, termed composition histogram images "CHI," which are presented as images [33]. The construction of a CHI from two single band images proceeds as follows. The corresponding pixels in the two compositional maps, A and B, are interrogated and the values of the concentrations, C_A and C_B, are used to find the location in a new data matrix whose dimensions are defined by concentration ranges for A and B. These ranges may be 0%–100% or they may be selected between any two arbitrarily chosen endpoints, e.g., 45%–65% and 38%–48%. The address in the matrix which is selected by C_A and C_B is augmented by one unit, and the procedure is repeated for the next pixel. When all of the pixels in images A and B have been interrogated, the resulting data matrix can be converted into a new image in

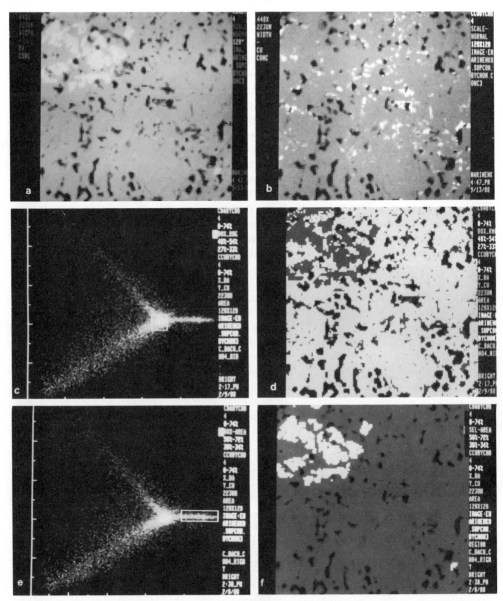

Figure 23. Concentration Histogram Image analysis applied to the analysis of a "123" high-T_c superconductor: (a) barium map; (b) copper map; (c) CHI from barium (horizontal axis) and copper (vertical axis); (d) traceback location of pixels selected by boxed area in (c), highlighted in green with barium map in red, locating pixels associated with the majority superconducting phase; (e) selection of horizontal lobe in CHI; (f) traceback location of pixels selected by boxed area in (e) highlighted in green with barium map in red, locating pixels associated with minority phase; (g) selection of 45° lobe in CHI; (h) traceback location of pixels selected by boxed area in (g), highlighted in green with barium map in red, locating pixels at edge of voids.

which the numerical values at each location in the matrix are encoded with the thermal color scale. The thermal scale permits the recognition of features spanning a wide range of intensity in the CHI. If the original compositional maps depicted a single phase, then ideally all of the pixel counts would be found in one location in the CHI. In fact, the practical influence of counting statistics causes the counts and the calculated concentration to be distributed in a disc whose center gives the composition of the phase. In a map of a multi-phase structure, there would be a disc corresponding to each distinct phase composition. In real systems, more complex structures are observed in the CHI. To relate regions of interest in the CHI to structures in the original compositional maps, a "traceback" feature has been designed into the computer algorithm. Traceback permits the location of each pixel in the original maps which is defined by its location in a selected concentration region of the CHI. The CHI technique can be readily applied to simultaneous analysis of three constituents, and in principle can be extended to higher dimensions.

The CHI procedure is illustrated in figure 23. The single band images for copper and barium in a "123" high-T_c superconductor, figures 23(a) and (b), are converted into the CHI shown in figure 23(c). The CHI shows one distinct region of high intensity, corresponding to the superconducting phase, and three connected bands of low intensity. By selecting the intensely populated concentration region outlined in figure 23(c) and applying the trace-back function, the corresponding pixels in the original images can be located. As shown in the color overlay of figure 23(d), the matrix of superconducting phase is delineated. When the horizontal lobe in the CHI is selected, figure 23(e), corresponding to variable barium concentration at constant copper concentration, the traceback function locates the minority phase shown in figure 23(f). Note the small region of the minority phase which is detected in the lower right-hand corner of the image, demonstrating the sensitivity of the technique to locating rare structures in the compositional maps. When the lobe in the CHI is selected which runs at 45°, shown in figure 23(g) and corresponding to copper and barium concentrations changing commensurately, the edges of the voids in the microstructure are highlighted by the traceback function, figure 23(h). The edges of the voids represent areas where the analytical total falls because of the deviation from the condition of a bulk solid due to beam penetration through the thin material.

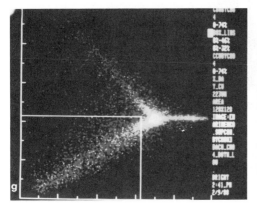
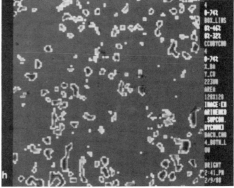

IV. Summary

Quantitative compositional mapping can overcome many of the limitations of the classical dot mapping or area scanning technique. Elemental count rate data is collected at each point in a digitally controlled scan. A complete quantitative analysis is performed at each picture element with special attention to correcting the instrumental artifacts such as wavelength spectrometer defocusing and to x-ray artifacts such as the bremsstrahlung background. From this data through proper standardization and matrix correction, matrices of concentration data can be obtained. The compositional maps are then generated with a digital image processor by assigning various gray scales or color scales to the concentration data. The images are readily amenable to manipulation by a wide range of image processing algorithms. Certain color scales such as the thermal scale are especially effective as an aid to the analyst in recognizing features of interest in the compositional maps. The use of multidimensional concentration histogram images coupled with an algorithm which can trace the pixel correspondence provides a powerful analysis tool. An image in the form of a compositional map, in which viewers can readily recognize structures of interest, which is supported at every pixel by complete elemental concentration data, provides a powerful tool for microstructural analysis.

V. References

[1] Cosslett, V. E. and Duncumb, P., Nature **177** (1956) 1172.
[2] Heinrich, K. F. J., Scanning Electron Probe Microanalysis, in Advances in Optical and Electron Microscopy, Barer, R. and Cosslett, V. E., eds., **6** (1975) 275.
[3] Yakowitz, H. and Heinrich, K. F. J., J. Res. Natl. Bur. Stand. Sec. A, **73** (1969) 113.
[4] Piccone, T. J., Butrymowicz, D. B., Newbury, D. E., Manning, J. R., and Cahn, J. W., Scripta Met. **16** (1982) 839.
[5] Heinrich, K. F. J., Rev. Sci. Inst. **33** (1962) 884.
[6] Birks, L. S. and Batt, A., Anal. Chem. **35** (1963) 778.
[7] Marinenko, R. B., Heinrich, K. F. J., and Ruegg, F. C., Micro-Homogeneity Studies of NBS Standard Reference Materials, NBS Research Materials, and Other Related Samples, Natl. Bur. Stand. (U.S.) Spec. Publ. 260-65 (1979).
[8] McCarthy, J. J., Fritz, G. S., and Lee, R. J., Microbeam Analysis (San Francisco Press, 1981) 30.
[9] Statham, P. J. and Jones, M., Scanning **3** (1980) 168.
[10] Fiori, C. E., Swyt, C. R., and Gorlen, K. E., Microbeam Analysis (San Francisco Press, 1984) 179.
[11] Marinenko, R. B., Myklebust, R. L., Bright, D. S., and Newbury, D. E., Microbeam Analysis (San Francisco Press, 1985) 159.
[12] Fiori, C. E., Leapman, R. D., and Gorlen, K. E., Microbeam Analysis (San Francisco Press, 1985) 219.
[13] Marinenko, R. B., Myklebust, R. L., Bright, D. S., and Newbury, D. E., J. Microscopy **145** (1987) 207.
[14] Goldstein, J. I., Newbury, D. E., Echlin, P., Joy, D. C., Fiori, C. E., and Lifshin, E., Scanning Electron Microscopy and X-Ray Microanalysis (Plenum, New York, 1981).
[15] Heinrich, K. F. J., Electron Beam X-Ray Microanalysis (Van Nostrand Reinhold, New York, 1981).
[16] Mayr, M. and Angeli, J., X-Ray Spect. **14** (1985) 89.
[17] Cameca Instruments, Instruction Manual for Camebax Dynamic Focusing Device for X-Ray Spectrometers, Cameca, Courbevoie Cedex, France.
[18] Swyt, C. R. and Fiori, C. E., Microbeam Analysis (San Francisco, Press, 1986) 482.
[19] Myklebust, R. L., Newbury, D. E., Marinenko, R. B., and Bright, D. S., Microbeam Analysis (San Francisco Press, 1986) 495.
[20] Marinenko, R. B., Newbury, D. E., Myklebust, R. B., and Bright, D. S., J. Microscopy, in press.
[21] Newbury, D. E., Joy, D. C., Echlin, P., Fiori, C. E., and Goldstein, J. I., Advanced Scanning Electron Microscopy and X-Ray Microanalysis (Plenum, New York, 1986) 181.
[22] Kramers, H. A., Phil. Mag. **48** (1923) 836.
[23] Fiori, C. E., Myklebust, R. L., Heinrich, K. F. J., and Yakowitz, H., Anal. Chem. **48** (1976) 172.
[24] Myklebust, R. L., Newbury, D. E., Marinenko, R. B., and Bright, D. S., Microbeam Analysis (San Francisco Press, 1987) 25.

[25] Myklebust, R. L., Newbury, D. E., Marinenko, R. B., Anal. Chem. **61** (1989) 1612.

[26] Newbury, D. E., Anal. Chem. **54** (1982) 1059A.

[27] Myklebust, R. L., Fiori, C. E., and Heinrich, K. F. J., FRAME C: A Compact Procedure for Quantitative Energy-Dispersive Electron Probe X-Ray Analysis, Natl. Bur. Stand. (U.S.) Tech. Note 1106 (1979).

[28] Fiori, C. E., Microbeam Analysis (San Francisco Press, 1986) 183.

[29] Butrymowicz, D. B., Newbury, D. E., Turnbull, D., and Cahn, J. W., Scripta Met. **18** (1984) 1005.

[30] Heinrich, K. F. J., J. de Physique **45** C2 (1984) C2-201.

[31] Duncumb, P. in X-Ray Microscopy and Microradiography, Cosslett, V. E., Engstrom, A., and Pattee, H. H., eds. (Academic, New York, 1957) 617.

[32] Butrymowicz, D. B., Cahn, J. W., Manning, J. R., Newbury, D. E., and Piccone, T. J., Diffusion-Induced Grain Boundary Migration, in Character of Grain Boundaries, eds., Yan, M. F., and Heuer, A. H. (American Ceramics Society, Columbus, OH, 1983) 202.

[33] Bright, D. S., Newbury, D. E., and Marinenko, R. B., Microbeam Analysis (San Francisco Press, 1988) 18.

QUANTITATIVE X-RAY MICROANALYSIS IN THE ANALYTICAL ELECTRON MICROSCOPE

D. B. WILLIAMS AND J. I. GOLDSTEIN

Department of Materials Science and Engineering
Lehigh University
Bethlehem, PA 18015

I. Introduction

The analytical electron microscope (AEM) is related to the transmission electron microscope (TEM) in a manner closely analogous to the relationship between the electron probe microanalyzer (EPMA) and the scanning electron microscope (SEM). While the term "analytical electron microscopy" is ill-defined, it is generally taken to mean quantitative elemental and structural analysis of thin (electron-transparent) specimens. This analysis is invariably performed in a probe-forming, scanning TEM, or STEM, at an accelerating voltage in the range 100–400 kV, substantially higher than in an SEM/EPMA. In this paper we will emphasize only the aspect of the AEM concerned with quantitative analysis of x rays using an energy-dispersive spectrometer (EDS) interfaced to a STEM. However, it is important to recognize that there are many other signals generated when an electron beam traverses a thin sample, as shown in figure 1. In particular, the electron energy-loss signal and the convergent-beam electron diffraction pattern are important analytical tools, which in many ways complement the x-ray spectral information. In addition, of course, the AEM is also capable of generating the many forms of images traditionally associated with a conventional TEM (and SEM). In fact, it is arguable that the development of the AEM has prevented the TEM from declining in importance as a tool for materials characterization.

By the early 1970s, TEM studies of thin crystals had reached a mature state of development. Aberration-limited image resolution was being approached in the first structure images showing the arrangements of atomic groups in complex oxides. Dynamical theory was able to predict, with remarkable accuracy, the diffraction contrast from all known crystalline defects. The principal imaging techniques of bright and dark field and their several variants were fully exploited. The average microscopist was generally happy to live with the rather limited technique of selected area electron diffraction. Specimen preparation techniques had advanced to the stage where all types of inorganic specimens could be thinned to electron transparency. The usual fate of a technique that has approached this level of maturity is to become merely an experimental tool which no longer attracts researchers with an interest in developing the technique. Future development is often confined to such aspects as "user friendliness," "computer control," etc.

Fortunately for materials scientists, TEM did not stagnate into an unchanging research tool. Considerable efforts were directed into improving the point-to-point resolution in phase contrast mode (so-called "high-resolution" or "atomic-resolution" microscopy), but the major changes that have occurred to the TEM have been in the development of analytical electron microscopy.

There were several developments necessary to create an AEM from a conventional TEM. First the symmetrical condenser-objective lens was required since this permitted the formation of fine (<10 nm) convergent electron probes necessary for high spatial resolution signals while retaining the ability to create much larger parallel electron beams for conventional TEM. Second, high-brightness electron sources (e.g., LaB_6 and field emission

Electron Probe Quantitation, Edited by K.F.J. Heinrich and
D. E. Newbury, Plenum Press, New York, 1991

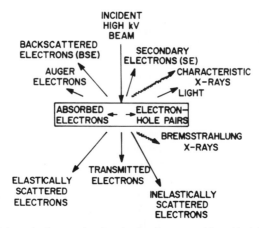

Figure 1. Schematic diagram showing the signals generated in a thin foil in the AEM.

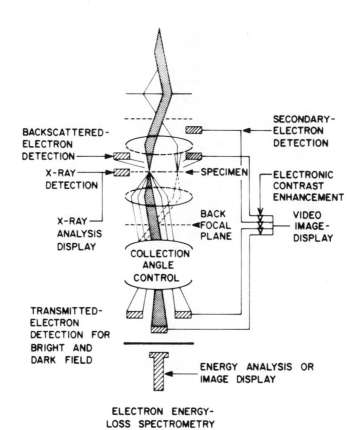

Figure 2. Schematic diagram of the scanning and imaging systems in a TEM/SEM and the various detectors that turn the instrument into an AEM.

guns (FEGs)) were needed to put sufficient current into sub-10 nm probes. Third, efficient, x-ray and electron spectrometers were needed so that the weak signals generated from thin samples could be detected with statistical significance in reasonable time. Fourth, clean vacuum systems were required so that the desired signals were not immediately swamped by the accumulation of hydrocarbon contamination. All these developments had been realized and commercial AEMs were a reality around 1978. Improvement has continued since then and some aspects of AEM are now in danger of becoming mature techniques. Figure 2 shows, schematically, the construction of a typical AEM.

As we shall see in the subsequent sections, the major advantage of x-ray analysis in the AEM compared with the EPMA is the enormous improvement in spatial resolution from ~1 μm to <0.01 μm. This arises from the combination of thin samples and high accelerating voltages. There are also accompanying drawbacks, notably in analytical sensitivity, but the progress in spatial resolution has permitted spectacular advances in the fine-scale analysis of materials. A simple example serves to demonstrate the advantages of the high spatial resolution of AEM compared with EPMA.

Figure 3a is an optical micrograph of a two-phase structure in an Fe-Ni meteorite. The resolution of the image is of the order of a few micrometers and so compares well with the

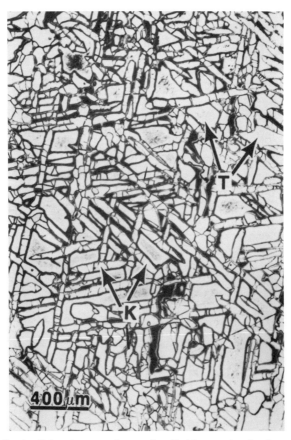

Figure 3a. Low magnification light microscope image of an Fe-Ni meteorite showing ostensibly a two-phase structure of α plates (labeled K) and γ matrix (labeled T) (courtesy K. B. Reuter).

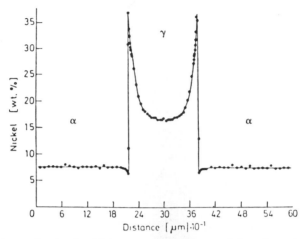

Figure 3b. EPMA trace showing the nickel concentration across an $\alpha/\gamma/\alpha$ region such as in figure 3a.

spatial resolution of microanalysis in the EPMA. Figure 3b shows an EPMA trace across two of the two-phase interfaces; the nickel content is observed to be low in the bcc α phase and high in the fcc γ phase. From such data, especially the values of the nickel content at the α/γ interface, it was possible to determine the Fe-Ni phase diagram down to ~ 500 °C as shown in figure 3c. If we now take the same sample, thin it to electron transparency and examine it in the AEM then, as shown in figure 3d, the microstructure reveals a far more complex nature and the γ region, instead of appearing to be a single phase, is seen to have transformed on a fine scale to several different regions. Using the AEM, the nickel composition profile was measured across the α/γ interface as shown in figure 3e. What is immediately apparent is that the nickel contents at the α/γ interface are much lower in the α phase and much higher in the γ phase than in the EPMA trace in figure 3b. Also the fine-scale breakdown of the γ phase is reflected in the oscillating nature of the nickel content across the region labeled 2 in figure 3d. Without going into detail (see Reuter et al. [1]), it is possible to extract enough information in the AEM to produce the phase diagram shown in

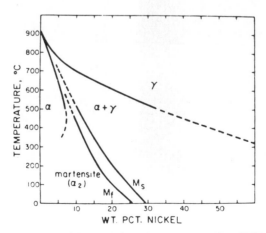

Figure 3c. The high temperature portion of the Fe-Ni phase diagram generated from EPMA data such as in figure 3b.

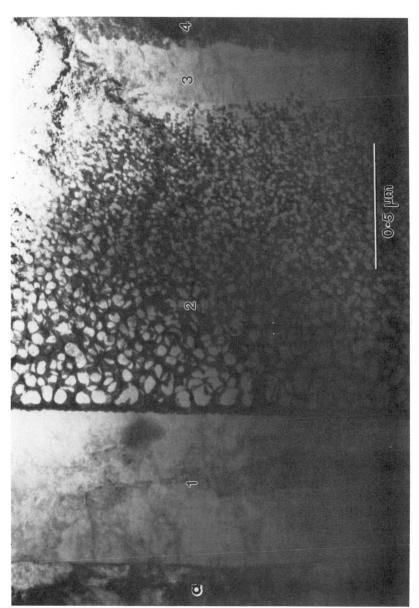

Figure 3d. High magnification TEM image of the α (a) region and the γ region of a meteorite. The γ region has transformed on a submicron scale into four regions labeled 1-4 (from Reuter et al. [1], courtesy Pergamon Press).

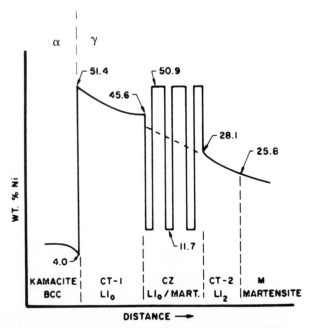

Figure 3e. Schematic AEM trace summarizing the typical nickel concentration variations across a microstructure such as that shown in figure 3d. Note the difference between the composition details here and in figure 3b due to the improved spatial resolution (from Reuter et al. [1], courtesy Pergamon Press).

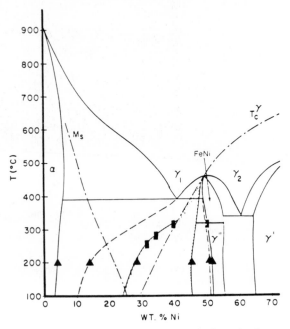

Figure 3f. The Fe-Ni phase diagram generated from the AEM data in figure 3e. Compare with the diagram in figure 3c.

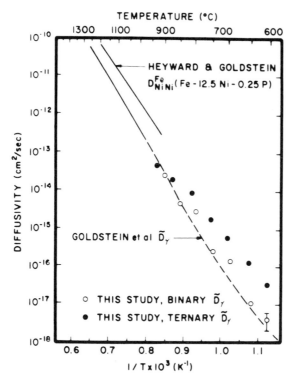

Figure 3g. Comparison of nickel diffusion data obtained in the EPMA (down to temperature $T = 900$ °C) and in the AEM (down to $T = 600$ °C). Note that the low temperature data lie on a line extrapolated from the high temperature values (from Dean and Goldstein [2], courtesy ASM International).

figure 3f. Comparing the two phase diagrams (figs. 3c and 3f) is a compelling demonstration of the power of the AEM. Furthermore, the ability to measure nickel diffusion on a scale of a few tens of nanometers in the AEM has permitted the extension of the diffusion coefficient (D) determination from ~ 1000 °C down to ~ 600 °C, which represents six orders of magnitude change in D, as shown in figure 3g (Dean and Goldstein [2]).

Having demonstrated the power of x-ray analysis in the AEM, we will now go on to describe in some detail the procedure for quantitative analysis of EDS spectra. At all times we will endeavor to draw parallels between the thin-sample analysis procedure and the analysis of spectra from bulk samples, which occupies most of this book.

A. INTRODUCTION TO QUANTITATIVE ANALYSIS

The method of quantitative x-ray analysis of thin specimens using the analytical electron microscope (AEM) is similar in many ways to the various methods of quantitative x-ray analysis of solid samples using the electron probe microanalyzer (EPMA). As in the approach outlined by Castaing [3] we keep instrumental settings (kV, beam current, specimen—x-ray spectrometer configuration) constant while the x-ray readings are taken. Considerations of atomic number, absorption and fluorescence, that is, matrix corrections, are also made. However, in AEM analyses a particular x-ray line from a specimen is not compared directly with a reference standard of known composition, as in the EPMA.

Because of the thin nature of the sample a much simplified approach can be used in the AEM, which we will now develop.

II. The Thin-Foil Analysis Approach

In the x-ray analysis of element A in a bulk or thin specimen containing elements A, B, C, the intensity of the generated primary x-ray emission $I_A^{SPEC^*}$ from the specimen is given by Scott and Love [4]:

$$I_A^{SPEC^*} = \phi_A^{\Delta \rho t} \int_0^\infty \phi_A(\rho t) \exp(-\chi \rho t) d\rho t \tag{1}$$

where ρ is the specimen density, $\phi_A^{\Delta \rho t}$ corresponds to the generated x-ray emission in counts per second from element A in an isolated thin film of the specimen of mass thickness $\Delta \rho t$, and $\phi_A(\rho t)$ corresponds to the ratio of the x-ray emission from a layer of element A, of thickness $\Delta \rho t$ at a depth t in the specimen to the x-ray emission from element A in an isolated thin film, of thickness $\Delta \rho t$. The x-ray emission at depth t is partially absorbed before leaving the specimen. The amount of absorption is calculated by the term:

$$\exp(-\chi \rho t) \tag{2}$$

and $\chi = \mu/\rho|_{SPEC}^A \csc \alpha$, where $\mu/\rho|_{SPEC}^A$ is the mass absorption coefficient for element A in the specimen and α is the x-ray take-off angle. The total generated intensity including any fluorescence contributions (δ_A) is then:

$$I_A^{SPEC^*} = \phi_A^{\Delta \rho t} \int_0^\infty \phi_A(\rho t) \exp(-\chi \rho t) d\rho t \cdot (1 + \delta_A) . \tag{3}$$

The intensity of the generated x-ray emission of element A from an isolated thin film $\phi_A^{\Delta \rho t}$ in a specimen of mass thickness $\Delta \rho t$ can be given as:

$$\phi_A^{\Delta \rho t} = N(\omega_A a_A Q_A / A_A) C_A \Delta \rho t \tag{4}$$

where N is Avogadro's number, A_A is the atomic weight of A, Q_A is the ionization cross section, ω_A is the fluorescence yield for the characteristic K or L lines from element A, C_A is the weight fraction of element A in the specimen, and "a" is the relative transition probability, i.e., the fraction of the total K or L line intensity that is measured as $K\alpha$ or $L\alpha$ radiation.

Equation (4) is applicable to electron-transparent thin films in which electrons lose only a small fraction of their energy in the film (~ 5 eV/nm). Q_A is considered not to vary significantly with electron beam energy over the small change in energy which results from passing through a thin foil and must be evaluated at E_0, the operating voltage of the AEM. Therefore the generated x-ray emission intensity for element A in a thin foil sample is obtained by substituting eq (4) into eq (3) and limiting the integral in eq (3) to the foil thickness t, thus:

$$I_A^{SPEC^*} = N(\omega_A a_A Q_A / A_A) C_A \Delta \rho t \int_0^t \phi_A(\rho t) \exp(-\chi \rho t) d\rho t \cdot (1 + \delta_A) . \tag{5}$$

In bulk (i.e., not electron transparent) specimens, the measured intensity of a particular characteristic x-ray line from the specimen is compared with that from a reference standard of known composition, often a pure element. In thin-specimen analysis, it is difficult to analyze a reference standard of known composition for two reasons. First of all the "ZAF" or the "matrix" corrections are not well developed for bulk sample standards at the high operating voltages ($\geqslant$100 keV) used in the AEM. Second, because of the specimen holder design and the limited volume of the stage region, it is difficult to insert a standard specimen (either thick or thin), in the AEM and recreate the exact same conditions as for the unknown sample. The AEM illumination system is not designed to reproduce accurately on the standard sample the same beam current used for the thin, unknown specimen.

The obvious spatial resolution advantages of thin-foil microanalysis were recognized many years ago and originally led to the development of the so-called Electron Microscope Micro-Analyzer (EMMA) pioneered by Duncumb in England in the 1960s [5]. Unfortunately the EMMA was ahead of its time, mainly because a wavelength dispersive spectrometer (WDS) was the only x-ray detector system available. The WDS was handicapped by its serial operation, poor collection efficiency, relatively cumbersome size and slowness of operation. However, Cliff and Lorimer [6] refitted an EMMA with an EDS system and developed a "parallel-collection" mode, which had a greater collection efficiency and improved stability compared with WDS. The ability to measure intensities of several elements simultaneously with EDS resulted in a major breakthrough in quantitative analysis.

III. The Cliff-Lorimer Ratio Technique

Cliff and Lorimer [6,7] proposed that, if the x-ray intensities (I_A, I_B) of two elements A and B in the same thin specimen can be measured simultaneously, the procedure for obtaining the concentrations of elements A and B can be greatly simplified. Using eq (5) we can take the ratio of generated intensities I_A/I_B, in the specimen as:

$$\frac{I_A^{SPEC^*}}{I_B^{SPEC^*}} = \frac{C_A}{C_B} \cdot \frac{\omega_A a_A Q_A/A_A}{\omega_B a_B Q_B/A_B} \cdot \frac{\int_0^t \phi_A(\rho t)\exp(-\chi\rho t)d\rho t}{\int_0^t \phi_B(\rho t)\exp(-\chi\rho t)d\rho t} \cdot \frac{1+\delta_A}{1+\delta_B} \tag{6}$$

The isolated thin-film mass thickness $\Delta\rho t$ drops out in the ratio making the measurement of $\Delta\rho t$ unnecessary. In this equation, the absorption calculation is carried out from the surface of the specimen, $t=0$, to the full thickness of the specimen, t.

In eq (6) we can think of the term:

$$\frac{\omega_A a_A Q_A/A_A}{\omega_B a_B Q_B/A_B}$$
as the atomic-number correction Z,

$$\frac{\int_0^t \phi_A(\rho t)\exp(-\chi\rho t)d\rho t}{\int_0^t \phi_B(\rho t)\exp(-\chi\rho t)d\rho t}$$
as the absorption correction A,

and $\dfrac{1+\delta_A}{1+\delta_B}$ as the fluorescence correction F.

The measured intensity I_A^{SPEC} from the EDS may be different from the generated intensity $I_A^{SPEC^*}$ because the x rays may be absorbed as they enter the EDS detector in the beryllium window, gold surface layer, and silicon dead layer. If the incoming x rays are very energetic, they may not be totally absorbed in the active area of the detector and may penetrate through the detector. (This is a particular problem in 300–400 keV AEMs where sufficient overvoltage is available to generate high-energy K lines from all elements in the periodic table. Under these circumstances it is possible to use intrinsic-germanium detectors in the AEM. These detectors offer very attractive properties such as high energy resolution (~ 120 eV) and the ability to detect x rays from B $K\alpha$ to U $K\alpha$ (Cox et al. [8]). It would not be possible to obtain these properties with a Si(Li) detector in an EPMA.)

We can relate the measured intensity, I_A^{SPEC}, to the generated intensity, $I_A^{SPEC^*}$, through the term ϵ_A (Goldstein et al. [9]). Therefore:

$$\frac{I_A^{SPEC}}{I_B^{SPEC}} = (ZAF) \cdot \frac{\epsilon_A}{\epsilon_B} \cdot \frac{C_A}{C_B} \tag{7a}$$

and ϵ (for element A) is given by:

$$\epsilon_A = \exp\left\{ -\left[\mu/\rho \big|_{Be}^A \, \rho_{Be} \, t_{Be} + \mu/\rho \big|_{Au}^A \, \rho_{Au} \, t_{Au} + \mu/\rho \big|_{Si}^A \rho_{Si} \, t_{Si} \right] \right\} \cdot$$
$$\left\{ 1 - \exp\left[-\mu/\rho \big|_{Si}^A \rho_{Si} \, t'_{Si} \right] \right\} \tag{7b}$$

where μ/ρ is the mass absorption coefficient of x rays from element A in various detector components comprising the beryllium window, the gold contact layer, the silicon dead layer and the active silicon layer; ρ is the density and t the thickness of the same detector components; t' is the thickness of the active silicon layer. A similar expression can be written for ϵ_B.

As in a solid specimen the matrix correction for ZAF is a function of the specimen composition, $(C_A, C_B \ldots)$. Since $C_A + C_B + \ldots = 1$, the composition of each element in the specimen can be calculated using an iterative procedure. This approach contrasts with the ZAF technique for bulk specimens which does not assume that all the concentrations sum to unity. Therefore the thin specimen analysis technique relies on the investigator to include all elements present in the specimen in the calculation.

If the A and F corrections are negligible and approach unity, the measured intensity ratio of elements A and B in the specimen can be given as:

$$\frac{I_A}{I_B} = (Z) \cdot \frac{\epsilon_A}{\epsilon_B} \cdot \frac{C_A}{C_B} = \frac{\omega_A \, a_A \, Q_A / A_A}{\omega_B \, a_B \, Q_B / A_B} \cdot \frac{\epsilon_A}{\epsilon_B} \cdot \frac{C_A}{C_B} \cdot \tag{8}$$

In this equation, none of the terms ω, a, Q, A or ϵ are a function of composition. Therefore:

$$\frac{C_A}{C_B} = \frac{1}{Z} \cdot \frac{\epsilon_B}{\epsilon_A} \cdot \frac{I_A}{I_B} = \left[\frac{\omega_B \, a_B \, Q_B \, \epsilon_B / A_B}{\omega_A \, a_A \, Q_A \, \epsilon_A / A_A} \right] \cdot \frac{I_A}{I_B} \cdot \tag{9}$$

The term in the square brackets in eq (9) is a constant at a given operating voltage, and is referred to as the k_{AB} factor or Cliff-Lorimer factor. Equation (9) is usually given in a simplified form as:

$$\frac{C_A}{C_B} = k_{AB} \cdot \frac{I_A}{I_B} \tag{10}$$

This relationship was applied initially by Cliff and Lorimer [7] and has gained great popularity because of its simplicity. If A or F corrections must be applied, then the complete matrix correction technique in eq (7) must be used.

The ratio or Cliff-Lorimer method is often referred to as a standardless technique. However, this description is strictly true only when k_{AB} is determined by the calculation of the ϵ, Q, ω, etc., terms given in eq (9). In some cases the specimen can be used as its own standard if its composition is known in some areas and only unknown in certain regions where solute migration has occurred. More often, the k_{AB} factor is determined using thin-specimen standards where the concentrations C_A, C_B, etc., are known. In this case, the characteristic x-ray intensities are measured and the Cliff-Lorimer k_{AB} factors, which include the atomic number correction Z and instrumental factors, ϵ_A, ϵ_B, are determined directly by using eqs (9) and (10). The standards approach although tedious is often more accurate, particularly because ϵ_A and ϵ_B vary from one instrument to another and even over a period of time for one EDS detector. The standards and standardless k_{AB} factor approach are summarized in many papers and textbooks, e.g., Williams [10], Williams et al. [11], Joy et al. [12], Wirmark and Nordén [13], and will not be developed further in this paper. Since atomic number and instrumental factors are well understood, our objective here will be to discuss more fully the absorption and fluorescence corrections since many thin-foil specimens require the full matrix correction. With the increasing availability of 300–400 kV AEMs, the ability to see through thicker foils will mean that absorption and fluorescence effects may become more prevalent.

IV. Absorption Correction

The absorption correction involves evaluating the expression:

$$\int_0^t \phi_A(\rho t) \exp[-\mu/\rho \,|_{\text{SPEC}}^A \csc \alpha \, (\rho t)] d\rho t \tag{11}$$

for all the elements present in the sample. It is critical therefore to know $\phi_A(\rho t)$ for the thin specimen under investigation. The measurement of $\phi(\rho t)$ curves for bulk specimens is a well-established procedure in the EPMA [14]. In contrast, only two simplified measurements of $\phi(\rho t)$ have been made for thin specimens [15,16]. In the work of Stenton et al. [15], the tracer layer, $\Delta\rho t$, was placed at the bottom of matrix foils of various thicknesses to measure $\phi(\rho t)$. This was a simplified technique since it did not account for scattering in the foil below the tracer. In essence what was measured was $\Delta\phi(\rho t)_{\text{max}}$ not the values of $\phi(\rho t)$ vs. t. In these measurements, normal incidence was used and foils were of intermediate atomic number. Tilted specimens and high atomic number specimens would allow for more multiple scattering, but were not considered by the authors.

The effects of tilted specimens and high atomic number specimens have been studied by Newbury et al. [16], both experimentally and by Monte-Carlo calculations. Figure 4 shows $\phi(\rho t)$ curves for gold tilted at 45°. The beam energy was 100 keV and the foils were 50, 100, 150, 200, 250, and 300 nm thick. In the thickest specimen $\phi(\rho t)$ increased from 1.2 at the surface to a maximum of 1.5 within the specimen. For specimen thicknesses typically used in the AEM ($\leqslant 100$ nm), $\phi(\rho t)$ increased from ~ 1.05 to 1.20. The major points from this work are first that $\phi(\rho t)$ in thin films will vary with atomic number, film thickness, tilt and beam energy and second that Monte-Carlo techniques can be used to calculate quite reasonable $\phi(\rho t)$ curves for these variables.

Wirmark and Nordén [13] have measured average, $\phi_{AV}(\rho t)$, values for iron in an austenitic stainless steel. The $\phi_{AV}(\rho t)$ value was obtained by dividing the measured x-ray intensity by the sample thickness t_0 at each point of analysis. The data indicated a linear increase in $\phi_{AV}(\rho t)$ with mass thickness, as might be expected since the amount of elastic single scattering increases with mass thickness. The $\phi_{AV}(\rho t)$ iron data were also consistent with calculated results of a model for $\phi(\rho t)$ derived by Jones and Loretto [17] based on

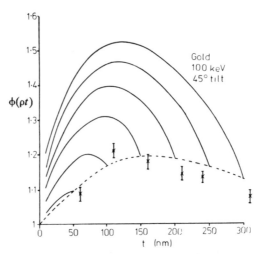

Figure 4. $\phi(\rho t)$ curves for gold foils tilted at 45°. The accelerating voltage was 100 keV and the foils were 50, 100, 150, 200, 250, and 300 nm thick (from Newbury et al. [16], courtesy San Francisco Press).

single-scattering theory. The model of Jones and Loretto gives the depth distribution of x rays, $\phi(\rho t)$, from a thin specimen normal to the electron beam as a function of atomic number Z, atomic weight A, specimen thickness t_0, specimen depth below the surface t, and beam energy E_0. The relation is:

$$\phi(\rho t) = 1 + \left[\frac{Z^2}{E_0^2(A/\rho)}\right] \cdot (at_0 + bt) \ . \tag{12}$$

The relevant units are E_0 (eV), A/ρ (cm^3), and t_0 and t (nm). For silicon the constants a, b are $a = 7,000$ and $b = 16,000$. The variation, if any, of a and b with atomic number is unknown.

Using eq (12), the variation of $\phi(0)$ and $\phi(\rho t)_{max}$ can be calculated. These terms represent the values of $\phi(\rho t)$ at the surface and at the bottom of the foil respectively. Table 1 shows the values of $\phi(0)$ and $\phi(\rho t)_{max}$ for silicon, nickel, and gold at 100 keV and various values of t_{max}. The calculated values of $\phi(\rho t)$ agree quite well with the Monte-Carlo calculations of Newbury [18] for nickel samples with thicknesses of 100 and 350 nm and Newbury et al. [16] for gold samples 50 and 100 nm thick.

Table 1. X-ray distribution function at 100 keV vs. atomic number and specimen thickness

Element	Atomic number, Z	Foil thickness, t	$\phi(0)$	$\phi(\rho t_{max})$
Si	14	67 nm	1.0008	1.0025
		*335 nm	1.0038	1.013
Ni	28	25 nm	1.002	1.0068
		*100 nm	1.0083	1.027
		350 nm	1.029	1.096
Au	79	28.5nm	1.012	1.04
		*50 nm	1.021	1.07
		100 nm	1.04	1.14

* Typical specimen thickness at 100 kV

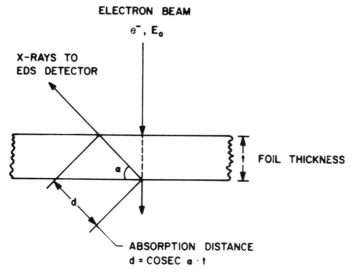

Figure 5. Idealized thin foil geometry in the AEM showing the definition of the take-off angle α and the absorption path length $t \csc \alpha$.

Within the range of typical specimen thicknesses used for x-ray analysis in the AEM at 100 keV, $\phi(\rho t)$ varies no more than 5 percent relative throughout the film and $\phi(0)$ is no greater than 1.03. At the current state of the development of x-ray analysis in the AEM one can then assume that x-ray production is essentially uniform throughout the film and that:

$$\phi_A(\rho t) \simeq \phi_B(\rho t) \simeq 1.0 \ . \tag{13}$$

This assumption was made in the initial approach to the absorption correction by Goldstein et al. [19] in which the integrals in eq (6) were evaluated to give the absorption correction factor A:

$$A = \frac{\mu/\rho\,|_{\text{SPEC}}^{A}}{\mu/\rho\,|_{\text{SPEC}}^{B}} \cdot \frac{1-\exp\left[-\mu/\rho\,|_{\text{SPEC}}^{B}\csc(\alpha)\rho t\right]}{1-\exp\left[-\mu/\rho\,|_{\text{SPEC}}^{A}\csc(\alpha)\rho t\right]} \ . \tag{14}$$

To calculate the absorption correction factor, the values of the mass absorption coefficient, μ/ρ, the specimen density ρ, and the absorption path length must be accurately known. Figure 5 shows the configuration of the specimen, electron beam and EDS detector that must be considered, in order to determine the absorption path length $t \csc \alpha$.

The absorption correction is usually not applied if A is between 0.97 and 1.03 (a correction $\leqslant \pm 3\%$ rel.) since the Z factor, ϵ_A/ϵ_B and the counting statistics usually result in errors of $\geqslant \pm 3\%$ rel. However, the specimen thickness at which the A correction must be applied is often much smaller than the specimen thickness which is obtained as a result of the preparation process. A few examples of calculated thickness above which the A corrections must be made ($A < 0.97$ or > 1.03) are 9 nm for Al $K\alpha$ in NiAl, 22 nm for P $K\alpha$ in Fe₃P and 25 nm for Mg $K\alpha$, O $K\alpha$ in MgO. These thicknesses are not often achieved during the specimen preparation process. As the trend to higher voltage AEMs continues, it will be possible to "see" through thicker samples and perform analyses from thicker areas. Under these circumstances, the "typical" thicknesses in table 1 will increase by factors of 2–3 and the need for absorption corrections will increase.

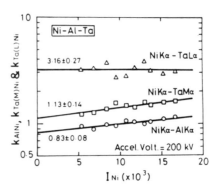

Figure 6. Data from Horita et al. [25] showing intensity ratios from three different nickel-base alloys as a function of the sample thickness. The thickness was measured in terms of the Ni $K\alpha$ intensity. By extrapolating to zero nickel intensity, the true k_{AB} factor can be obtained and it is noted above each set of data (courtesy North-Holland Physics Publishing).

The various factors which limit the accuracy of the absorption correction, mass absorption coefficient, specimen density, absorption path length and specimen geometry have been discussed by various authors [9,10] and will not be discussed in this paper. By far the most difficult parameter to measure is the specimen thickness t, which can vary substantially in a wedge-shaped thin foil. Among the methods used to obtain t are contamination-spot separation Lorimer et al. [20], the spacing of Kossell-Möllenstedt fringes in convergent-beam electron diffraction (CBED) patterns (Kelly et al. [21]), the K to L intensity ratio as a function of t for a given element (Morris et al. [22]), the EDS spectrum variation caused by using different tilt angles at the same analysis point [22], the EELS log-ratio technique (Malis et al. [23]), and the standards technique (Porter and Westengen [24]).

The popular contamination-spot-separation method unfortunately tends to overestimate the thickness [15] and may not be available as sample preparation techniques are better controlled and vacuums are improved in the specimen area. The CBED technique is accurate but occasionally the required diffraction conditions may not be obtainable at precisely the region of interest. The method involving the K/L ratio is only suitable if both peaks are visible in the spectrum. Each of the other methods have their own limiting factors as well.

The standards technique, however, has clear promise. A thin foil or region of thin foil of known mass thickness (usually determined by CBED techniques) and of known composition is used to calibrate x-ray intensity in terms of thickness. In this case one usually measures the intensity of the whole EDS x-ray spectrum entering the detector. The intensity of the EDS x-ray spectrum is a direct function of thickness until absorption effects become significant. One can therefore use a linear calibration of EDS x-ray intensity versus thickness as calibrated by the standard specimen. A standard which is most useful is a specimen of similar composition to the specimen of interest. This technique should be quite useful as it does not rely on the direct measurement of thickness at each analysis point. However, the technique requires that the beam current remains constant as measurements of sample and standard are made, and this is difficult to achieve in an AEM.

As CBED and particularly parallel EELS techniques become more popular and computer control of the stage, combined with digital imaging becomes available, the possibility of on-line CBED and EELS analysis will become a reality. Under those circumstances, these approaches will become the accepted on-line methods to determine the thickness of crystalline samples.

Figure 6 shows the results of a technique in which the absorption correction can be eliminated (Horita et al. [25]). In this simplification of the x-ray absorption correction, the ratios of I_A/I_B, I_B/I_C, etc. in a homogeneous phase are plotted vs. x-ray intensity from one of the elements (e.g., I_A or I_B or ...) which is not absorbed significantly as the specimen increases in thickness. An extrapolation of the nonabsorbed x-ray intensity to zero provides an I_A/I_B, I_B/I_C, etc. ratio in which the absorption A factor is 1.0 and I_A/I_B, I_B/I_C equals k_{AB} C_A/C_B, $k_{BC} \cdot C_B/C_C$, etc. This technique holds great promise. However, it cannot be used in some of the more important analyses, where C_A, C_B ... vary throughout a phase or region of the sample.

The assumption that the specimen is a parallel-sided thin film (fig. 5), is rarely if ever the case in practical x-ray microanalysis because of the ways in which materials specimens are thinned. Electropolishing and ion-beam thinning both give rise to wedge-shaped specimens which complicate the analysis considerably. In addition, for multiphase specimens, one must orient the interphase interface parallel to the electron beam and parallel to the EDS detector axis if absorption effects are to be minimized. Specimen geometry effects as outlined above have been described in detail by other authors (Maher et al. [26], Zaluzec et al. [27]) and will not be discussed here, although other sample-related problems will be described later in the paper.

V. Fluorescence Correction

X rays produced by electron ionization within the specimen may themselves be sufficiently energetic to excite and ionize other atoms. This effect, known as x-ray fluorescence, is intimately associated with the absorption process, because the primary cause of fluorescence is the absorption of another x ray. Figure 7 shows the measured variation in the fluorescence factor δ_A, as expressed by the percentage increase of chromium as a function

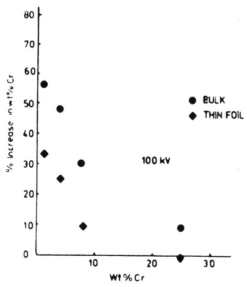

Figure 7. Measured variations in the fluorescence enhancement ratio δ_A at 100 keV as a function of the chromium content in a series of Fe-Cr alloys (from data given in Nockolds et al. [28], published in Goldstein et al. [9], courtesy Plenum Press).

of chromium concentration for Fe-Cr specimens at 100 kV. The Cr $K\alpha$ is excited by Fe $K\alpha$ and δ_A is roughly half that experienced in bulk specimens (Nockolds et al. [28]).

Equations for δ_A have been developed by Philibert and Tixier [29] and Nockolds et al. [28]. The major difference between the equations of the two authors is that Nockolds et al. consider the effect of x-ray generation throughout the foil while Philibert and Tixier assumed that all x rays originated from the middle of the foil on the beam axis. The equation of Nockolds et al. fits the data shown in figure 7 and it would appear that their expression is theoretically more suitable than the earlier expression of Philibert and Tixier. The equation for δ_A by Nockolds et al. is:

$$\delta_A = C_B \omega_B \cdot \frac{r_A - 1}{r_A} \cdot \frac{A_A}{A_B} \cdot \mu/\rho|_A \cdot \frac{E_C^A}{E_C^B} \cdot \frac{(\ln E_0/E_C^B)}{(\ln E_0/E_C^A)} \cdot \frac{\rho t}{2} \cdot$$

$$[0.932 - \ln \mu/\rho|_{SPEC} \rho t] \cdot \sec \beta \ . \qquad (15)$$

Additional terms used here are:

r_A = absorption edge jump ratio of element A

E_C^A, E_C^B = critical excitation energies for the characteristic x rays from A and B

β = tilt angle of the thin foil

It is important to note that, as in the absorption correction, the calculation of δ_A depends on the accurate measurement of the foil thickness t. The full fluorescence correction F involves the measurement of δ_A as well as δ_B for the B element in the foil. There are a few measurements of δ_A, δ_B in the literature indicating that the F correction is small and not often needed. It must be considered, however, when C_A is small and the exciting element B is present in large concentrations. In practice the only case where fluorescence effects in thin foils have been examined, is the fluorescence of Cr $K\alpha$ by Fe $K\alpha$ in samples of Fe$\sim$10wt%Cr. In this case, the fluorescence effect was minor, and the maximum correction factor was only $\sim$5 percent [28].

VI. Quantitative Analysis Procedures

The full quantitative analysis procedure is given by eq (7) in which ZAF and ϵ_A/ϵ_B must be calculated or measured. In the early development of the analysis procedure we would consider an "infinitely" thin foil where the effects of x-ray absorption, A, and fluorescence, F, could be neglected. Procedures for deciding whether a full analysis must be made, for example by calculating if A would be close to 1.0 (e.g., between 0.97 to 1.03)[9], have been developed. In these cases, the Cliff-Lorimer ratio method, eqs (9) and (10), could be applied and all that is required is the determination of k_{AB}. Some k_{AB} values are available in the literature (e.g., [30,31]) and a comparison of values obtained in several investigations is listed in Williams [10]. In many cases k_{AB} can be measured from specimens of known composition and, since it is not a function of composition, k_{AB} can be applied directly to an unknown specimen. The direct calculation of k_{AB} factors (standardless analysis) is not as accurate since ϵ_A, ϵ_B etc., are usually not well known and there are significant errors in calculating ionization cross sections Q_A, Q_B especially when L and M shell radiation is considered [11]. We estimate that errors in calculating k_{AB} factors exceed $\pm$5 percent relative. It should be noted that $k_{AB} = (Z \cdot \epsilon_A/\epsilon_B)^{-1}$; therefore, to perform a full quantitative analysis k_{AB} must be measured or calculated to obtain the atomic number correction and the instrumental (EDS), correction.

The critical factor in applying the A and F corrections is the need either to measure the foil thickness, t, or to eliminate the A correction by extrapolation methods of the type suggested by Horita et al. [25]. The standards technique in which the EDS x-ray spectrum intensity is calibrated against standard thickness specimens holds great promise. This technique may only measure t to an accuracy of ± 5 percent relative. However, the practical effect in terms of errors in the calculated absorption correction, A, is less than ± 1 percent relative unless the sample is unusually thick or the mass absorption coefficient exceeds 1,000 cm^2/g. Measurement errors in I_A, I_B at the 95 percent confidence level usually exceed ± 2 percent. Therefore one can tolerate the inaccuracy in measured specimen thickness at each point in the analysis and perform a perfectly acceptable full quantitative analysis. The major caveat in suggesting that one can apply the standards technique for specimen thickness, on line, is that the electron beam current must be measured and held constant as specimen and standard thickness specimens are analyzed. Modern AEM instruments must be designed with this feature in mind, and we discuss this point at the end of the paper.

VII. Crystallographic Effects Unique to the AEM

Because microanalysis in the AEM is usually carried out on a scale $\ll 1$ μm, the analyzed volume is almost invariably a single crystal. The crystal orientation and defect structure can be determined by conventional TEM imaging and diffraction and so it becomes possible in the AEM to examine the effect of crystallography on the microanalysis process. To our knowledge, any such effects are largely ignored in the measurement of composition in the EPMA.

The crystal orientation affects x-ray emission through electron channeling, sometimes referred to as the "Borrmann effect" in TEM terminology. This phenomenon was first observed by Duncumb [32] and quantified by Cherns et al. [33] using an EMMA. It was shown that anomalous x-ray intensities are generated close to strong Bragg diffraction orientations where the amount of channeling changes rapidly with orientation. A simple-minded interpretation is that where strong channeling occurs x-ray generation decreases because of the reduced electron-atom interactions. Conversely, where strong Bragg diffraction occurs, x-ray generation is enhanced. Because the presence of strong diffraction is easily discerned in either the TEM image or diffraction pattern, it is a simple matter to avoid the conditions where anomalous effects may occur.

However, there is another positive side to this channeling artifact, which results in a unique aspect of x-ray microanalysis in thin foils. Spence and Taftø [34] made use of the variation of x-ray intensity due to channeling effects to locate the positions of certain atoms in the crystal structure of complex compounds. Spence, with typical antipodean phraseology, described the process as "atom location by channeling enhanced microanalysis" because it results (approximately!) in the memorable acronym ALCHEMI.

In ALCHEMI, x-ray spectra are acquired from crystals tilted to two specific channeling orientations at which Bloch wave interactions are strongest with two different sets of planes. Under these circumstances, if a certain atom is preferentially located on one set of planes, then its x-ray peak intensity is different at each orientation, and the atom fraction on certain sites can be easily determined. This process has been applied to many problems (for a summary see Spence et al. [35]) and its only limitation is that the change in x-ray emission is often very small, so very good statistics are essential if a significant result is to be obtained.

VIII.   Spatial Resolution and Minimum Detectability

As we described in the introduction, a major driving force for the development of the AEM was the improvement in spatial resolution compared with the EPMA. This improvement arises for two reasons. First, the use of thin foil samples means that less elastic and inelastic scatter occurs as the beam traverses the sample. Second, the higher beam energy ($>$100–400 keV in the AEM compared with 5–30 keV in the EPMA) further reduces beam scatter because the mean free path for both elastic and inelastic collisions increases with the electron energy. The net result is that increasing the accelerating voltage when using thin samples decreases the total beam-specimen interaction volume thus giving a more localized x-ray signal source and a higher spatial resolution. With bulk samples in the EPMA, increasing the voltage increases the interaction volume thus degrading spatial resolution. In the EPMA, the physics of beam-specimen interactions, at the voltages used, mean that spatial resolution never improves below ~0.5–1 μm. Therefore, there is little interest in the theory of spatial resolution and little or no effort is routinely made to optimize this parameter. By contrast, much theoretical and experimental work has been carried out to both define and measure the spatial resolution in the AEM.

In the AEM field it is generally agreed that R comprises some combination of the incident electron probe size (d) and the spreading of the probe (b) as it traverses the specimen. Goldstein [36] proposed simply that:

$$R = b + d, \tag{16}$$

while Reed [37], arguing that the probe was Gaussian, proposed that:

$$R = (b^2 + d^2)^{1/2} . \tag{17}$$

There are many theories governing the probe size [38] and the beam broadening [9,39] but the most consistent approach is when both d and b are defined in terms of the diameter of the Gaussian electron beam that contains 90 percent of the electrons, i.e., the full width at tenth maximum (FWTM). Usually b is governed by the beam energy (E_0), sample thickness (t) and density (ρ) such that:

$$b \sim \frac{\rho^{1/2} \cdot t^{3/2}}{E_0} . \tag{18}$$

This simple relationship between b and t appears to hold for most of the available theories, from simple "single-scattering" models to full Monte-Carlo simulations, as shown in figure 8.

It is obvious from eq (17) that to improve spatial resolution, both d and b must be minimized. Minimizing d reduces the input probe current and for thermionic sources, if $d < 10$ nm, count rates are unacceptably low. However, with an FEG sufficient current (~1 nA) can be generated in a 1 nm probe to permit quantitative analyses. With a thermionic source, sample thicknesses have to be such that sufficient counts are generated for quantification but the net result is usually that b dominates in eq (17). Under these circumstances, much attention has been paid to increasing the accelerating voltage since, from eq (18), that is the only variable left for a given sample. The development of 300–400 keV instruments owes much to this perceived need for improved resolution. Unfortunately, as discussed in the section on x-ray absorption, higher voltages mean increased penetration and often the practical result of using an intermediate voltage AEM is that analyses are actually performed in thicker areas than at 100 kV and spatial resolution degrades. There is

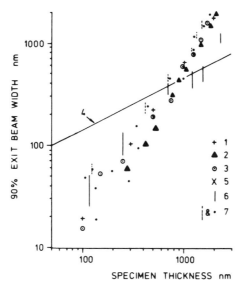

Figure 8. Beam spreading as a function of specimen thickness predicted by a variety of theoretical treatments. Most models describe a $t^{3/2}$ relationship, as expressed in eq (18) (from Goldstein et al. [9], courtesy Plenum Press).

still debate as to whether the expected linear improvement in spatial resolution occurs at 300–400 keV [38,40].

As high brightness FEG sources become available and higher voltage AEMs become commonplace, it will be easier to get quantitative data from thinner samples and then d will play a more dominant role in eq (17), and the expected combination of 300 keV and FEG will result in spatial resolutions $\ll 5$ nm.

Comparison of figures 3b and 3e in the introduction shows most clearly the improved spatial resolution of the AEM compared with the EPMA. However, spatial resolution and minimum detectability are intimately related. It is a feature of any microanalysis technique that an improvement in spatial resolution is balanced by a worsening of the detectability limit (all other factors being equal). This is because at higher spatial resolution the analyzed volume is smaller and, therefore, the signal intensity is reduced. The reduction in signal intensity means that the acquired spectrum will be noisier and therefore small peaks from trace elements will be less detectable. Therefore, in the AEM, the price that is paid for improved spatial resolution is a relatively poor minimum detectability.

The minimum mass fraction (MMF) represents the smallest concentration of an element that can be measured in the analysis volume. The MMF can be decreased or improved by increasing the peak x-ray intensity (P) for the element of interest, by increasing the peak to background ratio (P/B) for the element of interest, and by increasing the analysis time (τ). (See Ziebold, [41].) In general:

$$\text{MMF} \sim \frac{1}{(P \times P/B \times \tau)^{1/2}} \, . \tag{19}$$

To increase P, we can increase the current in the electron probe and increase the thickness (t) of the specimen, while to increase P/B we can increase the operating voltage (E_0) and decrease instrumental contributions to the background. Improvements in AEM instrument design such as using a high brightness source, a high voltage gun, and a higher collection angle for EDS can increase P. To increase P/B, a stable instrument with a clean vacuum environment is needed to minimize or eliminate specimen deterioration and contamination. Improved AEM stage design, to minimize stray electrons and bremsstrahlung radiation, both of which contribute to the detected spectrum, could also help to increase P/B. The P/B ratio is a strong variable from instrument to instrument [42], and there is great need to create a standard, acceptable level of the P/B ratio in AEMs, in order that optimum performance is always achieved.

An alternative approach is to define the MMF by a purely statistical criterion. The characteristic x-ray intensity for the solute, above background, may be assumed to be statistically significant at a 95 percent confidence level if the peak is greater than three times the standard deviation of the counts in the background under the peak. Combining this criterion with the Cliff-Lorimer equation and assuming Gaussian statistics, Romig and Goldstein [43] showed that the detectability (in wt%) of element B in element A may be expressed as:

$$C_B(\text{MMF}) = \frac{3(2I_B^b)^{1/2}}{I_A - I_A^b} \cdot C_A \cdot k_{AB}^{-1} \tag{20}$$

where I_A^b and I_B^b are background intensities for elements A and B; I_A is the raw integrated intensity of peak A; C_A is the concentration of A (in wt%); and k_{AB}^{-1} is the reciprocal of the Cliff-Lorimer k-factor k_{AB}. However, by expressing the Cliff-Lorimer equation as:

$$\frac{C_A k_{AB}^{-1}}{I_A - I_A^b} = \frac{C_B}{I_B - I_B^b} \tag{21}$$

and substituting into eq (20), Michael [44] found the detectability limit for B to be:

$$C_B(\text{MMF}) = \frac{3(2I_B^b)^{1/2}}{I_B - I_B^b} \cdot C_B \ . \tag{22}$$

The value of $C_B(\text{MMF})$ calculated with eq (22) has been used by Lyman [45] as a measure of analytical sensitivity, when comparing several different experimental approaches.

The best compromise in terms of improving MMF and x-ray spatial resolution is to use high operating voltages (300–400 kV) and thin specimens to minimize beam broadening. The loss of x-ray intensity, [P in eq (19)], a consequence of using thin specimens, can be compensated in part by the higher voltages and/or by using a high brightness FEG source where a small spot size of 1 to 2 nm can still be maintained. In summary, optimum MMF and spatial resolution can be obtained by using a high brightness, intermediate voltage source with thin foils, perhaps of the order of $t \sim 10$ nm. Under these circumstances, MMF values <0.1 wt% will become routine. This of course does not compare well with the EPMA, but a more favorable aspect of detectability limits in the AEM is evident when we consider the absolute sensitivity rather than the mass fraction.

Of interest is the ultimate ability of the AEM to detect the presence of only a few atoms in the analysis volume. For the EPMA with 1 μm^3 excitation volume and a 0.01 wt% MMF, $\sim 10^7$ atoms are detected in the analysis volume. Using data for the minimum mass fraction of chromium [46] in a 304L stainless steel measured in a VG HB-501 AEM, Lyman and Michael [46] obtained an MMF of 0.069 wt% chromium in a 164 nm foil with a spatial resolution of 44 nm and a 200 s counting time. The electron probe size was 2 nm (FWTM) with a beam current of 1.7 nA. In this analysis, an estimated 2×10^4 atoms were detected. If

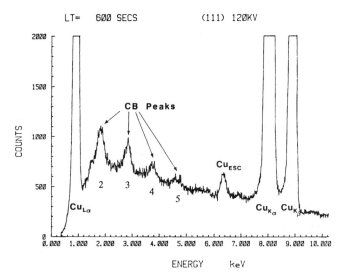

Figure 9. X-ray spectrum from pure copper showing the normal characteristic and escape peaks, and the small Gaussian coherent bremsstrahlung peaks labeled CB (courtesy K. S. Vecchio).

the counting time is increased by a factor of 10 and if the operating voltage is increased to 300 kV, the spatial resolution would improve to ~15 nm and MMF would improve to ~0.01 wt%. We estimate that about 300 atoms could be detected. For a foil thickness of 16 nm (1/10th the above measured thickness), the MMF would increase to ~0.03 wt%. However, the spatial resolution would improve to about 2 nm. For this case, we estimate that about 20 atoms would be detected. Therefore, in ~10 nm thick specimens, with a spatial resolution approaching the beam diameter d, of 1 to 2 nm, one will be able to detect the presence of 10 to 100 atoms in the analysis volume (10^{-8} μm^3) with x-ray emission spectroscopy.

A further limitation to detecting small spectral peaks in the AEM is the phenomenon of coherent bremsstrahlung. The bremsstrahlung or background in the x-ray spectrum is sometimes referred to as the "continuum" because the intensity is assumed to be a smooth, slowly-varying function of energy. This assumption is perfectly reasonable with the bremsstrahlung generated in bulk samples by electrons with energies $< \sim 30$ keV, such as in an SEM/EPMA. However, in thin samples illuminated by high-energy electrons, it is possible to generate a bremsstrahlung x-ray spectrum that contains small, Gaussian-shaped peaks known as "coherent bremsstrahlung" (CB). The phenomenon of CB is well known in high-energy physics experiments, but no one thought it would occur at AEM voltages until it was clearly demonstrated by Reese et al. [47]. Figure 9 shows a portion of an x-ray spectrum from a thin foil of pure copper taken at 120 keV. The primary peaks, as expected, are the Cu $K\alpha/\beta$ and the $L\alpha$ lines. In addition, the Cu $K\alpha$ escape peak is identified. The other small peaks, labeled 2, 3, 4, 5, are the CB peaks. They arise from the regular nature of the Coulombic interaction process with the nuclei that occurs as the beam electron traverses the crystal lattice, close to a row of atoms. The regular interactions result in x-ray photons of similar energy E_{CB} given by:

$$E_{CB} = 12.4\beta/[d \cos(90+\alpha)] \tag{23}$$

where β is the electron velocity (v) divided by the velocity of light (c), d is the lattice spacing in the beam direction and α is the detector take-off angle defined in the same manner as in figure 5 (i.e., from the plane of the sample, normal to the optic axis of the microscope). Obviously, there is great danger that these CB peaks will be mistakenly identified as characteristic peaks arising from the presence of a small amount of some element in the sample. Fortunately, it is easy to distinguish CB peaks from characteristic peaks because, as predicted by eq (23), the CB peaks will move depending on the accelerating voltage (which will alter v and hence β) as well as the specimen orientation, which will change the value of d. Of course, characteristic peaks show no such behavior, and are dependent only on the elements present in the sample. The numbers on the CB peaks in figure 9 indicate the Laue zones from which the x rays are being generated. While the CB peaks are a nuisance, it may be possible to use them to advantage. There is some evidence (Vecchio and Williams [48]) that the true bremsstrahlung intensity is low in the regions between the CB peaks. Therefore, if you are seeking to detect a very small amount of segregant, e.g., sulfur segregated to grain boundaries in copper, then it is possible to "tune up" the CB peaks by careful choice of kV and orientation to insure that the S $K\alpha$ line will appear between two CB peaks and not be masked by them.

IX. Specimen Preparation Problems

The preparation of thin foil specimens is, in many ways, more of an art than a science. Despite the publication of two textbooks (Thompson-Russell and Edington [49] and Goodhew [50]) that list in extraordinary detail the various methods reported in the literature, it is still usual to have to undergo extensive individual experimentation to find out the best method to prepare a given sample. It is rare that one method works the first time it is tried, and even rarer that a method works reproducibly all the times it is used. This variability is

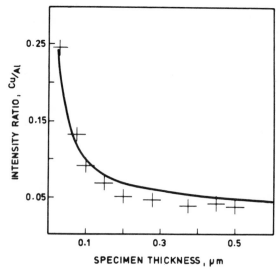

Figure 10. A typical effect of sample preparation is the apparent composition variation with foil thickness in a homogeneous sample. A thin surface solute-enriched layer results in the observed composition "profile." The solid line represents a calculation for a 15 nm thick surface layer of pure copper. Crosses represent experimental measurements (from Thompson et al. [51], courtesy Taylor and Francis).

probably a result of the drastic measures that are required to produce electron-transparent samples. The two major methods, jet electropolishing and ion beam thinning, subject the thin regions of the sample to either severe mechanical stress combined with rapid electro-chemical dissolution or intense ion-beam heating, differential sputtering and surface damage. Not surprisingly, the question is often raised as to whether the thin foil is still characteristic of the bulk sample.

For TEM imaging investigations, the question is whether the sample preparation has introduced crystal defects, and over the years it has become possible to identify noncharacteristic defect structures. For AEM investigations, the question is whether or not the sample chemistry has been altered during thinning. There is ample evidence that both electropolishing (Thompson et al. [51], Pountney and Loretto [52]) and ion-beam thinning (Morris et al. [22] and Njegic and Williams [53]) can change the surface chemistry and introduce severe heating effects. The thinner the sample, the more severe the problem and the easiest way to see the effect is to measure the composition variation with thickness in a supposedly homogeneous sample. If surface chemical changes have occurred then, instead of a constant composition, a change in one or more elemental composition is often detectable in the thinner regions of the sample, as shown in figure 10. If this effect is observed it is necessary to change the specimen preparation method, or clean the sample surface in some way. As with most specimen preparation problems, there is no single solution, and not all "solutions" work in a reproducible manner. It is probably safe to assume that the specimen surface chemistry is not characteristic of the bulk chemistry and a short "dusting off" in an ion beam thinner is recommended, prior to microanalysis. This dusting off should be carried out for only a couple of minutes at 1–2 keV, with the specimen tilted at 10°–15° and cooled to liquid nitrogen temperature to minimize further damage. Despite such precautions, great care should be taken, particularly when performing microanalysis in the thinnest portions of the specimen. If high spatial resolution is not required then thicker portions of the sample may be analyzed where surface artifacts do not affect the results.

In addition to the production of surface films, other sample effects arise which cause significant problems in AEM but are usually negligible in the EPMA. As we have already seen, the thin foil that results from sample preparation is not the ideal parallel-sided shape, but is almost invariably wedge-shaped, with local variations in thickness. The thickness changes make it practically impossible at present to carry out quantification using pure element standards and the ratio technique remains the only valid approach. Until accurate, on-line thickness determination becomes a reality, the sample shape will remain a major limitation. Furthermore, thickness measurement is a major source of error in both absorption and fluorescence corrections. The thickness change is often most severe around defects such as grain boundaries or interphase interfaces, because local chemistry changes affect the sample thinning rate. Unfortunately, of course, it is precisely such chemical variations that we are seeking to measure in the AEM.

In the EPMA it is customary to coat almost every sample to prevent charging. In the AEM a conducting film such as carbon is applied only in the case of ceramics, glasses and some minerals and semiconductors. The carbon film often adds to the problem of hydrocarbon contamination which still remains a major limitation in the AEM. While TEM/STEM vacuums have improved to the point where instrumental contributions to contamination are minimal, the contamination problem is still ubiquitous because of the sample-preparation procedures. Electropolishing is almost invariably performed in a hydrocarbon-based solvent, and the resulting foils are usually cleaned in ethanol. Ion-beam thinning is performed in instruments that are evacuated by oil-based mechanical and diffusion pumps. Carbon or other coatings are usually deposited in evaporators/sputter coaters with relatively poor vacuum systems. The result is often an alarming rate of hydrocarbon deposition at the precise area of interest where the electron probe is positioned. In the EPMA, oxygen gas jets are an easy solution to any hydrocarbon build-up, but no AEM has been similarly equipped. There is no alternative for the serious AEM analyst but to improve the specimen preparation process and invest in good clean vacuum equipment and avoid residual hydrocarbon films by prior infrared heating, or always operating with a liquid-nitrogen cooled sample holder.

The ultimate limitation to any procedure (imaging, diffraction or microanalysis) in the AEM is beam damage. In the EPMA, accelerating voltages are such that knock-on damage or sputtering effects do not occur, although sample heating may damage some polymers and other organic materials. However, in the AEM, particularly in 300–400 kV instruments, the very act of observing the sample may change its local chemistry. This analytical analog of Heisenberg's uncertainty principle represents the physical limits of the technique.

Voltages for knock-on damage of most elements are well known and tabulated from the high-voltage electron microscopy literature (e.g., [54,55]) and more recently the problem of surface atom sputtering has been studied (Zaluzec and Mansfield [56]). In either case, the obvious solution is to operate below the critical threshold voltages, although this will remove many of the attractive aspects of operating at the highest available voltage. However, it is the thermal damage that remains the ultimate problem for most samples, and this in fact is worse at lower voltages because of the reduced cross section for inelastic scatter—particularly the phonon vibrations which heat up the sample. As we seek more intense electron sources, the power-input capabilities of modern AEMs become impressive. For example, a Vacuum Generators HB501 100 keV FEG STEM can generate 1 nA in a probe of <2 nm (FWTM) [39]. If all the electrons deposited their energy in the sample, the power input would be ~ 10 GW/cm^2, which is equivalent to the output of several nuclear power stations being deposited into 1 cm^2 of the sample. Fortunately, in the AEM most electrons ignore the sample completely; otherwise it would be vaporized. Nevertheless, it is easy to see that we are rapidly approaching the limits of the technique in terms of the ability of the sample to remain unchanged in the microscope.

However, we are a long way from approaching the limits of the technique in other aspects, particularly the design of the AEM, and we will finish the paper by discussing the ways in which instrument design is moving to optimize the analytical performance.

X. Future Developments

The EPMA went through several design changes during its early history before an optimal design was agreed upon, after which most changes have been cosmetic and have not influenced the data generation process. From the standpoint of x-ray microanalysis, the AEM is still in the development stage, and at several points in the preceding text we have mentioned specific instrument-related problems which can be improved. It is somewhat ironic that the original analytical electron microscopes, the EMMAs, were in fact very well designed for microanalysis. This is not surprising since Duncumb, the designer, was primarily a microanalyst. However, the EMMAs suffered from electron optical (probe-forming) limitations as well as from the relatively primitive state of the WDS at the time. More modern AEMs are based on TEM/STEMs wherein the design has been driven primarily by electron optical considerations and the EDS has been attached to the column almost as an afterthought. Clearly, if the AEM is to be an optimized microanalytical tool, it must in effect become an "electron nanoprobe analyzer" based on the tried and true EPMA concepts. Where the AEM fails to meet these criteria is in the following six areas which we will discuss individually—(1) electron source stability, (2) in-situ probe-current measurement,

(3) multi-specimen capability, (4) high detector take-off angle and collection angle, (5) maximum peak to background ratio, (6) computerized stage control.

1. *Source stability:* In the EPMA, tungsten or LaB_6 sources are used as a matter of course, and high levels of probe-current stability are achieved by feedback mechanisms from isolated condenser aperture drives. There is absolutely no reason why this should not be incorporated into all AEMs. With FEG sources, particularly cold FEG systems, the emission behavior is notoriously unstable and a stabilizing feedback mechanism would appear imperative even for imaging purposes.

2. *On-line probe-current measurements:* This requires a Faraday cup in the column below the final probe-limiting aperture. No such option exists as part of any AEM column, although some sample holders do offer Faraday cups close to the sample.

3. *Multi-specimen capability:* Some AEMs offer two-sample holders and more recently four- and five-sample holders have become an option. However the majority of instruments can still only examine one sample per specimen exchange, and thus require that the beam be turned off between sample viewings.

If areas 1–3 were available and on-line thickness determination a possibility, then x-ray microanalysis in the AEM could be carried out using thin-film pure-element standards in entirely the same manner as in the EPMA. Ultimately this will provide more accurate analysis than the standardless or conventional Cliff-Lorimer ratio approach.

4. *High take-off angle and collection angle:* In the EPMA a high value of α is required to minimize the absorption correction, which is the major source of error in quantification. The same argument could be made for the AEM, but the confined stage region of the TEM-based system, essential to maintain reasonable imaging and diffraction performance, militates against this ideal configuration. Furthermore the smaller absorption correction in thin foils means that a high take-off angle is not as essential as in the EPMA. High take-off angles are possible if the detector "looks" through the upper objective pole-piece, but the resultant collection angle is so small (<0.05 sr) that count rates became unacceptably low even for relatively large probe sizes. A compromise take-off angle of $\sim 20°$ permits reasonable absorption path lengths and higher collection angles ($> \sim 0.1$ sr), but this is still far from ideal. With careful design, collection angles closer to 0.3 sr should be feasible when α is $\sim 20°$, and this aspect is most critical in the AEM where the low count rate from thin samples and fine probes limits the analytical sensitivity.

5. *Peak-to-background ratio:* Part of the problem with confined AEM stages, and very high voltage electron beams is that electron scatter and high-energy bremsstrahlung production within the illumination system and stage combine to lower the P/B ratio in the spectrum. Commercial AEMs show a wide range of P/B [42] and considerable work is required before the stage area is shielded sufficiently to produce the maximum, theoretical P/B ratio. (See Nicholson et al. [57].)

6. *Computerized stage control:* This is a standard feature of the EPMA, permitting unattended microanalysis. The AEM has not achieved this status yet, but part of the problem is that since the analysis is performed at much higher magnifications, specimen drift is large enough over long time periods to be the major limitation to unattended microanalysis. Recently, in combination with digital image-storage and image-matching programs, on-line

drift correction has become available, and unattended microanalysis is possible (Vale and Statham [58]). In combination with digital storage, computerized traversing and tilting, it now becomes feasible to project that all operations currently possible on the EPMA at the micron level will soon be reproducible on the AEM at the sub-10 nm level. If contamination does not occur, there is no reason why a combination of an FEG and large collection angle spectrometers on the AEM cannot match the counting statistics of a thermionic source and a bulk sample in the EPMA. The possibility of compact WDS systems on the AEM would further enhance the prospects of the instrument (Goldstein et al. [59]).

XI. Summary

We have tried to draw parallels between x-ray microanalysis in the EPMA and the AEM. The analysis procedure in thin samples is often very simple, but if absorption and fluorescence occur, then the quantification routines are very similar. To date, in the AEM it remains essential to use a ratio technique because of variations in both the sample thickness and the stability of the probe. However, it will soon be possible to pursue the EPMA approach of using pure element standards. Spatial resolution in the AEM is ~10 to 100 times better than in the EPMA but detectability limits are ~10 times worse (in terms of minimum mass fraction). Absolute sensitivity may, however, approach the level of a few atoms. Sample preparation is simple for the EPMA but remains a thorn in the side of thin-foil analysts, contributing many problems. The ultimate limit for AEM microanalysis is beam damage, and it is doubtful if there is anything to be gained in this respect by increasing accelerating voltages above the current levels of 300-400 keV. Finally we note that the AEM, unlike the EPMA, is still not optimally designed for x-ray microanalysis. Nevertheless the changes that remain to be made are all within our current technical capabilities. Then it will be possible to talk about the "Electron Probe Nano-Analyzer" (EPNA)!

XII. Acknowledgments

The authors wish to acknowledge stimulating discussions with C. E. Lyman and J. R. Michael. Many of the ideas presented in this chapter arose from work funded by NASA under Grant NAG9-45.

XIII. References

[1]   Reuter, K. B., Williams, D. B., and Goldstein, J. I. (1988), Geochim. Cosmochim. Acta. **52**, 617.

[2]   Dean, D. C. and Goldstein, J. I. (1986), Met. Trans. **17A**, 1131.

[3]   Castaing, R. (1951), Ph.D. Thesis, University of Paris.

[4]   Scott, V. D. and Love, G. (1983), Quantitative Electron Probe Microanalysis, Ellis Harwood Ltd., Chichester, U.K., 58.

[5]   Duncumb, P. (1968), J. de. Microscopie 7, 581.

[6]   Cliff, G. and Lorimer, G. W. (1972), Proc. 5th European Cong. on Electron Microscopy, The Institute of Physics, Bristol and London, 141.

[7]   Cliff, G. and Lorimer, G. W. (1975), J. Microsc. **103**, 203.

[8]   Cox, C. E., Lowe, B. G., and Sareen, R. A. (1988), IEEE Transactions on Nuclear Science **35.1**, 28.

[9]   Goldstein, J. I., Williams, D. B., and Cliff, G. (1986), Principles of Analytical Electron Microscopy, Joy, D. C., Romig, A. D. Jr., and Goldstein, J. I., eds., Plenum Press, New York, 155.

[10] Williams, D. B. (1984), Practical Analytical Electron Microscopy in Materials Science, Electron Optics Publishing Group, Mahwah, NJ.

[11] Williams, D. B., Newbury, D. E., Goldstein, J. I., and Fiori, C. E. (1984), J. Microsc. **136**, 209.

[12] Joy, D. C., Romig, A. D., and Goldstein, J. I., eds. (1986), Principles of Analytical Electron Microscopy, Plenum Press, New York.

[13] Wirmark, G. and Nordén, H. (1987), J. Microsc. **148**, 167.

[14] Goldstein, J. I., Newbury, D. E., Echlin, P., Joy, D. C., Fiori, C. E., and Lifshin, E. (1981), Scanning Electron Microscopy and X-Ray Microanalysis, Plenum Press, New York.

[15] Stenton, N., Notis, M. R., Goldstein, J. I., and Williams, D. B. (1981), Quantitative Microanalysis with High Spatial Resolution, Lorimer, G. W., Jacobs, M. H., and Doig, P., eds., The Metals Soc., London, Book 277, 35.

[16] Newbury, D. E., Myklebust, R. L., Romig, A. D., and Bieg, K. W. (1983), Microbeam Analysis—1983, Gooley, R., ed., San Francisco Press, San Francisco, CA, 168.

[17] Jones, I. P. and Loretto, M. H. (1981), J. Microsc. 124, 3.

[18] Newbury, D. E. (1981), private communication in Joy et al. (1986).

[19] Goldstein, J. I., Costley, J. L., Lorimer, G. W., and Reed, S. J. B. (1977), SEM 1977, Johari, O., ed., IITRI, Chicago, 1, 315.

[20] Lorimer, G. W., Cliff, G., and Clark, J. N. (1976), Developments in Electron Microscopy and Analysis, Venables, J. A., ed., Academic Press, London, 153.

[21] Kelly, P. M., Jostsons, A., Blake, R. G., and Napier, J. G. (1975), Phys. Stat. Sol. 31, 771.

[22] Morris, P. L., Ball, M. D., and Statham, P. L. (1979), Electron Microscopy and Analysis, Mulvey, T., ed., Inst. of Physics, Bristol and London, 413.

[23] Malis, T., Cheng, S. C., and Egerton, R. F. (1988), J. Electron Microscopy Technique 8, 193.

[24] Porter, D. A. and Westengen, H. (1981), Quantitative Microanalysis with High Spatial Resolution, Lorimer, G. W., Jacobs, M. H., and Doig, P., eds., The Metals Soc., London, Book 277, 94.

[25] Horita, Z., Sano, T., and Nemoto, M. (1987), Ultramicroscopy 21, 271.

[26] Maher, D. M., Ellington, M. B., Joy, D. C., Schmidt, P. H., Zaluzec, N. J., and Mochel, D. E. (1981), Analytical Electron Microscopy—1981, Geiss, R. H., ed., San Francisco Press, San Francisco, CA, 29.

[27] Zaluzec, N. J., Maher, D. M., and Mochel, P. E. (1981), Analytical Electron Microscopy—1981, Geiss, R. G., ed., San Francisco Press, San Francisco, CA, 47.

[28] Nockolds, C., Nasir, M. J., Cliff, G., and Lorimer, G. W. (1980), Electron Microscopy and Analysis, Mulvey, T., ed., Inst. of Physics, Bristol and London, 417.

[29] Philibert, J. and Tixier, R. (1975), Physical Aspects of Electron Microscopy and Microbeam Analysis, Siegel, B. M. and Beaman, D. R., eds., J. Wiley, New York, 333.

[30] Wood, J. E., Williams, D. B., and Goldstein, J. I. (1984), J. Microsc. 133, 255.

[31] Graham, R. J. and Steeds, J. W. (1989), J. Microsc. 133, 275.

[32] Duncumb, P. (1962), Phil. Mag. 7, 2101.

[33] Cherns, D., Howie, A., and Jacobs, M. H. (1973), Z. Naturforschung A28, 565.

[34] Spence, J. C. H. and Taftø, J. (1983), J. Microsc. 130, 147.

[35] Spence, J. C. H., Graham, R. J., and Shindo, D. (1986), Materials Problem Solving with the Transmission Electron Microscope, Hobbs, L. W., Westmacott, K. H., and Williams, D. B., eds., Materials Research Society, Pittsburgh, PA, 139.

[36] Goldstein, J. I. (1979), Introduction to Analytical Electron Microscopy, Hren, J. J., Goldstein, J. I., and Joy, D. C., eds., Plenum Press, New York, 83.

[37] Reed, S. J. R. (1982), Ultramicroscopy 7, 405.

[38] Klein, C. F., Ayer, R. A., and Williams, D. B. (1987), Intermediate Voltage Microscopy and its Application to Materials Science, Rajan, K., ed., Electron Optics Publishing Group, Mahwah, NJ, 24.

[39] Michael, J. R. and Williams, D. B. (1987), J. Microsc. 147, 289.

[40] Rajan, K., McCaffery, J., Sewell, P. B., Leavens, C. R., and L'Esperance, G. (1987), Intermediate Voltage Microscopy and Its Application to Materials Science, Rajan, K., ed., Electron Optics Publishing Group, Mahwah, NJ, 11.

[41] Ziebold, T. O. (1967), Anal. Chem. 36, 322.

[42] Williams, D. B. and Steel, E. B. (1987), Analytical Electron Microscopy—1987, Joy, D. C., ed., San Francisco Press, San Francisco, CA, 228.

[43] Romig, A. D. Jr. and Goldstein, J. I. (1979), Microbeam Analysis—1979, Newbury, D. E., ed., San Francisco Press, San Francisco, CA, 124.

[44] Michael, J. R. (1982), M. S. Thesis, Lehigh University.

[45] Lyman, C. E. (1986), Microbeam Analysis—1986, Romig, A. D. Jr. and Chambers, W. F., eds., San Francisco Press, San Francisco, CA, 434.

[46] Lyman, C. E. and Michael, J. R. (1987), Analytical Electron Microscopy, Joy, D. C., ed., San Francisco Press, San Francisco, CA, 231.

[47] Reese, G. M., Spence, J. C. H., and Yamamoto, N. (1984), Phil. Mag. A49, 697.

[48] Vecchio and Williams (1987), J. Microsc. 147, 15.

[49] Thompson-Russell, K. C. and Edington, J. W. (1977), Electron Microscope Specimen Preparation Techniques in Materials Science, MacMillan Press, London.

[50]   Goodhew, P. J. (1985), Thin Foil Preparation for Electron Microscopy, Elsevier, New York.

[51]   Thompson, M. N., Doig, P., Edington, J. W., and Flewitt, P. E. J. (1977), Phil. Mag. **35**, 1532.

[52]   Pountney and Loretto (1980), Electron Microscopy 1980, Brederoo, P. and Cosslett, V. E., eds., 7th European Congress on Electron Microscopy Foundation, Lieden, Netherlands, **3**, 180.

[53]   Njegic, A. and Williams, D. B. (1981), J. Microsc. **123**, 293.

[54]   Bell, W. L. and Thomas, G. (1971), Electron Microscopy and Structure of Materials, Thomas, G., ed., University of California Press, Berkeley, CA, 23.

[55]   Cherns, D., Finnis, M. W., and Matthews, M. D. (1977), Phil. Mag. **35**, 693.

[56]   Zaluzec, N. J. and Mansfield, J. F. (1987), Intermediate Voltage Microscopy and Its Application to Materials Science, Rajan, K., ed., Electron Optics Publishing Group, Mahwah, NJ, 29.

[57]   Nicholson, W. A. P., Gray, C. C., Chapman, J. N., and Robertson, B. W. (1982), J. Microsc. **125**, 25.

[58]   Vale, S. H. and Statham, P. J. (1986), Proc. XI Int. Cong. on Electron Microscopy, The Japanese Society of Electron Microscopy, Kyoto, **1**, 573.

[59]   Goldstein, J. I., Lyman, C. E., and Williams, D. B. (1989), Ultramicroscopy **28**, 162.

INDEX

BC